NEUROMETHODS

Series Editor
Wolfgang Walz
University of Saskatchewan
Saskatoon, SK, Canada

For further volumes:
http://www.springer.com/series/7657

Neuromethods publishes cutting-edge methods and protocols in all areas of neuroscience as well as translational neurological and mental research. Each volume in the series offers tested laboratory protocols, step-by-step methods for reproducible lab experiments and addresses methodological controversies and pitfalls in order to aid neuroscientists in experimentation. *Neuromethods* focuses on traditional and emerging topics with wide-ranging implications to brain function, such as electrophysiology, neuroimaging, behavioral analysis, genomics, neurodegeneration, translational research and clinical trials. *Neuromethods* provides investigators and trainees with highly useful compendiums of key strategies and approaches for successful research in animal and human brain function including translational "bench to bedside" approaches to mental and neurological diseases.

Dopaminergic System Function and Dysfunction: Experimental Approaches

Edited by

José Antonio Fuentealba-Evans

Escuela de Química y Farmacia y Centro Interdisciplinario de Neurociencia, Facultad de Química y de Farmacia, Pontificia Universidad Católica de Chile, Santiago, Chile

Pablo Henny

Departamento de Anatomía Normal, Escuela de Medicina, Pontificia Universidad Católica de Chile, Santiago, Chile

Humana Press

Editors
José Antonio Fuentealba-Evans
Escuela de Química y Farmacia y Centro Interdisciplinario de Neurociencia, Facultad de Química y de Farmacia
Pontificia Universidad Católica de Chile
Santiago, Chile

Pablo Henny
Departamento de Anatomía Normal, Escuela de Medicina
Pontificia Universidad Católica de Chile
Santiago, Chile

ISSN 0893-2336 ISSN 1940-6045 (electronic)
Neuromethods
ISBN 978-1-0716-2798-3 ISBN 978-1-0716-2799-0 (eBook)
https://doi.org/10.1007/978-1-0716-2799-0

This Humana imprint is published by the registered company Springer Science+Business Media, LLC, part of Springer Nature.
The registered company address is: 1 New York Plaza, New York, NY 10004, U.S.A.

Preface to the Series

Experimental life sciences have two basic foundations: concepts and tools. The Neuromethods series focuses on the tools and techniques unique to the investigation of the nervous system and excitable cells. It will not, however, shortchange the concept side of things as care has been taken to integrate these tools within the context of the concepts and questions under investigation. In this way, the series is unique in that it not only collects protocols but also includes theoretical background information and critiques which led to the methods and their development. Thus, it gives the reader a better understanding of the origin of the techniques and their potential future development. The Neuromethods publishing program strikes a balance between recent and exciting developments like those concerning new animal models of disease, imaging, *in vivo* methods, and more established techniques, including, for example, immunocytochemistry and electrophysiological technologies. New trainees in neurosciences still need a sound footing in these older methods in order to apply a critical approach to their results.

Under the guidance of its founders, Alan Boulton and Glen Baker, the *Neuromethods* series has been a success since its first volume published through Humana Press in 1985. The series continues to flourish through many changes over the years. It is now published under the umbrella of Springer Protocols. While methods involving brain research have changed a lot since the series started, the publishing environment and technology have changed even more radically. Neuromethods has the distinct layout and style of the Springer Protocols program, designed specifically for readability and ease of reference in a laboratory setting.

The careful application of methods is potentially the most important step in the process of scientific inquiry. In the past, new methodologies led the way in developing new disciplines in the biological and medical sciences. For example, physiology emerged out of anatomy in the nineteenth century by harnessing new methods based on the newly discovered phenomenon of electricity. Nowadays, the relationships between disciplines and methods are more complex. Methods are now widely shared between disciplines and research areas. New developments in electronic publishing make it possible for scientists that encounter new methods to quickly find sources of information electronically. The design of individual volumes and chapters in this series takes this new access technology into account. Springer Protocols makes it possible to download single protocols separately. In addition, Springer makes its print-on-demand technology available globally. A print copy can therefore be acquired quickly and for a competitive price anywhere in the world.

Saskatoon, SK, Canada ***Wolfgang Walz***

Preface

The midbrain dopaminergic system, in fact more appropriately defined as the *mesodiencephalic* dopaminergic system given its embryonic origin and current brain parcellation [1–3], is one of the most studied brain neurotransmitter–specific neuronal groups, and its research has been highly influential outside the basic neuroscience field [4–6]. This is partly because it subserves functions that are essential to mammalian and otherwise vertebrate life, including movement and motor activity, motivation, and learning [7–9]. Pioneering pharmacological and anatomical studies in the late 1950s and early 1960s showed not only that dopamine was a genuine brain neurotransmitter but also that it had an indisputable role in movement and motor disorders [10–12], discoveries that later formed the basis of our, still current, basic conceptualization of basal ganglia function in movement [13]. Then, animal models and human studies in the following decades revealed the intimate relationship between *in vivo* dopamine neurons single-cell activity pattern, dopamine release and ongoing behavior [9, 14, 15]. These advances further developed into a strong framework that today allows us to interpret motivation, preference formation and contingency learning in terms of cellular and circuit-level processing [8, 9], in turn illuminating our understanding of addictive processes [16, 17], and neuropsychiatric diseases [18].

Obvious as it can be, these enormous advances owe their success to the use of a myriad of different methodologies and technical refinements through the years that allow us to understand and interpret in greater detail the mechanisms that underlie dopamine's role in health and disease. In this book, we humbly honour the importance of those earlier experimental approaches by offering a glimpse of current protocols and methodologies that allow us to, after more than 60 years, keep scrutinizing this system.

Chapters of this book are divided, more or less arbitrarily, in four main themes as follows: neuroanatomy (Chapters 1, 2, and 3), cell physiology (Chapters 4, 5, and 6), models of addiction and neuropsychiatric diseases (Chapters 7, 8, 9, and 10), and circuit level and behavior (Chapters 11, 12, 13, and 14), which we hope will guide the reader more directly to the methodologies she or he may be more interested in of course, research topics and methodologies intertwine, which is a reflection of the multifaceted influence of the dopamine system in the brain, as well as the diversity of technical approaches that we now have to address current problems in neuroscience. Thus, for instance, ultrastructural analysis of striatal dopaminergic synapses by Doig and Larsson and changes in multi-regional activity by Garcia-Nuñez et al. are both studied in the context of Parkinson's disease models. The methodological approach presented by Reynolds et al. is carried out in the context of dopaminergic system development and its relation to neuropsychiatric diseases, which in turn links to the developmental schizophrenia model of Zhu et al. or to the development of the dopaminergic system in children, adolescent and adult humans as studied by Parr et al. Similarly, the protocol by Awadallah et al. on dopamine transporter function is relevant to the development of addiction, a topic that is also studied with various methodological approaches by Pautassi et al., Scorza et al., and Perez-Valenzuela and Laviolette. Finally, single-cell-level firing pattern physiology from an *in silico* model perspective by Knowlton and Canavier can be linked to methodologies for *in vivo* electrophysiological characterization of dopaminergic neurons single-cell firing pattern, including waveform analysis, by

Aguilar-Rivera et al., or methods to study the structural determinants of single-cell activity by Gatica et al., which intimately relates to striatal dopamine release, as studied by Clark and España, and overall dopaminergic neurons' role in behavior by Naneix and Coutureau.

We would like to end this preface by thanking all the contributing authors who shared their work with us, especially acknowledging their enthusiasm and patience during this long but ultimately rewarding endeavour.

Santiago, Chile *José Antonio Fuentealba-Evans*
Pablo Henny

References

1. Bissonette GB, Roesch MR (2016) Development and function of the midbrain dopamine system: what we know and what we need to. Genes Brain Behav 15:62–73
2. Mesman S, Smidt MP (2020) Acquisition of the midbrain dopaminergic neuronal identity. Int J Mol Sci 21:1–20. https://doi.org/10.3390/ijms21134638
3. Puelles E, Martínez-de-la-Torre M, Watson C, Puelles L (2012) Midbrain. In: The mouse nervous system. Elsevier Inc., pp 337–359
4. Björklund A, Dunnett SB (2007) Fifty years of dopamine research. Trends Neurosci 30:185–187
5. Björklund A, Dunnett SB (2007) Dopamine neuron systems in the brain: an update. Trends Neurosci 30:194–202. https://doi.org/10.1016/j.tins.2007.03.006
6. versen SD, Iversen LL (2007) Dopamine: 50 years in perspective. Trends Neurosci 30:188–193
7. Reiner A, Medina L, Veenman CL (1998) Structural and functional evolution of the basal ganglia in vertebrates. Brain Res Rev 28:235–285. https://doi.org/10.1016/S0165-0173(98)00016-2
8. Wise RA (2004) Dopamine, learning and motivation. Nat Rev Neurosci 5:483–494
9. Schultz W (2007) Behavioral dopamine signals. Trends Neurosci 30:203–210
10. Fahn S (2008) The history of dopamine and levodopa in the treatment of Parkinson's disease. Mov Disord 23:S497–S508
11. Hornykiewicz O (2002) Dopamine miracle: from brain homogenate to dopamine replacement. Mov Disord 17:501–508. https://doi.org/10.1002/mds.10115
12. Abbott A (2007) Neuroscience: the molecular wake-up call. Nature 447:368–370. https://doi.org/10.1038/447368a
13. DeLong MR (1990) Primate models of movement disorders of basal ganglia origin. Trends Neurosci 13:281–285. https://doi.org/10.1016/0166-2236(90)90110-V
14. Schultz W, Dayan P, Montague PR (1997) A neural substrate of prediction and reward. Science (1979) 275:1593–1599. https://doi.org/10.1126/science.275.5306.1593
15. Grace AA, Bunney BS (1984) The control of firing pattern in nigral dopamine neurons: burst firing. J Neurosci 4:2877–2890. https://doi.org/10.1523/jneurosci.04-11-02877.1984
16. Fields HL, Margolis EB (2015) Understanding opioid reward. Trends Neurosci 38:217–225
17. Subramaniyan M, Dani JA (2015) Dopaminergic and cholinergic learning mechanisms in nicotine addiction. Ann N Y Acad Sci 1349:46–63. https://doi.org/10.1111/nyas.12871
18. Grace AA (2016) Dysregulation of the dopamine system in the pathophysiology of schizophrenia and depression. Nat Rev Neurosci 17:524–532

Contents

Contributors

MARCELO AGUILAR-RIVERA • *Department of Bioengineering, Stanford University, Stanford, CA, USA*

CAROLINA ASTUDILLO-VALENZUELA • *Department of Health Sciences, Faculty of Medicine, Pontificia Universidad Católica de Chile, Santiago, Chile*

NORA AWADALLAH • *Department of Molecular Pharmacology and Neuroscience, Loyola University Chicago School of Medicine, Chicago, IL, USA*

VICTOR AZÓCAR • *Laboratorio de Neuroquimica, Escuela de Química y Farmacia y Centro Interdisciplinario en Neurociencia, Faculta de Química y de Farmacia, Pontificia Universidad Católica de Chile, Santiago, Chile*

DARINKA BUC • *Laboratorio de Neuroquimica, Escuela de Química y Farmacia y Centro Interdisciplinario en Neurociencia, Faculta de Química y de Farmacia, Pontificia Universidad Católica de Chile, Santiago, Chile*

FINNEGAN CALABRO • *Department of Psychiatry and Department of Bioengineering, University of Pittsburgh, Pittsburgh, PA, USA*

CARMEN C. CANAVIER • *Department of Cell Biology and Anatomy, School of Medicine, Louisiana State Health Sciences Center-New Orleans, New Orleans, LA, USA*

PHILIP J. CLARK • *Department of Neurobiology and Anatomy, Drexel University College of Medicine, Philadelphia, PA, USA*

ETIENNE COUTUREAU • *University of Bordeaux, CNRS, INCIA, Bordeaux, France; Institut de Neurosciences Cognitives et Intégratives d'Aquitaine (INCIA), UMR 5287 CNRS/ Université de Bordeaux, Université de Bordeaux-Site Carreire, Bordeaux, France*

NATALIE DOIG • *MRC Brain Network Dynamics Unit, Division of NDCN, University of Oxford, Oxford, UK*

RODRIGO A. ESPAÑA • *Department of Neurobiology and Anatomy, Drexel University College of Medicine, Philadelphia, PA, USA*

MARÍA CAROLINA FABIO • *Facultad de Psicología, Universidad Nacional de Córdoba (UNC), Córdoba, Argentina; Instituto de Investigación Médica M. y M. Ferreyra, INIMEC-CONICET-UNC, Córdoba, Argentina*

SARA FABIUS • *Departamento de Neurofarmacología Experimental, Instituto de Investigaciones Biológicas Clemente Estable, Montevideo, Uruguay*

MACARENA S. FERNÁNDEZ • *Instituto de Investigaciones Psicológicas, IIPsi-CONICET-UNC, Córdoba, Argentina*

CECILIA FLORES • *Department of Psychiatry and Department of Neurology and Neurosurgery, McGill University, Douglas Mental Health University Institute, Montreal, QC, Canada*

JOSÉ ANTONIO FUENTEALBA-EVANS • *Escuela de Química y Farmacia y Centro Interdisciplinario en Neurociencia, Facultad de Química y de Farmacia, Pontificia Universidad Católica de Chile, Santiago, Chile*

ROMULO FUENTES-FLORES • *Departamento de Neurociencia, Facultad de Medicina de la Universidad de Chile, Santiago, Chile*

MARTÍN GALVALISI • *Departamento de Neurofarmacología Experimental, Instituto de Investigaciones Biológicas Clemente Estable, Montevideo, Uruguay*

Ximena Paz Garcia-Nuñez • *Laboratorio de Control Motor & Neuromodulación, Facultad de Medicina de la Universidad de Chile, Santiago, Chile*

Rafael Ignacio Gatica • *Laboratorio de Neuroanatomía, Departamento de Anatomía Normal, Escuela de Medicina, Pontificia Universidad Católica de Chile, Santiago, Chile*

A. A. Grace • *Departments of Neuroscience, Psychiatry and Psychology, University of Pittsburgh, Pittsburgh, PA, USA*

Pablo Henny • *Departamento de Anatomía Normal, Escuela de Medicina, Pontificia Universidad Católica de Chile, Santiago, Chile*

Christopher Knowlton • *Department of Cell Biology and Anatomy, School of Medicine, Louisiana State Health Sciences Center-New Orleans, New Orleans, LA, USA*

Dimitri N. Koutzoumis • *Department of Pharmacology and Therapeutics, University of Florida College of Medicine, Gainesville, FL, USA*

Bart Larsen • *Department of Psychiatry, Perelman School of Medicine, University of Pennsylvania, Philadelphia, PA, USA*

Max Larsson • *Department of Biomedical and Clinical Sciences, Division of Neurobiology, Linköping University, Linköping, Sweden*

Steven R. Laviolette • *Addiction Research Group, University of Western Ontario, London, ON, Canada; Department of Anatomy & Cell Biology, University of Western Ontario, London, ON, Canada; Lawson Health Research Institute, University of Western Ontario, London, ON, Canada; Department of Psychiatry, Schulich School of Medicine & Dentistry, University of Western Ontario, London, ON, Canada*

Guillermo Lazcano • *Laboratorio de Neuroanatomía, Departamento de Anatomía Normal, Escuela de Medicina, Pontificia Universidad Católica de Chile, Santiago, Chile*

Beatriz Luna • *Department of Psychiatry, University of Pittsburgh, Pittsburgh, PA, USA*

Del MacGowan • *Integrated Program in Neuroscience, McGill University, Montreal, QC, Canada; Department of Psychiatry and Department of Neurology and Neurosurgery, McGill University, Douglas Mental Health University Institute, Montreal, QC, Canada*

Colleen Manitt • *Department of Psychiatry and Department of Neurology and Neurosurgery, McGill University, Douglas Mental Health University Institute, Montreal, QC, Canada*

Fabien Naneix • *The Rowett Institute, University of Aberdeen, Aberdeen, UK*

Dominique Nouel • *Department of Psychiatry and Department of Neurology and Neurosurgery, McGill University, Douglas Mental Health University Institute, Montreal, QC, Canada*

Andrea Harée Pantoja-Urbán • *Integrated Program in Neuroscience, McGill University, Montreal, QC, Canada; Department of Psychiatry and Department of Neurology and Neurosurgery, McGill University, Douglas Mental Health University Institute, Montreal, QC, Canada*

Ashley Parr • *Department of Psychiatry, University of Pittsburgh, Pittsburgh, PA, USA*

Ricardo M. Pautassi • *Facultad de Psicología, Universidad Nacional de Córdoba (UNC), Córdoba, Argentina; Instituto de Investigación Médica M. y M. Ferreyra, INIMEC-CONICET-UNC, Córdoba, Argentina*

Enzo Perez-Valenzuela • *Addiction Research Group, University of Western Ontario, London, ON, Canada; Department of Anatomy & Cell Biology, University of Western Ontario, London, ON, Canada; Lawson Health Research Institute, University of Western Ontario, London, ON, Canada*

Jose A. Pino • *Department of Medicine, University of Atacama, Copiapó, Chile*

José Pedro Prieto • *Departamento de Neurofarmacología Experimental, Instituto de Investigaciones Biológicas Clemente Estable, Montevideo, Uruguay*
Lauren M. Reynolds • *Plasticité du Cerveau CNRS UMR8249, École Supérieure de Physique et de Chimie Industrielles de la Ville de Paris (ESPCI Paris), Paris, France; Neuroscience Paris Seine CNRS UMR 8246 INSERM U1130, Institut de Biologie Paris Seine, Sorbonne Université, Paris, France*
Cecilia Scorza • *Departamento de Neurofarmacología Experimental, Instituto de Investigaciones Biológicas Clemente Estable, Montevideo, Uruguay*
Brenden Tervo-Clemmens • *Department of Psychiatry, Massachusetts General Hospital, Harvard Medical School, Boston, MA, USA*
Gonzalo E. Torres • *Department of Molecular Pharmacology and Neuroscience, Loyola University Chicago School of Medicine, Chicago, IL, USA*
D. L. Uliana • *Departments of Neuroscience, Psychiatry and Psychology, University of Pittsburgh, Pittsburgh, PA, USA*
Aranza Wille-Bille • *Instituto de Investigación Médica M. y M. Ferreyra, INIMEC-CONICET-UNC, Córdoba, Argentina*
X. Zhu • *Departments of Neuroscience, Psychiatry and Psychology, University of Pittsburgh, Pittsburgh, PA, USA; Department of Psychiatry and Behavioral Sciences, University of California San Francisco, San Francisco, CA, USA*

Part I

Neuroanatomy

Chapter 1

Ultrastructural Study of Dopaminergic Axon Terminals

Natalie Doig and Max Larsson

Abstract

The way in which dopamine is released in the striatum is critical to its function, and a wide range of disorders are characterized by altered dopamine signaling, from Parkinson's disease to an array of neuropsychiatric disorders. To fully comprehend the pathogenic pathways of dopamine disorders, it is essential that we investigate the subcellular modifications which underpin them. Research into the genetic variants of Parkinson's disease has identified common pathways mediated by the mutated genes and their proteins, many of which have a role in vesicular function. A growing body of evidence suggests that neurodegeneration in Parkinson's disease starts at the axon terminal and propagates backwards, ultimately resulting in the death of neurons in the substantia nigra pars compacta. Indeed, some of the early symptoms of Parkinson's disease, preceding cell death, can be attributed to impaired synaptic transmission caused by alterations in proteins expressed at the axon terminal. In order to examine this, electron microscopy can be used, as it allows us to perform detailed analyses of the ultrastructure of dopaminergic terminals. The aim of this chapter is to provide a comprehensive protocol for the study of the ultrastructural features of dopaminergic axon terminals in the striatum, including terminal area, the length of the active zone, the number of vesicles, and vesicle distribution. This protocol provides detailed methods covering the preparation of the tissue, immunohistochemistry using gold labeling, preparation of sections for electron microscopy, examination of ultrathin sections using a transmission electron microscope, acquisition of digital electron microscope images, and detailed analyses of ultrastructure using freely available open-source software.

Key words Dopamine, Transmission electron microscopy (TEM), Electron microscopy (EM), Substantia nigra pars compacta (SNc), Ventral tegmental area (VTA), Vesicle, Synapse, Ultrastructure, Immuno-EM, Immunogold, Quantification, ImageJ

1 Introduction

A relatively small group of neurons in the midbrain gives rise to complex and massive axons that innervate the forebrain and utilize the neurotransmitter dopamine. The importance of dopamine neurons in normal brain function is exemplified by the devastating symptoms of Parkinson's disease (PD), and their role in an extensive array of brain processes means that small changes in the precise temporal or spatial coding of dopamine release can result in severe functional impairments [1–4]. In addition to PD, altered dopamine

José Antonio Fuentealba-Evans and Pablo Henny (eds.), *Dopaminergic System Function and Dysfunction: Experimental Approaches*, Neuromethods, vol. 193, https://doi.org/10.1007/978-1-0716-2799-0_1,

signaling is linked to depression, schizophrenia, Tourette's syndrome, ADHD, and addiction, to name but a few [1, 2, 5–7].

PD and other disorders characterized by altered dopamine homeostasis are multifaceted in their onset, presentation, and progression; in order to understand and treat these disorders, we need to examine the underlying ultrastructural changes that occur. Although some dopaminergic disorders are mediated by postsynaptic changes, others are characterized by the altered release of dopamine from the presynaptic axon terminal. In PD there is evidence that some of the known disease-causing mutations, such as alpha-synuclein, can alter synaptic vesicle functioning and thus dopamine release [8–12].

To investigate structural alterations at the level of the axon terminal, it is necessary to use electron microscopy (EM). Here we describe, in detail, a protocol that allows the experimenter to assess potential changes in a variety of ultrastructural features of dopaminergic axon terminals in the striatum including their size, incidence of synapses, length of the active zone/postsynaptic specialization, and the number and distribution of synaptic vesicles. This protocol is ideally suited to examining potential alterations in ultrastructure that may underlie functional changes/deficits following genetic manipulations or in genetic models of disease [13–16] but can also be used to examine changes as a result of other manipulations. Furthermore, the protocol could be applied to the study of axon terminals containing any other neurotransmitter.

The protocol involves five main stages: (i) preparation of the tissue, (ii) pre-embedding immunohistochemistry for tyrosine hydroxylase (TH) as a marker of dopaminergic terminals, (iii) preparation of the sections for EM, (iv) examining and imaging the ultrathin sections using the transmission electron microscope (TEM), and (v) analysis of the digital electron microscope images.

Preparing the tissue involves three main techniques; perfusion-fixation of the experimental animal, which is a key prerequisite for all the following steps, sectioning the fixed brain using a vibrating microtome, and freeze-thawing of the sections to increase the penetration of the immunochemicals. Pre-embedding immunohistochemistry is then performed on the sections, including the application of a primary antibody against TH, followed by a gold-conjugated secondary antibody for localization of dopaminergic terminals in the striatum. The colloidal gold conjugated to the secondary antibody is enhanced by silver intensification in order to make the axon terminals easier to localize at lower magnifications. Here we use a gold-conjugated secondary antibody to reveal dopaminergic terminals rather than a peroxidase reaction as it allows for better visualization of synaptic vesicles.

Once immunolabeled sections have been chosen for EM, several steps must be carried out in order to prepare them for examination in the electron microscope; the main aims of these

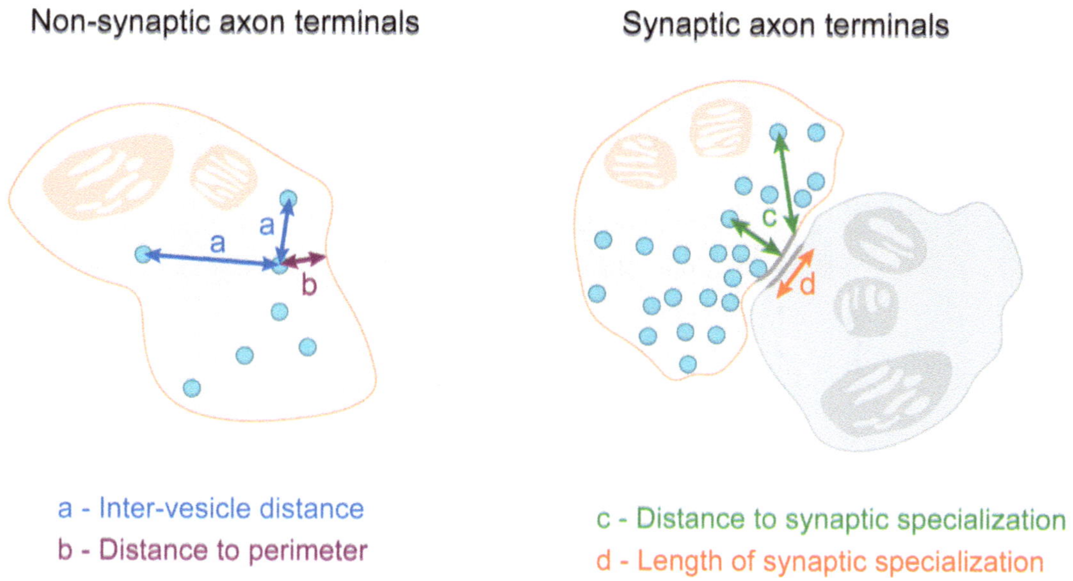

Fig. 1 Schematic showing vesicle distribution parameters that can be analyzed. *Left*: Vesicle distribution parameters in non-synaptic axon terminals; inter-vesicle distance (a), shown for one vesicle and its distance to two other vesicles for clarity, and the vesicle distance to perimeter (b) measured as the shortest distance to the plasma membrane. *Right*: Vesicle distribution parameters in synapse-forming axon terminals; the vesicle distance to synaptic specialization (c) measured as the shortest distance to the delineated active zone (gray thickening of membrane on presynaptic side of synaptic cleft) and the length of the synaptic specialization (d), shown on the postsynaptic side of the synaptic cleft for clarity

techniques are to maximize the preservation of the ultrastructure, to increase contrast for EM, and to acquire ultrathin sections that can be examined in the electron microscope. The next stage involves acquiring images in accordance with a sampling strategy, which is discussed.

Digital electron microscope images are then analyzed. Here we describe a method involving the use of freely available open-source software: *ImageJ* [17, 18] and custom-designed plugins for electron microscopic image analysis, *PointDensity* and *PointDensitySyn* [13–16, 19]. This software allows the researcher to analyze electron micrographs and obtain data on multiple variables including axon terminal size, incidence of synapses, number of vesicles, density of vesicles, distance of vesicles to the cell membrane, distance of vesicles to the active zone (if present), and the inter-vesicle distance as an indicator of clustering (Fig. 1).

2 Materials

Note that many of the chemicals used in the preparation of tissue for EM are toxic. We recommend discussing all protocols and chemicals with your Institute's safety officer before beginning.

2.1 Tissue Preparation

2.1.1 Perfusion-Fixation

- Paraformaldehyde (PFA; ***Safety Note 2.1.1***).
- Glutaraldehyde (***Safety Note 2.1.1***).
  - Normally supplied as a 25% aqueous solution.
  - Should be specially purified for EM.
- 0.5-M NaOH.
- 0.2-M phosphate buffer, pH = 7.4 (PB).
- 1 × phosphate-buffered saline, pH = 7.4 (PBS).
- Anesthetic.
- Peristaltic pump.
- Dissection instruments.
- Plastic or blunt needle to insert into the heart.

Safety Note 2.1.1
Paraformaldehyde and glutaraldehyde are toxic and harmful if swallowed, inhaled, or absorbed through the skin. Wear appropriate personal protective equipment. All manipulations should be carried out in a fume hood. Dispose of waste in accordance with your Institution's safety guidelines.

2.1.2 Vibrating Microtome Sections

- Vibrating microtome (e.g., Leica VT1000 S).
- Cyanoacrylate adhesive (superglue).
- PBS.
- Labeled glass vials.
- Paint brushes.
- Brain matrix (stainless steel, e.g., from Ted Pella Inc.).
- Agar (4% in dH_2O).
- 50-mL centrifuge tube(s).
- Razor blades (wipe clean with 70% ethanol prior to use).
- Filter paper.

2.1.3 Freeze-Thaw

- Cryoprotectant (25% sucrose w/v and 10% glycerol v/v in 0.05-M PB).
  - Sucrose.
  - Glycerol.
  - 0.2 M PB, pH = 7.4.
- Liquid nitrogen.
- Extra strong aluminum foil.

2.2 Immunohistochemistry

2.2.1 Primary and Secondary Antibody Incubation

- Normal goat serum (NGS; Vector Laboratories Cat# S-1000, RRID:AB_2336615).
- Rabbit anti-tyrosine hydroxylase antibody; Millipore Cat# AB152, RRID:AB_390204.
- Nanogold IgG goat anti-rabbit IgG (H + L) Antibody; Nanoprobes Cat# 2003, RRID:AB_2687591.

2.2.2 Silver Intensification of Gold Conjugated to Secondary Antibody

- HQ silver enhancement kit from Nanoprobes (#2012-45ML; use at RT; reagents are light sensitive, store at −20 °C, make sure to thaw completely before use).
- Sodium acetate buffer (0.1 M, pH 7.0–7.5). To make 250 mL: 3.4-g sodium acetate trihydrate dissolved in 250-mL dH_2O, adjust pH to 7.0–7.5 with dilute acetic acid.
- Glass embryo dishes (or similar) with foil to cover.
- Glass Pasteur pipettes.
- Microcentrifuge tubes.

2.3 Preparation of Sections for EM

2.3.1 Post-Fixation with Osmium Tetroxide

- 1% solution of osmium tetroxide in 0.1-M PB pH 7.4 (***Safety Note 2.3.1***).
- Glass petri dishes with lids.
- Glass Pasteur pipettes.

Safety Note 2.3.1

Osmium tetroxide is a very powerful oxidizing agent and is highly toxic. Osmium tetroxide is volatile and can cause irreversible damage to the lungs and eyes at low concentrations. All manipulations should be carried out in a fume hood and wearing appropriate personal protective equipment. Osmium tetroxide should be stored at low temperature (2–8 °C). It can penetrate plastics and packaging; therefore, stock solutions should be stored in accordance with your Institute's safety guidelines; we routinely use double ground glass stoppered bottles. Inactivate and dispose of waste according to your Institution's safety guidelines.

2.3.2 Dehydration and Embedding in Resin

- Graded series of v/v dilutions of ethanol in dH_2O (50%, 70%, 90%, 100%).
- Dry absolute ethanol (stored over anhydrous copper sulfate).
- Propylene oxide (***Safety Note 2.3.2a***).
- 1% solution of uranyl acetate in 70% v/v ethanol (***Safety Note 2.3.2b***). This solution should be covered or stored in the dark (or in an opaque container) and must be filtered before use.
- EM epoxy resin (Durcupan ACM; ***Safety Note 2.3.2c***).

Safety Note 2.3.2a

Propylene oxide is a volatile and extremely flammable liquid. It is highly toxic and a potential mutagen and carcinogen. It is toxic via ingestion, inhalation, or absorption through the skin. All manipulations must be carried out in a fume hood with appropriate personal protective equipment. Please note that propylene oxide requires specific gloves for handling and should not be handled with standard nitrile gloves. Store, handle, and dispose of propylene oxide in line with your Institution's safety guidelines.

Safety Note 2.3.2b

Uranyl acetate is toxic as it is both mildly radioactive and a heavy metal. Uranyl acetate is very toxic if ingested, inhaled as dust, or by skin contact if the skin is cut or abraded. Wear appropriate personal protective equipment and only handle in a fume hood. It must be handled, stored, and disposed of in line with your Institution's safety and radiation guidelines.

Safety Note 2.3.2c

Durcupan ACM epoxy resin contains toxic components. Wear appropriate personal protective equipment and only handle in a fume hood. Polymerize all excess resin and all contaminated materials at 60 °C for 48 h before disposal.

2.3.3 Re-embedding for EM

- Dissection microscope.
- Double- and single-edged razor blades.
- Scalpel.
- Pre-made resin blocks (these are made with the same Durcupan resin as the tissue is embedded in; the resin is cured in a suitable mold/capsule to create a cylindrical block with a flat top).
- Cyanoacrylate adhesive (superglue).
- Cocktail sticks.
- Hot plate.

2.3.4 Re-sectioning for EM

- Ultramicrotome (e.g., Leica EM UC6).
- Pioloform-coated single-slot copper grids.
- Glass knives.
- Diamond knife (Diatome).

2.3.5 Staining of Ultrathin Sections with Reynold's Lead Citrate

- Lead nitrate (***Safety Note 2.3.5***)
- Trisodium citrate.
- 4-M NaOH.
- Petri dish.
- Parafilm.

- Filter paper.
- NaOH pellets.
- dH_2O (use double distilled if you can) [always use fresh water; do not allow it to stand for long periods as CO_2 will dissolve into it. If you have issues with lead precipitate, you can boil your distilled water to remove more air from it].
- Hair dryer.
- Pasteur pipettes.

Safety Note 2.3.5
Lead salts are extremely toxic so exercise care and wear suitable personal protective equipment. Store, handle, and dispose of lead salts in accordance with your Institution's safety guidelines.

2.4 Examination of Sections in the Electron Microscope

2.4.1 Image Acquisition

- TEM fitted with digital camera.
  - In our case we use a Hitachi HT7800 TEM with a XAROSA CMOS camera (EMSIS GmbH).

2.5 Analysis of Electron Micrographs

2.5.1 Software

- Fiji distribution of *ImageJ* [18]. https://imagej.net/software/fiji/.
- *ImageJ* plugins for image analysis *PointDensity* and *PointDensitySyn* [19] https://github.com/maxdl.

3 Methods

3.1 Tissue Preparation

With the increased use of genetically modified animals, it is essential to make sure that appropriate controls are used; this must be established when setting up the breeding. Furthermore, wherever possible the experimenter(s) should be blinded to genotype and/or other experimental factor(s) throughout the whole process. Animal use should be carried out in accordance with the institutional and national guidelines on the use of animals in research.

3.1.1 Perfusion-Fixation

A successful perfusion is not only key to ensuring good ultrastructure at the electron microscopic level but is a requirement for successful sectioning, staining, and embedding. The following protocol is written for an adult mouse but can be adapted for a rat [20]. Here we briefly outline perfusion-fixation; *see* refs. 21, 22 for more details. The main adaptation for EM, compared to light microscopy (LM), is that glutaraldehyde should be added to the paraformaldehyde (PFA) fixative solution for optimal morphology (*see* **Note 1**).

Protocol for Preparation of PFA-Glutaraldehyde Fixative Solution

This protocol makes 1 L of 4% w/v PFA in 0.1-M phosphate buffer, pH 7.4 with 0.1% glutaraldehyde (*see* **Note 1**). All procedures should be carried out in a fume hood. Note that you can purchase methanol-free ready-made 4% PFA.

(i) Heat approximately 400 mL of dH_2O to 65–70 °C on a hot-plate stirrer in a fume hood.

(ii) Weigh and add 40-g PFA and stir continuously.

(iii) Add 0.5-M NaOH dropwise until the solution clears; the solution will clear slowly at about 60 °C. There is a lag, so pause after adding NaOH.

(iv) Filter the solution.

(v) Add 500-mL 0.2-M PB. Make up to 1 L with dH_2O.

(vi) Just before starting the perfusion procedure, add the glutaraldehyde to make up to a final concentration of 0.1%. For 1 L of fixative solution, to achieve a final concentration of 0.1% glutaraldehyde (starting with a solution of 25% glutaraldehyde), you would need to add 4.0 mL.

Protocol for Perfusion-Fixation of an Adult Mouse

(i) Deeply anaesthetize the animal. We induce anesthesia using isoflurane (4% v/v in O_2) and then give an overdose of injectable anesthetic, e.g., pentobarbitone (1.5 g/kg, i.p.).

(ii) Open up the thorax and put a cannula into the left ventricle and gently push into the left atrium toward the ascending aorta.

(iii) Using a peristaltic pump, exsanguinate the mouse by transcardial perfusion with 20–50 mL of PBS. Once all blood has been removed, switch to the PFA-glutaraldehyde fixative solution. It is beneficial to limit the buffer rinse time to ~100 s or less to avoid significant changes to synaptic structure [23]. A good indicator of when to switch to fixative is when the liver becomes clear of blood and is a pale cream color.

(iv) Run through the fixative solution for a fixed time or volume; an average would be about 100 mL over 20 min, but this should be determined experimentally. Rapid hardening following muscle twitches is often an indicator of a successful perfusion.

(v) Remove brain carefully to avoid damage to your area of interest. The brain should be firm to the touch and a creamy white color. Any pinkness is generally an indicator of a poor perfusion.

(vi) Immerse brain(s) in the fixative solution for an additional 12 h at 4 °C.

3.1.2 Vibrating Microtome Sections

The most common method for preparing sections for EM is a vibrating microtome; the reason being that frozen sections made with a cryostat or a freezing microtome have poor ultrastructure [20]. The aim of this protocol is to obtain coronal sections of the mouse brain from anterior to the striatum to posterior of the midbrain, in order to allow the experimenter to examine axon in all parts of the striatum and preserve sections containing the midbrain dopaminergic cell bodies for any other investigations. For EM, it is beneficial to process the tissue as soon as possible.

Protocol for Sectioning a Mouse Brain with a Vibrating Microtome

(i) In the fume hood, rinse the brain thoroughly in PBS to get rid of any excess fixative solution. Trim/remove any meninges.

(ii) Block the most anterior part of the brain, i.e., the olfactory bulbs, making sure to leave the entire striatum intact. Trim the very caudal part of the cerebellum to create an even, flat surface. Use a brain matrix to help keep cuts straight and perpendicular to the direction of cutting. It is a common error to trim unevenly which results in the hemispheres within one section being at different planes (in relation to bregma); this is especially problematic if one hemisphere is the control for the other hemisphere.

(iii) When using the vibratome to section the majority of the mouse brain, we recommend embedding in agar to improve support of the tissue.

(iv) To embed in agar, make a 4% (w/v) solution of agar in dH_2O, heat the solution in a glass beaker until it boils, and then allow to cool until it is safe to the touch but not yet thickened.

(v) While the agar is cooling, dry the brain with filter paper and place it flat, cerebellum side down within a container (we use a trimmed 50-mL centrifuge tube on a plastic surface secured by Blu Tack®).

(vi) Pour the cooled agar over the brain ensuring that the brain remains in the center and does not float. Once the agar has set, remove it from the container and trim a safe distance around the brain to create a block for sectioning.

(vii) Dry the bottom of the agar-embedded brain with filter paper and glue it to the vibratome stage with superglue.

(viii) Set up the vibratome and fill the water bath with PBS.

(ix) Cut sections; a thickness of 50–70 μm is recommended.

(x) Start collecting sections at about the level where the forceps minor of the corpus callosum appears, approximately 2 mm anterior to bregma [24].

(xi) Using a paint brush, carefully collect sections into a one in six series into glass vials containing PBS. Keep collecting sections until you have cut past the midbrain.

(xii) Wash sections 3 × 10 min in PBS.

3.1.3 Freeze-Thaw

Freeze-thawing is a technique used to increase the penetration of immunochemicals by mechanically disrupting the tissue. It is preferred to the use of detergents for EM as it is less detrimental to the ultrastructure. The aim is to produce very small ice crystals that cause restricted damage to the tissue [20]. There are multiple methods to freeze-thaw tissue; here we describe one that we routinely use.

Protocol for Freeze-Thawing of Mouse Brain Sections

(i) Select the sections to be processed for EM and place these in a separate container.

(ii) To make 100-mL cryoprotectant: add 25-g sucrose, 10-mL glycerol, and 25-mL 0.2-M PB into a glass bottle and make up to 100 mL with dH_2O.

(iii) Place chosen sections into cryoprotectant for a minimum of 4 h. Gently shake until the sections have sunk. You can leave them overnight on a shaker at 4 °C.

(iv) Prepare liquid nitrogen in a Dewar flask (take safety measures when using liquid nitrogen including the use of appropriate personal protective equipment).

(v) The aim is to freeze the sections very rapidly on the surface of the liquid nitrogen. To enable this, the sections are placed in a holder that can be readily applied to the nitrogen. You can create such a holder out of extra strong aluminum foil and then make one end into an "envelope" to hold the sections and use the other end as the handle. A well plate or other holder that will not break when frozen rapidly can also be used.

(vi) Remove section(s) from cryoprotectant using a paintbrush. Place the section(s) on your foil envelope and remove all excess cryoprotectant using filter paper; no more than three sections should be processed at any one time. Fold the foil over lengthways so that your section(s) are carefully covered and secured.

(vii) Hold the long end of the foil and place the section end onto the top of the liquid nitrogen for 30 s, in the vapor. You can place a drop of PBS on top of the folded foil to act as a guide; the PBS should rapidly freeze during the 30 s period.

(viii) Remove the foil from the liquid nitrogen; the PBS drop should thaw.

(ix) Repeat the freeze and thaw process twice more—to give a total of 90 s of freezing.

(x) Once the PBS drop has fully thawed, begin to open the foil envelope; you can dribble a few drops of PBS into the envelope to help loosen the sections. Open the rest of the foil and collect the sections using a clean brush.

(xi) Pour off the PBS and dry the foil—try not to create any creases or indentations in the central area where the sections go—so it can be used again.

(xii) Repeat for all your sections.

3.2 Immunohistochemistry

In this protocol we use a primary antibody for tyrosine hydroxylase (TH) as a marker of dopamine in the striatum, followed by a gold-conjugated secondary antibody (Figs. 2 and 3). These antibodies have been tried and tested [15] but could be replaced by other suitable antibodies if needed. Silver intensification is then used to increase the size of the colloidal gold particles so that they can be seen at the LM level and more easily located at lower magnifications at the EM level [20].

3.2.1 Primary and Secondary Antibody Incubation

Protocol for Incubating Sections in Primary and Secondary Antibodies

(i) Wash sections thoroughly in PBS.

(ii) Block sections with 10% NGS in PBS for 1 h. All incubations should be carried out with constant gentle shaking.

(iii) Incubate in primary rabbit anti-TH antibody (1:1000) in PBS with 2% NGS overnight at RT. Alternatively, sections can be incubated for 72 h at 4 °C.

(iv) Wash sections 3 × 5 min in PBS.

(v) Incubate in gold-conjugated goat anti-rabbit secondary antibody (1:400) in PBS with 1% NGS for a minimum of 4 h at RT (can be incubated overnight at 4 °C).

3.2.2 Silver Intensification of Gold Conjugated to Secondary Antibody

Silver intensification is the process by which growing shells of silver are deposited over the gold particles that are conjugated to the secondary antibody [20]. We use commercial kits as they are easy to use and reliable. Here we use the HQ silver enhancement kit from Nanoprobes (Nanoprobes #2012-45ML), but there are other commercial kits available. Detailed instruction protocols come with the kits; here we describe our routinely used method (Figs. 2 and 3).

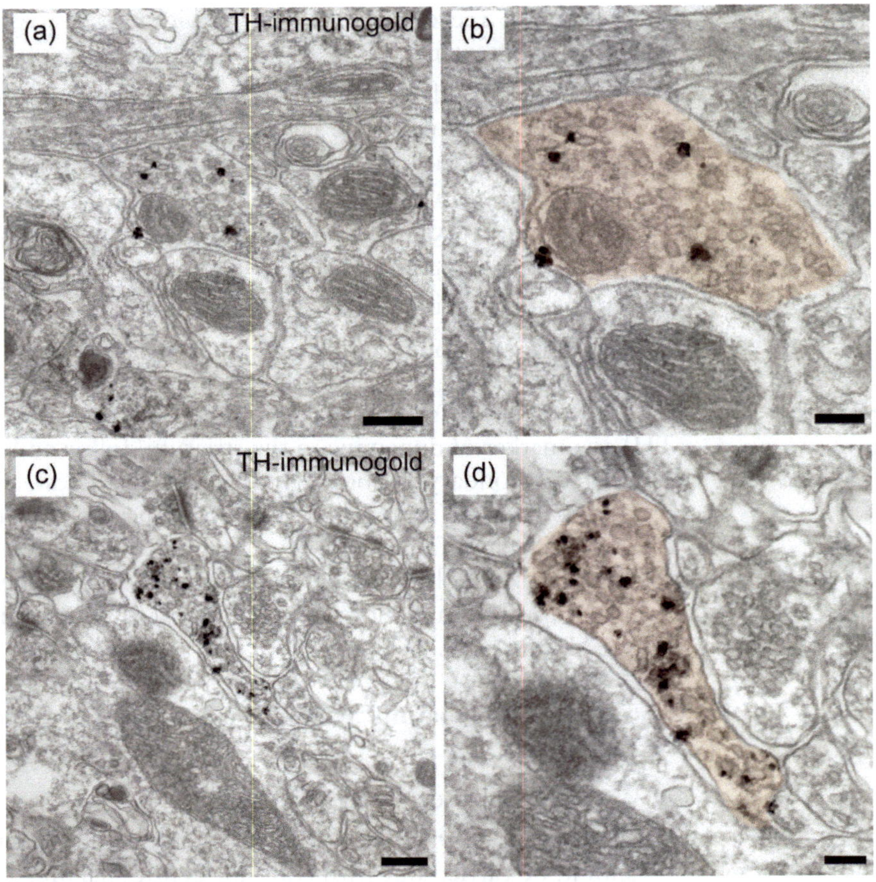

Fig. 2 Example electron micrographs of dopaminergic axon profiles in the striatum. (**a**) Axon terminal showing enriched immunogold labeling for tyrosine hydroxylase (TH) as a marker of dopaminergic structures. Note the presence of multiple vesicles in the axon terminal. (**b**) Enlarged view of (**a**) pseudocolored for illustration. (**c**) Axon terminal showing TH immunogold labeling. (**d**) Enlarged view of (**c**) pseudocolored for clarity. These axon profiles were acquired and analyzed as part of a previous study [15]. Scale bars: (**a**, **c**) 200 nm; (**b**, **d**) 100 nm

Protocol for Silver Intensification of Colloidal Gold

(i) Wash sections 2 × 10 min in PBS.

(ii) Wash in dH_2O (2 × 5 min). Sections must be thoroughly rinsed with deionized water before silver enhancement reagents are applied. This is because the buffers used for antibody incubations and washes contain chloride ions and other anions which form insoluble precipitates with silver [20].

(iii) Wash in sodium acetate buffer (2 × 5 min). This is optional; sodium acetate can speed up the reaction and reduce background.

(iv) Place a section into glass embryo dish (or similar), remove excess sodium acetate, and make sure section is flat.

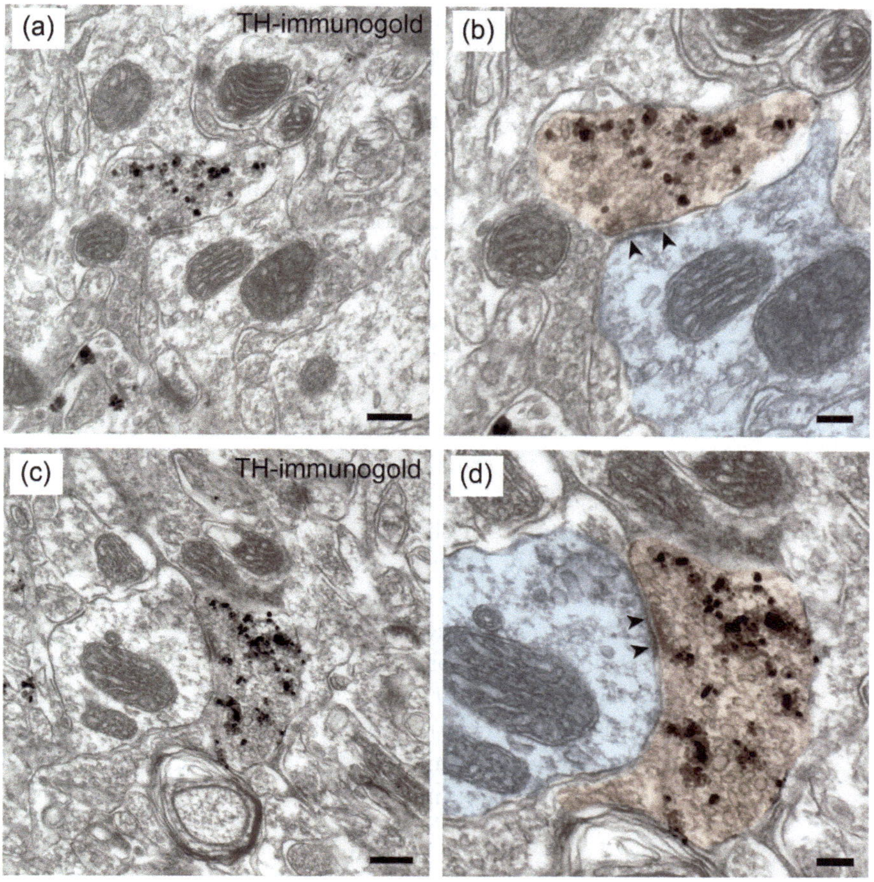

Fig. 3 Example electron micrographs of dopaminergic axon profiles in the striatum forming synapses. (**a**) Synapse-forming axon terminal showing enriched immunogold labeling for TH. Note the pre- and postsynaptic specializations. (**b**) Enlarged view of (**a**) pseudocolored for illustration; dopaminergic axon terminal (red) forming a synapse (arrowheads) with a dendrite (blue) containing two mitochondria. Note the presence of vesicles in the axon terminal and the darkening within the synaptic cleft. (**c**) Axon terminal showing TH immunogold labeling forming a synapse with a dendrite. (**d**) Enlarged view of (**c**) pseudocolored for clarity; a dopaminergic axon terminal (red) forms a synapse (arrowheads) with a dendrite (blue). These axon profiles were acquired and analyzed as part of a previous study [15]. Scale bars: (**a**, **c**) 200 nm; (**b**, **d**) 100 nm

(v) Mix solutions A and B from HQ silver enhancement kit in equal volumes (i.e., five drops of each) in a microcentrifuge tube. Add an equal volume of C into tube and mix.

(vi) Remove any remaining buffer from the dish and add the mixed silver enhancement solution to start the intensification process.

(vii) React in the dark for 2–10 min. Stop reaction with acetate buffer.

(viii) Wash with dH_2O (2×5 min). Wash with PBS (1×10 min).

(ix) More or less time controls particle size. You will need to determine the optimal reaction timing dependent on the antibody and tissue; to do this react and wash the first section and check it under a microscope to assess whether reaction time needs to be altered. It can be useful to react a section containing midbrain dopaminergic neurons as you will very clearly see if the reaction has worked if the TH cells become a dark brown color. To assess the extent of the reaction in the striatum, focus on the border between striatum and cortex; the cortex should remain a very pale cream color, and the striatum should look yellow gold in comparison. Note that the gold particles are on the surface of the tissue and do not penetrate into the middle of the section.

(x) Repeat for remaining sections.

3.3 Preparation of Sections for EM

3.3.1 Post-Fixation with Osmium Tetroxide

After immunohistochemical procedures have been carried out, sections that have been chosen for EM are post-fixed in a buffered osmium tetroxide solution. Osmium, a heavy metal, will embed directly into the lipid membranes and myelin and thereby create necessary contrast for EM [20, 21]. During the osmium treatment, the sections will turn dark brown, almost black (depending on length of treatment), and will become very brittle. Any folds in the tissue before osmium treatment will become fixed irreversibly; it is therefore very important to make sure your sections are very flat and even before applying the osmium solution.

Protocol for Post-fixation with Osmium Tetroxide

All manipulations involving osmium tetroxide must be carried out in a fume hood.

(i) Wash sections 2 × 10 min in 0.1-M PB.

(ii) Transfer sections to a small glass petri dish containing 0.1-M PB (make sure that dish has a suitable glass lid).

(iii) Remove the buffer slowly, using a paintbrush, to ensure that the sections remain flat on the bottom of the dish. Pipette off as much of the PB as possible.

(iv) Carefully pipette 1–2 mL of the osmium tetroxide solution onto the sections using a glass Pasteur pipette. Replace lid. Make sure section is well covered but not floating.

(v) Incubation time depends on the thickness of the sections. The osmium reaction time is also reduced when it is done following silver intensification as osmium tetroxide can cause some loss of silver. For 50-μm sections, we use between 7 and 12 min depending on the level of silver-intensified gold labeling.

(vi) Pipette off the osmium solution and place in a waste container (dispose according to your Institution's safety guidelines).

(vii) Wash section 3 × 15 min in 0.1-M PB.

(viii) It is preferable to dehydrate and embed the sections in resin immediately following osmium treatment.

3.3.2 Dehydration and Embedding in Resin

It is necessary to embed the sections in an electron microscopic resin before ultrathin sectioning. The resin provides support for re-sectioning but also allows high-resolution LM prior to sectioning. Most resins are not miscible with water; therefore, the tissue must be fully dehydrated with ethanol before embedding. Once all water has been removed, the tissue sections are equilibrated in a solution that is miscible with both the ethanol and the resin; in our protocol, this is propylene oxide [20]. During the dehydration, the tissue sections are also exposed to a solution of uranyl acetate, another heavy metal used to increase contrast.

Protocol for Dehydration and Embedding in Resin

(i) Remove the PB from the petri dishes and replace with 50% ethanol. Wash 2 × 15 min. Use a large volume to remove phosphate, as phosphate can react with uranyl acetate to form an electron dense precipitate.

(ii) Replace 50% ethanol solution with 1% uranyl acetate dissolved in 70% ethanol. Incubate for 20–45 min in the dark (cover with foil).

(iii) Replace with 95% ethanol for 15 min.

(iv) Replace with 100% ethanol for 10–15 min.

(v) Incubate in dry absolute ethanol 2 × 10 min.

(vi) Carefully replace with propylene oxide (2 × 10 min). Note that propylene oxide is extremely volatile, so ensure sections do not dry out when changing solutions. There are safer alternatives to propylene oxide commercially available.

(vii) During the dehydration, prepare the resin according to the manufacturer's instructions. For Durcupan ACM epoxy resin, we mix the components A:B:C:D in the ratio: 10:10:0.3:0.2, by weight. Mix the resin in a plastic beaker with a wooden spatula. Note that small changes to these ratios can result in drastic changes to the polymerization of the resin.

(viii) Transfer a few milliliter of resin into a shallow container such as a weigh boat. Make sure containers are well labeled (note that ethanol and propylene oxide will wash away permanent marker pen).

(ix) Using a paintbrush, carefully transfer the sections from the propylene oxide into the resin. Gently make sure that the sections are covered in resin and leave overnight in the fume hood.

(x) Gently warm boats on a hot plate in the fume hood, to reduce the viscosity of the resin. Using your resin paintbrush and some forceps, carefully transfer the sections to uncleaned slides. Allow the sections and resin to settle on the slide.

(xi) Carefully manipulate the sections using cocktail sticks or a paintbrush to make sure they are orientated correctly and not overlapping; it is helpful to do this using a dissection stereomicroscope (protected from resin using Parafilm).

(xii) Gently place a coverslip over the sections and take care not to press the coverslip down too hard. Try and remove any bubbles that are on or near the sections using a cocktail stick, but do not press down too hard on top of the sections as they are still brittle and liable to shatter.

(xiii) Getting the amount of resin just right is done by trial and error; enough resin is needed so that there is no air under the coverslip but not so much that it is seeping out the sides of the coverslip. Absorb excess resin with filter paper and add extra resin using the end of a cocktail stick and place small drops along the edge of the coverslip. If it is not necessary to do high-resolution LM prior to EM, you can use acetate foils to cover the sections rather than a glass coverslip; the benefit of this is that the acetate is easier to remove than a coverslip. If using coverslips, it is helpful to carefully rub them on your skin so that they have a thin layer of grease on them; this aids greatly when trying to remove them at a later stage.

(xiv) Place slides in a container lined with foil. To polymerize the resin, place the container with the slides in it into an oven at 60 °C for 48 h. Make sure the container is level. All materials that are contaminated with resin can be placed in the oven to polymerize and then be disposed of safely.

(xv) The slides and sections are now suitable for LM prior to EM and can be stored indefinitely.

3.3.3 Re-embedding for EM

Once you have examined your slides in the light microscope (LM), you can select specific regions of the sections to be further processed for examination in the electron microscope. It is very useful to take some LM images of these areas prior to re-sectioning the tissue for EM. We recommend taking a few low- and high-magnification images to help with the next few stages. In the re-embedding protocol, superglue is used to affix the dissected-out resin-embedded tissue to the top of a pre-made resin block; we use this method as it is quick, but the drawback is that you cannot take high-magnification LM images of the tissue in the block once re-embedded.

Procedure for Re-embedding Tissue for EM

(i) Examine your slides and sections. Identify which section (s) from each slide you will want to examine tissue from. It is useful to take pictures of the whole slides for reference (you can do this using a regular scanner or photocopier). Using a light microscope, identify the region(s) of the striatum to be examined at the EM level; take LM photos for correlation and orientation. The region of interest from each section will then be cut out and attached to a resin cylinder suitable for sectioning for the electron microscope.

(ii) If you are using acetate instead of glass coverslips, you can remove the acetate prior to imaging but take care not to damage the underlying tissue. If you need higher-magnification images, you can apply a drop of immersion oil and temporarily place a glass coverslip over the top of the tissue.

(iii) If using a glass coverslip, remove your coverslip from the slide; to do this, remove all the resin from around the coverslip using a single- or double-edged blade. Then gently insert a double-edged razor blade between the coverslip and the resin and gently push the blade under the coverslip until it becomes free. Wear eye protection.

(iv) Warm the slide on a hot plate; the polymerized resin is brittle at room temperature and will crack or split if not warmed enough.

(v) Make guide cuts using a scalpel or double-edged razor blade and then warm again and make cuts under dissecting microscope to free the piece of tissue in your region of interest. Keep warming the slide between cuts to prevent cracking of the tissue.

(vi) Place a small drop of superglue on top of a resin block and gently remove your section of resin-embedded tissue from the slide using forceps and place onto the superglue, adjusting with a cocktail stick if needed. You can flatten the piece of tissue on a razor blade on the hotplate if needed.

(vii) Label the resin block and allow the glue to set for 24 h before re-sectioning.

3.3.4 Re-sectioning for EM

To examine tissue in the transmission electron microscope (TEM), ultrathin sections need to be in the region of 50–70 nm thickness; this process is carried out using an ultramicrotome (we use Leica EM UC6). It is beyond the scope of this protocol to describe ultramicrotomy in detail (and the method is dependent on the model of ultramicrotome), but we will point the reader to some useful resources and give our most useful tips.

Resources

- Videos are a helpful way to learn and improve skills. Suppliers of ultramicrotomes often have videos of their models [25]. There are also several instructional videos on YouTube [26]. Some protocol papers also have supplemental movies [27].

Useful Tips

- To get a good ribbon of serial sections (e.g., trapezoid-shaped sections joined together), the top and bottom edges must be clean and parallel. You can achieve good parallel edges to your trapezoid block using a glass knife. This can also be achieved with a "trim tool"; *see* ref. 28.
  - Orientate and line up the block with the long edges at the top and bottom (as usual) and cut one to two semi-thin (500–1000 nm) full sections. Rotate the block 90° (either clockwise or counterclockwise), line up the glass knife with the surface of the block, and cut until you get a whole section.
  - Change the knife block angle (and only this angle) so that it is at the maximum (~30°) in one direction (e.g., left).
  - Set your cutting window and start cutting until you have cut a clean edge. Move the knife to the other extreme, i.e., 30° to the right if you started with the left. Repeat the cutting. You can then examine the block under a microscope—you should have a very neat pyramid with clean parallel edges at the top and bottom of your block. This method is also detailed here [29]. The bonus to having straight edges, apart from sections cutting well and forming a nice ribbon, is that the ribbon will not curve.
- A smaller block will cut and form ribbons better. If you are having any issues cutting, then the first troubleshooting step is to make your block smaller [20].
- Keep everything meticulously clean during ultramicrotomy, clean everything with ethanol prior to use, and filter your dH_2O. Carefully clean the diamond knife before cutting and replace the water in the bath often to avoid the buildup of any dirt or dust.

3.3.5 Staining of Ultrathin Sections with Reynold's Lead Citrate

Although sections can be examined in the electron microscope as soon as they have been collected onto copper grids, you can enhance contrast by staining with lead citrate. With microscopes that are equipped with a digital screen camera instead of eye pieces, this step is not essential but may be beneficial, nonetheless. You can purchase "ready-made" lead citrate, but it is relatively easy and economical to make it yourself; if you do purchase commercial lead citrate solutions, make sure that they are "EM grade."

Here we describe preparing and using Reynold's lead citrate which is formed as a product from the reaction between sodium citrate and lead nitrate [30]. Lead citrate will precipitate easily in the presence of CO_2 in the air or in water; therefore, we recommend not staining all your grids at once—stain a third or half of them to be sure that you do not have any issues with the lead stain precipitating. Note that it is very important when you are using immunogold particles to make sure that there is no lead precipitate as it can be confused with silver-intensified gold particles.

To Make the Lead Citrate

(i) Weigh out 0.27 g of lead nitrate into a glass vial. Weigh 0.35-g trisodium citrate in a separate vial or weigh boat.

(ii) Add 9.6-mL dH_2O to the lead nitrate. Put cap on and shake to dissolve.

(iii) Add the trisodium citrate to the lead nitrate. Shake to dissolve. The solution will become milky in appearance.

(iv) Add 0.4 mL of 4-M NaOH. Replace the lid quickly and shake. It is important to prevent contact with the air as much as possible as the lead citrate reacts with atmospheric carbon dioxide. The solution should be clear.

(v) You can aliquot and store the solution in syringes—make sure that all air is expelled from the syringe. The solution can be stored at 4 °C for a few weeks (store in appropriate toxic fridge).

Staining of Ultrathin Sections

(i) Line a large petri dish with Parafilm (clean side up). Carefully place some sodium hydroxide pellets into one section of the petri dish to absorb CO_2.

(ii) Carefully take up some of the lead citrate into a Pasteur pipette, discard the first drop, and then place separate drops onto the Parafilm.

(iii) Place a grid, sections down, onto each drop. Take care not to leave the petri dish lid off for long.

(iv) Leave sections to stain for 1–5 min.

(v) Wash each grid in distilled water; have approximately four vials of filtered distilled water and carefully dunk the grid several times in each before removing excess water with filter paper.

(vi) Completely dry grid with a hairdryer, ensuring that there is no fluid between the tips of the forceps, before replacing in the grid box. It is preferable to carefully dry off the drop of fluid on the grid with filter paper to avoid any contaminants in the water drying onto the sections.

(vii) Dispose of lead waste in accordance with your Institute's safety guidelines.

(viii) Check your grids in the EM to see if the lead staining has worked well and that there is not an issue with lead precipitate.

3.4 Examination of Sections in the Electron Microscope

3.4.1 Sampling

One of the challenges for EM is to design a sampling strategy to ensure your sample is representative, reflecting the true distribution/occurrence of a particular structure and to ensure that it will be possible to detect effects of experimental manipulation. The striatum is so densely packed with DA terminals that you could get 100s of images from a very small volume, but this would not be representative of the striatum as a whole and could easily bias your data, if, for example, all your axons studied were confined to the matrix or to a patch. Here we describe a sampling strategy based on previous work [15]. An example sampling strategy is laid out in Fig. 4. If you are examining DA terminals in the dorsolateral striatum, we would recommend re-embedding tissue from at least two sections in the rostrocaudal plane per animal. From each of these sections, you can re-embed tissue from at least two regions of interest, making two blocks. From each block you would then sample from at least two grids, and then from each grid, take images from at least two ultrathin sections (Fig. 4). This is merely a guideline; the optimum strategy (where possible) would be to carry out a pilot study to acquire some initial data so that you can perform a power analysis and estimate the sample size required.

3.4.2 Image Acquisition

This protocol describes a random systematic sampling strategy to acquire digital images for analysis using a TEM fitted with a digital camera.

Procedure for Image Acquisition

(i) Place grids into the specimen holder and insert into electron microscope.

(ii) Image the whole grid and choose two sections to sample from. Pick sections that are as distant as possible from each other (opposite ends of the ribbon) or examine different regions on the two different sections; this will decrease the likelihood of imaging the same terminals in both sections.

(iii) Determine a magnification at which you can locate positively labeled profiles, but you cannot see individual synapses; this is your "searching magnification." Define a criterion for "immunopositive" structures depending on the level of background labeling, e.g., a minimum of five gold particles/structures.

(iv) Immunogold labeling has limited penetration into the tissue but is also nonspecific at the very surface. In order to avoid

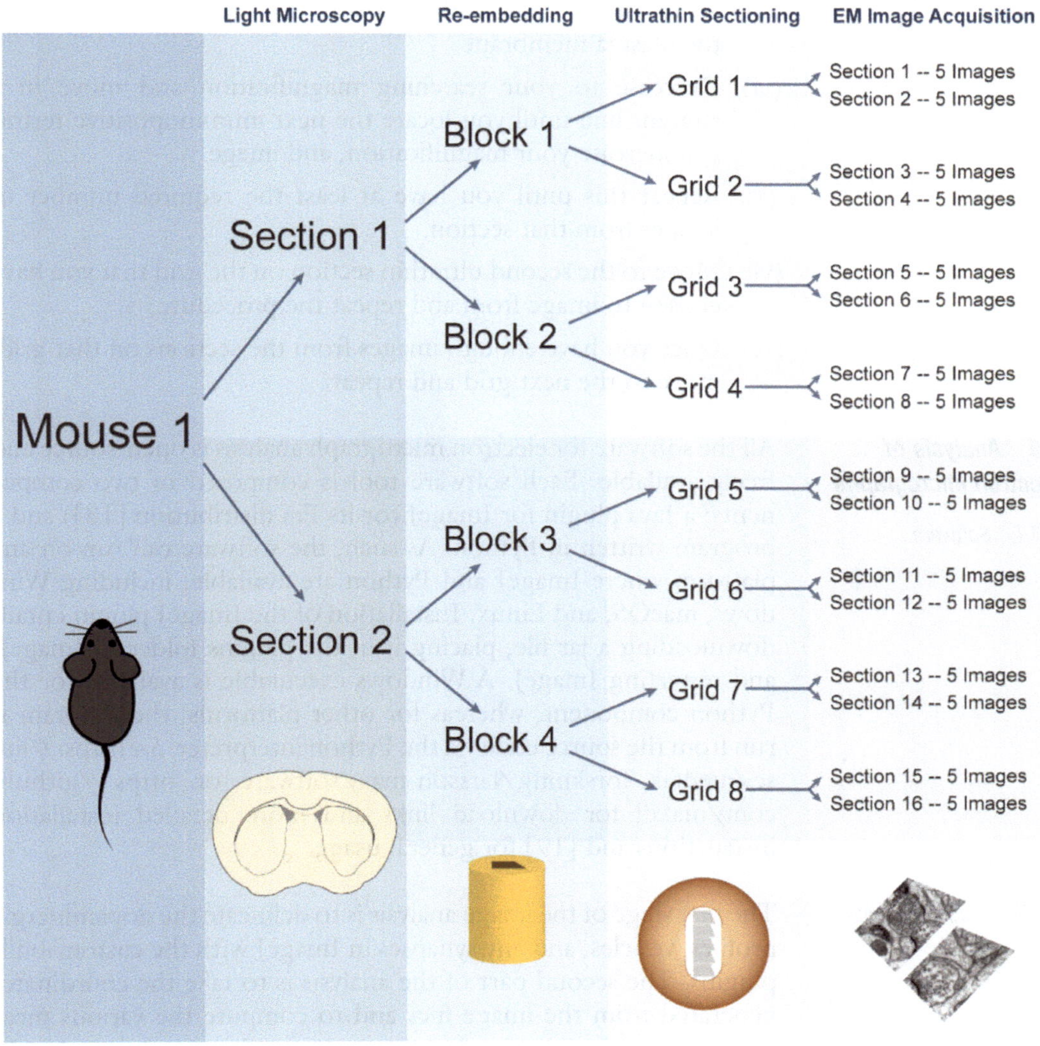

Fig. 4 A guideline sampling strategy for analysis of the ultrastructure of TH-positive axonal structures in the striatum. For each mouse, two sections in the rostrocaudal plane are examined with light microscopy for regions of interest. Two regions are then re-embedded from each section (to make blocks). Tissue on each block is then re-sectioned using an ultramicrotome. Ribbons of ultrathin sections are collected onto multiple grids. Grids are then analyzed in the electron microscope, and digital images are taken from two sections from each grid for further analysis

imaging nonspecific staining, locate the surface of the tissue (where tissue meets the covering resin) and move a set distance (e.g., 2–5 μm) from there before starting imaging.

(v) Once a safe distance from the surface, pick a random starting point and move in a straight line until you locate an immunopositive terminal, increase the magnification so that it is possible to visualize synapses and vesicles, and image the profile. When imaging, keep your immunopositive terminal in the center of the frame and make sure you do not crop off any of the plasma membrane.

(vi) Decrease to your searching magnification and move in a straight line until you locate the next immunopositive terminal, increase your magnification, and image.

(vii) Repeat this until you have at least the required number of images from that section.

(viii) Move to the second ultrathin section on the grid that you have chosen to image from and repeat the procedure.

(ix) Once you have enough images from the sections on that grid, move to the next grid and repeat.

3.5 Analysis of Electron Micrographs

3.5.1 Software

All the software for electron micrograph analysis is open-source and freely available. Each software tool is composed of two components: a Java plugin for ImageJ (or its Fiji distribution [18]) and a program written in Python. As such, the software can run on any platform where ImageJ and Python are available, including Windows, macOS, and Linux. Installation of the ImageJ plugin entails downloading a jar file, placing it in the plugins folder of ImageJ, and restarting ImageJ. A Windows executable is available for the Python component, whereas for other platforms, the program is run from the source code via the Python interpreter. *See* https://liu.se/medfak/forskning/larsson-max/software or https://github.com/maxdl for download links and more detailed installation instructions and [19] for general usage.

3.5.2 Image Analysis

The first stage of the image analysis is to delineate the dopaminergic profiles, vesicles, and any synapses in ImageJ with the custom-built plugins. The second part of the analysis is to take the coordinates generated from the image files and to compute the various measurements of interest, such as axon terminal cross-sectional area, inter-vesicle distances, density of vesicles, distances of vesicles to profile borders, etc. (*see* Fig. 1). All dopaminergic profiles are analyzed with the *PointDensity* tools, and any profiles that are forming synapses can be analyzed with the *PointDensitySyn* tools. The order in which you analyze the images should be considered before beginning analysis (*see* **Note 2**).

Analysis of Non-synaptic Profiles

(i) Open the first EM image for analysis in *ImageJ* and load the *PointDensity* plugin.

(ii) Trace the plasma membrane of the immunopositive profile using the *ImageJ* polygon tool. Close the border. Designate the selection as "Profile border" in the *PointDensity* panel (Fig. 5).

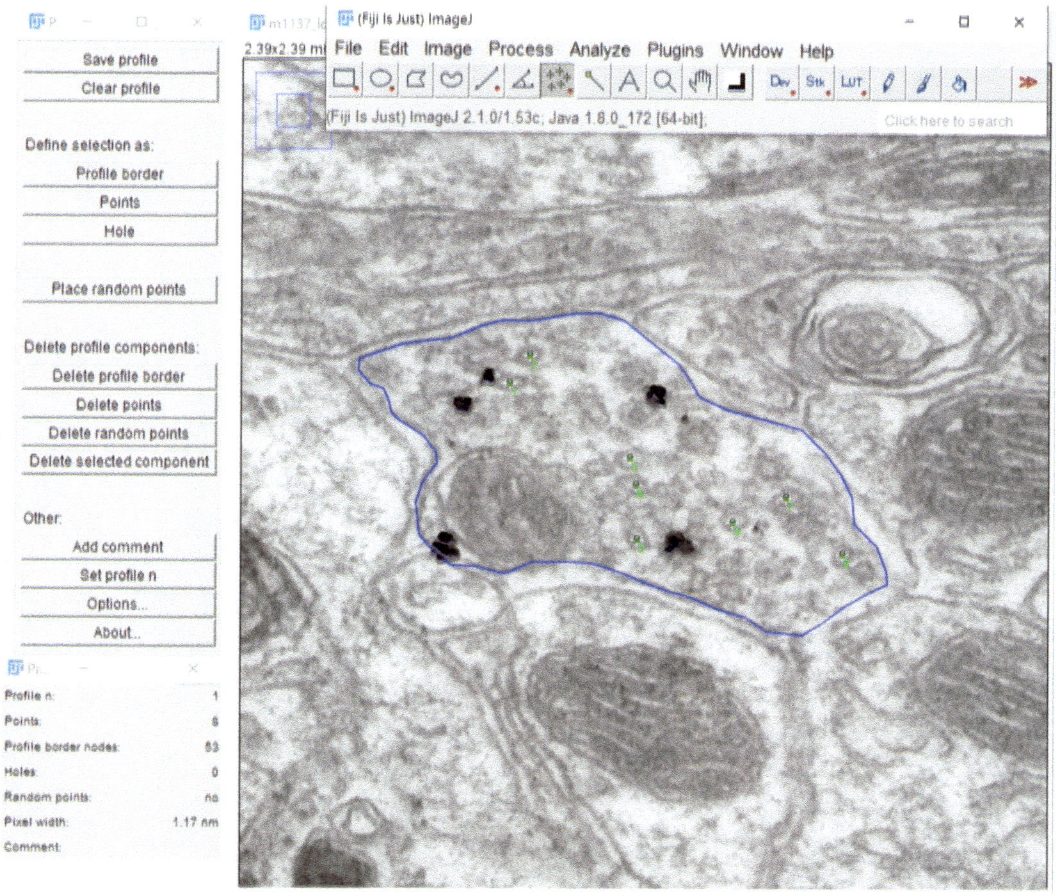

Fig. 5 Analysis of non-synaptic dopaminergic axon terminal using *ImageJ* and *PointDensity* plugin. An axon terminal (same terminal as shown in Fig. 2a) is delineated (blue line), and the outline is defined as "Profile border" using the *PointDensity* interface (left). Vesicles are then marked using the multipoint selection tool and defined as "Points"; a few vesicles are marked here for illustration (green)

(iii) Place a point marker in the center of each vesicle using the *ImageJ* multipoint tool. Vesicles should be counted if at least 50% of the vesicle membrane is visible. Define the selection as "Points" on the *PointDensity* panel (Fig. 5).

(iv) Carefully name and save all data.

Analysis of Synapse Forming Profiles

All axon terminals that are forming synapses can be analyzed with the *PointDensitySyn* tool. Dopaminergic terminals in the striatum normally form symmetric synapses (Gray's type II) which are characterized by the presence of presynaptic and postsynaptic membrane specializations, a widened synaptic cleft and cleft material (Figs. 3 and 6).

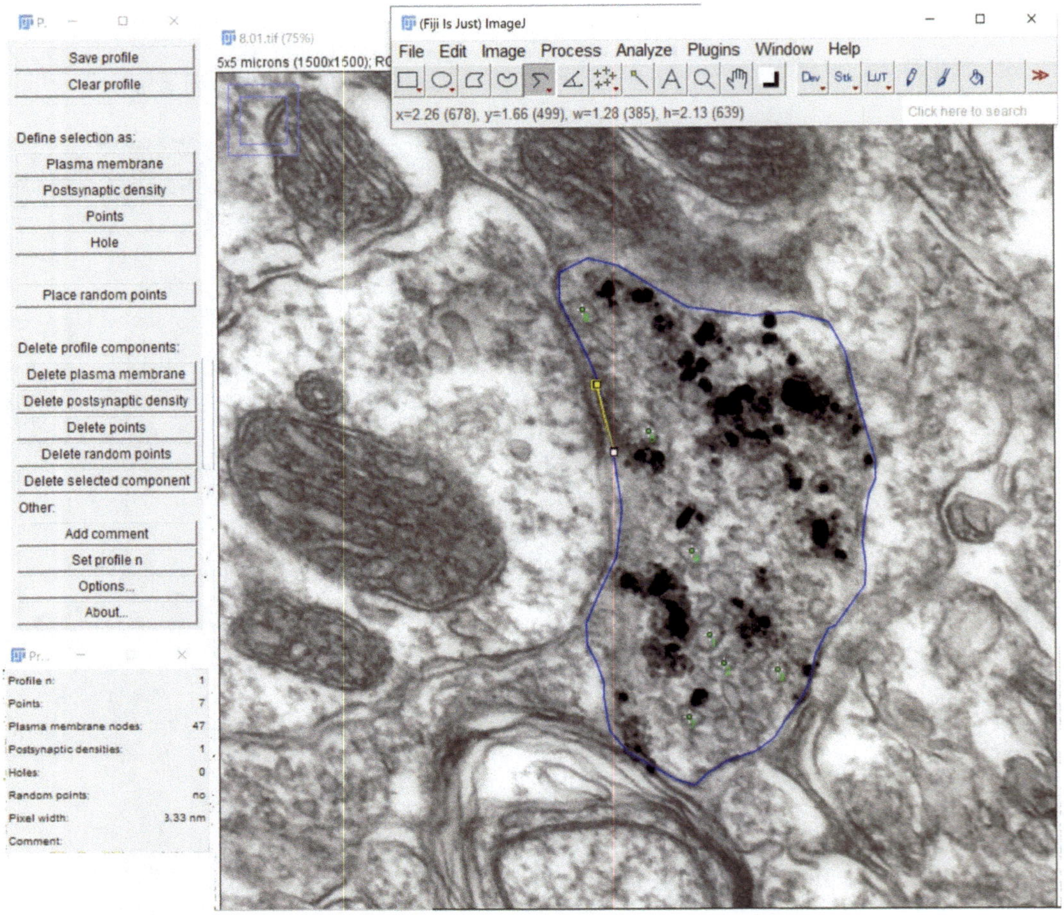

Fig. 6 Analysis of a synapse-forming dopaminergic axon terminal using *ImageJ* and *PointDensitySyn* plugin. An axon terminal (as shown in Fig. 3a) is delineated (blue line) and defined as "Plasma membrane" using the *PointDensitySyn* interface (left). Vesicles are then marked using the multipoint selection tool and defined as "Points"; a few vesicles are marked here for illustration (green). The presynaptic specialization is then delineated (yellow) and defined as "Postsynaptic density" using the interface

(i) Open EM image in *ImageJ* and load the *PointDensitySyn* plugin.

(ii) As before, draw around and define the profile border and then mark and define all vesicles.

(iii) Trace the length of the active zone(s) with the *ImageJ* segmented line tool. The active zone is defined as the length of plasma membrane directly opposing the postsynaptic density, on the presynaptic side of the synaptic cleft (Fig. 6).

(iv) Name and save all data.

Processing Coordinate Files

To compute the values of interest from the coordinates saved for each analyzed EM image, the text files need to be processed with the second (*Python*) component [19].

(i) Once all images have been analyzed in *ImageJ*, with either the *PointDensity* or *PointDensitySyn* plugins, you can run the related *Python* component of the tool, i.e., *PointDensity* or *PointDensitySyn*.

(ii) Add all profile coordinate files to the input file list, check your options, and run [19].

(iii) Values can be exported to Excel or tabulated text files depending on your preference.

4 Notes

1. Glutaraldehyde is essential for preserving the ultrastructure for EM, as it is a more effective cross-linker than PFA, but it is slower to penetrate the tissue, and thus a combination of both aldehydes is needed. There is, however, a balance to be found between immunogenicity and tissue preservation—higher concentrations of aldehydes will give you better preservation but may block antigenic sites; therefore, you might need to reduce the concentration of glutaraldehyde if you do not get good immunolabeling with your chosen antibodies. We routinely use 0.1% glutaraldehyde; the working range is from 0.01% up to 1%; if no immunolabeling is to be carried out at all (i.e., exanimation of ultrastructure alone), you can use a much higher concentration. The object of the protocol described here is to carry out immunohistochemistry for TH so ~0.1% is a good balance.
2. It is inevitable that you will become better and faster at delineation of axon terminals and marking vesicles over time and with practice. It is important to account for this effect, for example, if you were to begin analysis with an animal in your control group, you may falsely get the impression that there are fewer vesicles in this dataset simply because you got better at marking vesicles over time. Therefore, a method may be to "train" yourself with a random set of images before starting your analyses and then to randomize which experimental animal you begin with and the order you analyze them in. As the analyzed images are saved as copies of the original micrographs with the recorded coordinates visible, it is also possible to go back and reevaluate the accuracy of the analysis at a later time. All analyses should be performed blinded wherever possible.

Acknowledgments

Thank you to Lisa Conyers for writing and maintaining all our lab protocols so well. Thank you to Paul Bolam for instruction and editing.

References

1. Iversen SD, Iversen LL (2007) Dopamine: 50 years in perspective. Trends Neurosci 30: 188–193
2. Beaulieu J-M, Gainetdinov RR (2011) The physiology, signaling, and pharmacology of dopamine receptors. Pharmacol Rev 63: 182–217
3. Schultz W (2007) Multiple dopamine functions at different time courses. Annu Rev Neurosci 30:259–288
4. Grace AA (2016) Dysregulation of the dopamine system in the pathophysiology of schizophrenia and depression. Nat Rev Neurosci 17: 524–532
5. Nestler EJ, Carlezon WA (2006) The mesolimbic dopamine reward circuit in depression. Biol Psychiatry 59:1151–1159
6. Berke JD, Hyman SE (2000) Addiction, dopamine, and the molecular mechanisms of memory. Neuron 25:515–532
7. Nutt DJ, Lingford-Hughes A, Erritzoe D et al (2015) The dopamine theory of addiction: 40 years of highs and lows. Nat Rev Neurosci 16:305–312
8. Lotharius J, Brundin P (2002) Pathogenesis of Parkinson's disease: dopamine, vesicles and α-synuclein. Nat Rev Neurosci 3:932–942
9. Singleton AB, Farrer M, Johnson J et al (2003) [alpha]-Synuclein locus triplication causes Parkinson's disease. Science 302:841
10. Esposito G, Clara FA, Verstreken P (2012) Synaptic vesicle trafficking and Parkinson's disease. Dev Neurobiol 72:134–144
11. Wong YC, Krainc D (2017) α-Synuclein toxicity in neurodegeneration: mechanism and therapeutic strategies. Nat Med 23:1–13
12. Ebanks K, Lewis PA, Bandopadhyay R (2020) Vesicular dysfunction and the pathogenesis of Parkinson's disease: clues from genetic studies. Front Neurosci 13:1381
13. Anwar S, Peters O, Millership S et al (2011) Functional alterations to the nigrostriatal system in mice lacking all three members of the synuclein family. J Neurosci 31:7264–7274
14. Janezic S, Threlfell S, Dodson PD et al (2013) Deficits in dopaminergic transmission precede neuron loss and dysfunction in a new Parkinson model. Proc Natl Acad Sci 110:E4016–E4025
15. Kosillo P, Doig NM, Ahmed KM et al (2019) Tsc1-mTORC1 signaling controls striatal dopamine release and cognitive flexibility. Nat Commun 10:1–19
16. Connor-Robson N, Booth H, Martin JG et al (2019) An integrated transcriptomics and proteomics analysis reveals functional endocytic dysregulation caused by mutations in LRRK2. Neurobiol Dis 127:512–526
17. Rueden CT, Schindelin J, Hiner MC et al (2017) ImageJ2: ImageJ for the next generation of scientific image data. BMC Bioinformatics 18:529
18. Schindelin J, Arganda-Carreras I, Frise E et al (2012) Fiji: an open-source platform for biological-image analysis. Nat Methods 9: 676–682
19. Larsson M, Bergersen LH, Gundersen V (2015) Immunogold electron microscopic quantification of small molecular compounds and proteins at synapses and other neural profiles. In: Merighi A, Lossi L (eds) Immunocytochemistry and related techniques. Springer, New York, pp 281–297
20. Bolam JP (1992) Experimental neuroanatomy: a practical approach. IRL Press at Oxford University Press, Oxford/New York
21. Zhang J, Xiong H (2014) Brain tissue preparation, sectioning, and staining. In: Xiong H, Gendelman HE (eds) Current laboratory methods in neuroscience research. Springer, New York, pp 3–30
22. Gage GJ, Kipke DR, Shain W (2012) Whole animal perfusion fixation for rodents. J Vis Exp (65):e3564
23. Tao-Cheng J-H, Gallant PE, Brightman MW et al (2007) Structural changes at synapses after delayed perfusion fixation in different regions of the mouse brain. J Comp Neurol 501: 731–740
24. Paxinos G, Franklin KBJ (2019) Paxinos and Franklin's the mouse brain in stereotaxic coordinates. Academic Press
25. Leica Microsystems Sectioning of Aluminum with Ultramicrotome Leica EM UC7.

https://www.youtube.com/watch?v=p9vaiXq8T7Y
26. Denise Gautreau How to Cut Thin Sections Using an Ultramicrotome. https://www.youtube.com/watch?v=Nl-pE667M0s
27. Harris KM, Perry E, Bourne J et al (2006) Uniform serial sectioning for transmission electron microscopy. J Neurosci 26:12101–12103
28. SynapseWeb, Kristen M. Harris, PI. http://synapseweb.clm.utexas.edu/. Serial Sectioning | SynapseWeb. https://synapseweb.clm.utexas.edu/serial-sectioning
29. Yamaguchi M, Chibana H (2018) A method for obtaining serial ultrathin sections of microorganisms in transmission electron microscopy. J Vis Exp (131):e56235
30. Reynolds ES (1963) The use of lead citrate at high pH as an electron-opaque stain in electron microscopy. J Cell Biol 17:208–212

Chapter 2

Quantifying Dopaminergic Innervation in Rodents Using Unbiased Stereology

Lauren M. Reynolds, Andrea Harée Pantoja-Urbán, Del MacGowan, Colleen Manitt, Dominique Nouel, and Cecilia Flores

Abstract

Quantitative microscopy techniques are valuable tools for neuroanatomical and psychiatric research. Qualitative methods are limited due to inherent biases and an inability to provide an absolute number of markers throughout an entire region of interest. Design-based stereology provides an unbiased alternative, as it uses mathematical and statistical principles to form accurate predictions about the distribution of objects in three-dimensional brain tissue from two-dimensional images while avoiding assumptions about the distribution, orientation, size, or shape of objects. Stereological analysis requires careful a priori planning to provide accurate estimates and avoid bias. Here, we outline, from tissue collection to image processing, a complete stereology protocol to estimate the total number and density of varicosities and/or neurons throughout the mesocorticolimbic dopamine circuitry using Stereoinvestigator Software (MicroBrightField). These dopamine pathways continue to undergo profound development during postnatal life, and disruptions to their maturation are implicated in a wide variety of psychiatric illnesses, including depression, schizophrenia, and substance use disorder. Indeed, a stereological approach has allowed our group to identify and measure the impact of experiences in adolescence on dopamine neuroanatomy in adulthood. This protocol will prove useful for other groups studying the dopamine system and can be easily adapted to study countless other markers, cell types, and neurotransmitter systems.

Key words Stereology, Dopamine, Prefrontal cortex, Nucleus accumbens, Quantitative neuroanatomy

1 Introduction

The mesocorticolimbic dopamine (DA) system is involved in a broad variety of behaviors encompassing emotional, motivational, and cognitive domains [1–6]. Psychiatric patients often exhibit deficits in behaviors across these domains, and these symptoms are tightly associated with alterations to the anatomy of DA circuits. Mesocorticolimbic DA pathways continue to undergo exuberant development in adolescence [7, 8], when many psychiatric diseases emerge [9, 10], and can be dynamically shaped in response to

José Antonio Fuentealba-Evans and Pablo Henny (eds.), *Dopaminergic System Function and Dysfunction: Experimental Approaches*, Neuromethods, vol. 193, https://doi.org/10.1007/978-1-0716-2799-0_2,

environmental factors [11–15]. The architecture of the DA innervation to terminal regions is intimately linked to the establishment of the structure and function of postsynaptic neurons [16, 17], highlighting the evaluation of DA neuroanatomy as a critical research question in neurodevelopmental neuroscience and biological psychiatry.

The organization of DA neurons and their mesocorticolimbic terminal innervations have been heavily studied since the introduction of antibodies against DA or DA markers (e.g., tyrosine hydroxylase, TH). Despite this interest, the bulk of early studies on DA system anatomy were limited to qualitative imaging techniques. While these studies are informative, quantitative measurements are necessary to compare DA neuroanatomy between experimental groups. Exhaustive counting of DA neurons in the ventral tegmental area (VTA) or of axonal varicosities in terminal regions is not practical. Therefore, several methods to create estimations have been developed. The initial quantitative studies focused on counting objects in tissue sections, or on measuring "optical density" by using computer software to extract the number of pixels meeting a certain saturation threshold. This densitometric technique could then be used to compare pixel values (e.g., darker pixels indicate denser staining), or the area which saturated pixels occupy (e.g., considered a proxy for region volume or cell number) [18–23]. These studies are inherently limited by multiple sources of bias and generally can only provide measurements in the two-dimensional plane of section [24, 25].

To generate estimations of total number and/or density of DA neurons or DA axon varicosities, our laboratory has adopted the use of design-based stereology, a powerful, unbiased microscopy technique which allows for accurate interpretations about 3D brain tissue from 2D images [26–28]. Stereology uses geometric and statistical principles to accurately estimate the number particles within a given volume regardless of their physical characteristics and counters biases by avoiding assumptions about distribution, orientation, size, or shape of objects within samples [24, 28–30]. Using systematic sampling through the region of interest to eliminate sampling bias, objects (e.g., cells or terminals) are counted within stereological probes when they meet stereological assumptions. These counts are then used to determine the density of particles within each section. Since the ratio of each probe to the entire volume is known, estimates of the number of particles throughout the entire region of interest can be computed. The result provides an absolute number of particles contained within each region of interest, and densities or ratios can be calculated if needed [24, 28–30].

While stereology is an extremely powerful method for quantifying DA neurons or DA varicosities in three-dimensional space, stereological studies remain a minority of anatomical analyses due to barriers in their implementation. Stereological studies must be designed a priori to meet the necessary assumptions and avoid sampling bias, and they cannot be used on tissue that has already been prepared (*see* **Note 1**). Every step of a stereology experiment, starting with perfusion and brain sectioning, must be controlled in order to provide an accurate estimate free of bias. Once the assumptions and protocol are well understood, stereological experiments can be rapidly implemented to produce highly accurate estimates of a variety of cellular objects—including single- or double-labeled neurons and varicosities.

To meet the assumptions for stereology:

1. The object to be counted should be distributed relatively homogenously through the region of interest.
2. The exact (a) thickness of the sections to be counted from, (b) the series of sections you will count from (e.g., every fourth section, a 1:4 series), and (c) the resulting distance between each section to be counted must be defined at the start of the experiment.
3. Objects should be counted only within the probe.
4. The sampling grid must be determined by the stereological parameters.

Using the optical fractionator probe in *Stereoinvestigator* (MicroBrightField), we have assessed the number and density of DA varicosities (presynaptic sites) labeled with TH within the different subregions of the medial and orbital prefrontal cortex (PFC), the nucleus accumbens (NAc), and the dorsal striatum (STR) [16, 17, 31–35]. We have also quantified the number of DA neurons in the ventral tegmental area (VTA) and substantia nigra pars compacta (SNc) using the optical fractionator [17, 31] and used the nucleator probe to assess cell size in the VTA [31]. These stereological techniques have enabled us to efficiently assess the role of the Netrin-1/DCC guidance cue system in the postnatal structural development of DA circuitry [16, 17, 32, 35], as well as quantitatively evaluate how adolescent experiences alter adult PFC DA structure [14, 31, 34]. We have also evaluated the percentage of GFP-labeled DA neurons and varicosities using stereology either while counting live or from acquired image stacks [17]. While this protocol focuses on anatomical measurements of the mesocorticolimbic DA system, the principles described here can be easily adapted to other regions or markers of interest by first running a pilot study to verify the appropriate stereological parameters.

2 Materials

2.1 Tissue Processing

PFA (1003254182, Sigma-Aldrich). This should be made fresh and stored at 4 °C. While it is stable at 4 °C for up to 2 months, best results are obtained when made in the 24 h preceding the perfusion.

1× PBS—working solution for all steps. Dilute from 10× PBS (11666789001, Roche).

Vibratome (VT1000S, Leica).

24-well plates (229123, CELLTREAT) (*see* **Note 1**, Part iii).

Cryoprotectant

250-mL 1× PBS.

100-mL glycerol (G2025, Sigma-Aldrich).

150-mL ethylene glycol (E178, Thermo Fisher Scientific).

Blocking Solution

2% Bovine serum albumin (A7906, Sigma-Aldrich).

0.2% Tween-20 (P1379, Sigma-Aldrich).

Antibodies

See Table 1.

Mounting Media

Invitrogen SlowFade Gold Antifade Mountant medium (Thermo Fisher Scientific, Cat #S36936).

Clear nail polish (to seal edges of coverslip).

2.2 Stereology

Hardware

Leica microscope DM4000B.

Automated stage LUDL LEP, 99S100-LE, STG 3″ × 2″ STEPPER.

Light source X-Cite mini + Excelitas Technologies.

Hamamatsu Digital Camera C1440-42U ORCA-flash4.0LT.

DC1394 camera support.

Windows computer with:

Processor: Intel® Xeon® W-2245 CPU @ 3.90GHz on 16 Cores.

Installed Memory (RAM): 64 GB.

Video Card: NVIDIA GeForce RTX 2060.

Video Memory: 6 GB.

Software

Stereo Investigator version 2020.2.3 (MicroBrightField, St. Albans VT).

Table 1
Validated antibodies for stereological assessment of dopamine circuitry in mice

		Product number	Supplier	Host	Dilution	Notes
Primary antibodies	Polyclonal Anti-TH	AB152	Millipore	Rabbit	1:500–1:1000	
	Monoclonal Anti-TH	MAB318	Millipore	Mouse	1:300	Use M.O.M. blocking kit (Vector Labs #BMK-2202)
	Polyclonal Anti-GFP	NB100-1770	NOVUSBIO	Goat	1:1000	
Secondary antibodies	Alexa Fluor® 594 AffiniPure Donkey Anti-Rabbit	711-585-152	Jackson Immuno Research Labs	Donkey	1:500	
	Alexa Fluor® 633 Goat Anti-Mouse	A21052	Life Technologies	Goat	1:500	
	Alexa Fluor® 488 AffiniPure Donkey Anti-Goat	705-543-003	Jackson Immuno Research Labs	Donkey	1:500	

Other

MBF calibration grid slide (Part # 94-425 rev.1.05, MicroBrightField).

The Mouse Brain in Stereotaxic Coordinates (Third Edition, Academic Press).

3 Methods

3.1 Tissue Processing

3.1.1 Perfusion (See Note 2)

I. Perfuse mice with cold saline or 1× PBS followed by 4% PFA, brains postfix overnight by staying in 4% PFA.

II. Extract the brain from the skull. For the best sectioning results, care should be taken to remove the brain without damaging it (e.g., olfactory bulbs attached); however, once the brain is sectioned, superficial damage to the cortical surfaces will not affect the counting of objects in subcortical space.

3.1.2 Tissue Sectioning and Storage

I. Our preferred method of sectioning is to use the vibratome on fresh fixed tissue. In this case, brains are ideally sectioned the day after perfusion, and no more than 2 days later (*see* **Note 3**).

II. Our team uses a Leica vibratome (*see* Subheading 2) with the speed set to 6, vibration set to 7.5, and section thickness at 35 μm (*see* **Note 4**).

III. Section brain in a 1:6 series into a 24-well plate (i.e., section into six consecutive wells, keeping track of where each section belongs in the larger series, Fig. 1a) (*see* **Note 1**). Collect sections free-floating in 1× PBS.

! Pause point: tissue slices can be frozen at this step for later use (*see* **Note 3**).

3.1.3 Tyrosine Hydroxylase (TH) Immunofluorescence

Primary (*See* Note 5)

I. Select which wells of sections will be processed (*see* **Note 1**, Part iv).

II. Put sections into block for 60 min at room temperature on an orbital shaker set to a low frequency.

III. Change to primary antibody solution. 500 μL of solution per well is sufficient for processing in a 24-well plate. Place the plate with sections in primary on an orbital shaker set to a low frequency in a refrigerator or in a cold room (4 °C) for **48 h**.

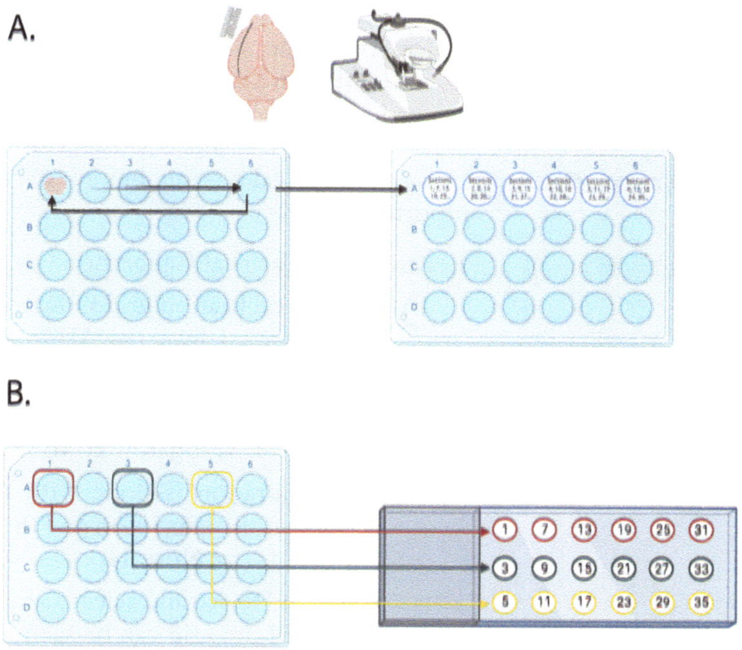

Fig. 1 Organization of brain sections for stereology experiments. (a) The brain is sectioned in a 1:6 series into a 24-well plate. *Top*: The brain should be marked with a cut from a razor or needle to indicate the side that will not be counted from and sectioned on a vibratome (or according to standard laboratory protocols). When sectioning or storing brains from different subjects together, marking or notching the brains differently before sectioning is strongly recommended in order to avoid mixing tissue. *Bottom*: Starting with the first section of the region of interest (ROI), sections 1–6 are stored in wells 1–6. The seventh section will be placed in well 1. Continue until sectioning through the ROI is finished; each well will then contain every sixth section through the entire anterior-posterior limits of the ROI. A 6- or 12-well plate can also be used, but the use of a 24-well plate allows tissue from four subjects to be stored together. (b) To mount a 1:2 series on the slide, three wells (1, 3, and 5 **OR** 2, 4, and 6) are used per slide, with each well containing a row. Sections are mounted in anterior to posterior order and will then constitute a 1:2 series by moving down columns and across rows on the slide

Secondary (*See* Note 6)

I. Wash sections 3× for 10 min in 1× PBS.

II. Put sections into secondary antibody, leave at room temperature on an orbital shaker set to a low frequency for 1 h, and cover with tin foil to protect from light.

III. Wash 3× for 10 min with PBS.

IV. Mount onto gel-coated or charged slides. Each well should be a row, mounted in anterior to posterior order (Fig. 1b). All sections that will be counted for a region of interest should be on the same slide. Coverslip slides with **aqueous** antifade mounting medium and seal the edges of the slide with clear nail polish (*see* **Note 7**).

3.2 Stereology (See Note 8)

Always check calibration before starting an experiment—most important step

3.2.1 Calibration

I. Parcentric/Parfocal Calibration (*see* **Note 9**)

(i) Use the calibration grid slide from MicroBrightField (MBF Bioscience). Clean the MBF slide with the glass cleaner liquid and put in on the stage. Use the smaller (25 μm) grid for the parcentric/parfocal calibration.

(ii) In Stereo Investigator, go to the Acquire Tab.

(iii) The calibration is done with transmitted light (TL), so verify that the lamp is on and the shutter for the fluorescent light is off (*see* **Note 10**).

(iv) Click on Device and select Lamp and then adjust the exposure until the grid can be seen.

(v) Bring the corner of the small (25 μm) grid into focus with the 5× objective and move through the objectives in increasing order (e.g., 10×, 20×, 40×, and then 100×) while keeping the corner of the grid in the center of the frame. **Do not use oil with the oil immersion objectives; you do not need it for the calibration**.

(vi) Click anywhere in the workspace to create a reference point.

(vii) Go to File → Calibration → Parcentric/Parfocal Calibration.

(viii) Choose all objectives to be used, from largest to smallest (e.g., 100×, 40×, 20×, 10×, 5×).

(ix) Bring the corner of the small (25 μm) grid into focus using the largest objective. **Remember that using the oil for high-power objectives for this calibration is not needed. Do NOT move the stage**. Follow the instructions on screen, bringing **the same point into focus** at each objective.

(x) When finished, make sure both parcentric and parfocal are checked on the menu and click OK.

Do NOT check ENABLE PARFOCAL if your microscope has an internal motorized Z axis.

II. Grid Tune (*see* **Note 11**)

(i) Go to *Tools → Grid Tune Current Lens* **for each objective, one at a time**.

(ii) At 5×, use the larger (250 μm) grid; for all others, use the 25-μm grid—make sure this is reflected in the "Box Size" area of the setup window.

(iii) A perforated white grid with an anchor will appear on the screen. When working with a 5× objective or lower, drag the anchor to an intersection at the center of the screen so that the **dotted lines are in the center of the black lines**. When working with a 10× or higher-power objective, drag the anchor to an intersection at the center of the screen so that the white dotted lines rest **on the side of the black grid lines**.

(iv) Hover the mouse over the perforated grid lines at the side of the screen until the cursor changes to a movement arrow ↔. Use this movement arrow to overlay the white dotted lines with the black grid lines on the calibration grid slide. Align the vertical lines and then the horizontal lines as consistently as possible (*see* **Note 12**).

(v) Right click and select "Finish calibrating current lens."

(vi) Repeat for all other objectives.

*3.2.2 Tracing the TH+ Innervation (See **Note 13**)*

I. Make **ONE file for each SUBJECT** (*see* **Note 14**).

II. Go to Probes → Optical Fractionator.

III. EITHER open a new file and click to create a reference point **OR** start your optical fractionator workflow to begin tracing (*see* **step 3** below for more information on the optical fractionator workflow) (*see* **Note 15**).

IV. Identify which sections will be counted from, keeping in mind that for a 1:4 series every other section on the slide will be counted if wells 1, 3, and 5 are mounted on the slide (1: 2 series) (*see* **Note 1**). If there is a missing a section within the region of interest, *see* **Note 16**.

For the PFC:

(i) Trace the dense DA innervation to the inner layers (V–IV) of the cingulate (Cg1), prelimbic (PrL), and infralimbic (IL) subregions of the PFC. **The PFC measurements are always defined primarily by this dense DA innervation**. The cingulate starts on the top of the curve where the TH+ staining is most intense. Use the atlas (we use *The Mouse Brain Atlas in Stereological Coordinates* from Paxinos and Franklin) as a guide as to where the boundaries between the sections fall and where the bottom boundary of the IL is. The lateral boundary is the white matter, while the medial boundary is the edge of the dense TH+ innervation (Fig. 2a). Close the shutter once the tracings are done, so the slide

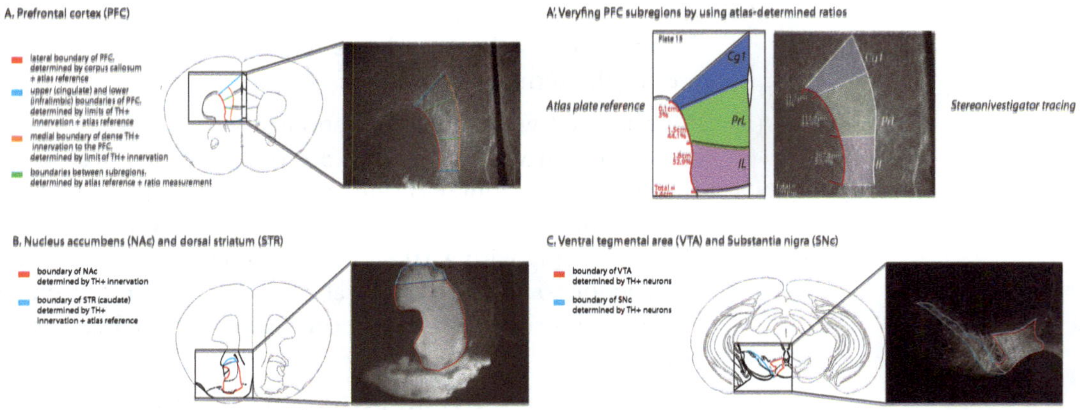

Fig. 2 Tracing dopaminergic innervation using TH staining. (**a**) *Left*: Diagram of a representative coronal atlas plate containing the PFC with color-coded guide to the *Stereoinvestigator* contour boundaries demarcating the dense TH+ innervation to layers V and VI. *Right*: Fluorescence image of the corresponding coronal brain section stained for TH at 5× magnification. (**a′**) When determining boundaries between PFC subregions, the proportion of each subregion's lateral boundary to the entire lateral span of the medial PFC (e.g., *right*) should be consistent with ratios derived from the atlas (e.g., *left*). (**b**) Dopaminergic innervation to the NAc and STR can be traced according to the dense areas of TH+ staining evident in fluorescent images from the region at 5×. *Left*: Example atlas coronal plate. *Right*: Example contours to a corresponding image. (**c**) The VTA and SNc can be traced and distinguished according to dense regions of TH+-stained neurons. *Left*: Example atlas coronal plate. *Right*: Example contours to a corresponding image

is not exposed to too much light; always keep the shutter closed if it is not necessary to have the light on.

(ii) Do the math: measure with the tool "quick measurement" the length of the lateral boundaries of your tracings. The proportion of each subregion tracing to the whole PFC is compared to the atlas and the proportions you calculated (*see* **Note 17**).

(iii) If the proportions are different than what would be expected from the atlas, correct your contours until you are close to the atlas proportions.

For other regions:

For regions with boundaries that can be easily identified by their TH staining, such as the NAc, STR, and VTA, it is sufficient to trace the densely innervated nuclei to make your contours, keeping in mind the same principles of section organization (e.g., keeping the 1:4 series) as you would in the PFC. *See* Fig. 2b, c for examples.

V. When you move to the next section, always use Joy Track to conserve the position of your tracing in the program relative to the stage. This way, when coming back to the file, only the first tracing needs to be aligned with Joy Free, and then it should be possible to snap to other sections using the "Go to" tool in the Macro View window (*see* **Note 9**).

3.2.3 Using the Cavalieri Estimator to Calculate Region Volume

I. *At least the first three steps of the optical fractionator workflow need to be done*—the contours must be all made and assigned to the correct section. When **step 3** is finished, close the optical fractionator workflow in order to use the Cavalieri estimator.

II. Go to Probes → Cavalieri Estimator and enter the grid size and section thickness where prompted (*see* **Note 18**).

III. Select the desired marker from the menu. Use 1 marker per subregion of interest in the file—for example, when counting from all three subregions of the PFC (Fig. 2a), the Cg1, PrL, and IL subregions will each require a distinct marker for the Cavalieri estimator (e.g., Cg1 = asterisk, PrL = dot, IL = cross). The names and colors of the markers can be saved in each user profile for ease of use.

IV. Right click in the workspace and select "Paint Cavalieri marker mode."

V. Right click in the workspace again and select "Paint markers into contour."

VI. A macro view of all of the contours in the workspace can now been seen. Click inside of the contour associated with your selected marker on **every section chosen for counting** (e.g., using the marker selected for the Cg1, click inside the Cg1 contour of each section on the slide to paint in the Cavalieri markers). Then select the next marker (e.g., the one chosen for the PrL) and repeat the process, painting the Cavalieri markers into their associated contour on every section spanning the region of interest. Continue until all regions of interest are filled.

VII. When done, uncheck Cavalieri Estimator from the Probe menu.

VIII. Go to the Probe run list. A probe run for each section should become visible with all the markers that were used (e.g., Cg1 & PrL & IL for PFC). Highlight all of these probe runs and click view sampling results. Confirm that sampling is done every fourth section by entering 4 when prompted for an interval.

IX. Export the counts to excel.

3.2.4 Optical Fractionator Workflow

I. Any time a new file is opened in Stereoinvestigator, click anywhere in the workspace to create a reference point.

II. Go to *Probes → Optical Fractionator Workflow.*

III. Start a new subject.

IV. Move through the workflow as prompted below:

Table 2
Validated stereological parameters for the assessment of dopamine circuitry in mice

	PFC	NAc/STR	VTA/SNc
Object to count	*DA varicosities*	*DA varicosities*	*DA neurons*
# of sections to count	5	5	4
Section's cut thickness	35 μm	35 μm	35 μm
Section evaluation interval	4	4	8
Starting section number	1	1	1
Grid size	175 × 175 μm	200 × 200 μm	150 × 150 μm
Counting frame size	25 × 25 μm	10 × 10 μm	60 × 60 μm
Optical disector height	10 μm	5 μm	10 μm
Guard zone	4–5 μm	4–5 μm	3 μm

Step 1. Set up the section.

Input the section details (number of sections to count, section cut thickness, section evaluation interval, starting section number; *see* Table 2 for the values we commonly use in the lab) (*see* **Note 19**).

Step 2. Set microscope to low magnification.

Select the low-magnification objective to trace the regions of interest (normally 5×).

Step 3. Trace the region(s) of interest.

If tracing is done within the workflow:

- Start with section 1 and trace all regions of interest on this section (Cg1, PrL, IL) using a designated contour for each separate region (*see* Subheading 3.2.2, for more information on how to make contours of TH+ regions of interest).
- Move to section 2 when all regions on section 1 are complete and repeat for each section you will count from. Ensure that **Joy Track** mode is used to move between sections (*see* **Note 9**) in order to register the position of each section on the slide. This will make it easier to move between the sections for counting; as once the contours are made, it is then possible to snap quickly to their position using *Macro View* → *Go to.*

If tracings have already been made before opening the workflow:

- The user will be prompted to assign each tracing to a section (*see* Subheading 3.2.2).

V. For live counting, use **steps 4–11**. For counting from IMAGE STACKS, *see* Part VI.

Step 4. Set the microscope to high magnification.

- Make sure to add oil if using an oil immersion objective!
- Make sure the contour is aligned properly over the section that will be counted at low (5×) and high (100×) magnification. *If it is necessary to adjust the contour over the section, use **Joy Free** mode to move the slide independently of the contour (e.g., the contour will stay as it is, and the slide will move "underneath" it; *see* **Note 9**).

Step 5. Measure mounted thickness.

- Check box "Refocus to top of section at each site."
- Under "Choose your method for measuring," check the box for "measure the mounted thickness while counting."
- Under "Measure mounted thickness while counting," check the box for "measure mounted thickness at sampling sites" and set the evaluation interval to 1.

Step 6. Define the counting frame size.

- For information on determining the counting frame settings, *see* **Note 19**.

Step 7. SRS grid layout settings.

- For information on determining your SRS (serial random sampling) grid settings, *see* **Note 19**.

Step 8. Define disector options.

- For information on determining the size of your disector and guard zone, *see* **Note 19**.

Step 9. Save sampling parameters.

- Here it is possible to choose to save the parameters for future experiments.

Step 10. Count objects.

- Select region of interest from menu. Press the blue arrow to begin a new probe run; the user will be directed to the first sampling site.
- The *Focus Top of Section* dialog box is displayed. Using the fine focus knob, focus to the top of the section and then click OK in the dialog box.

- The *Focus Bottom of Section* dialog box is displayed. Using the fine focus knob, focus to the bottom of the section and then click OK.
- Select a marker from the *Use Marker* drop-down menu in the workflow panel, or from the marker toolbar.
- Count the cells/varicosities following the counting rules (*see* Subheading 3.2.5).
- Click the "fast forward" button to go to the next site. Repeat step for all the regions in all the sections.
- Click the "I'm Finished Counting" button when done. Click the *Next Step* button to move to **step 11**.

Step 11. View the sampling results.

- Click the *Display Probe Run List* button.
- Highlight all of the entries for **one region of interest at a time** on the probe menu (e.g., PrL Section 1, PrL Section 2, PrL Section 3, PrL Section 4, PrL Section 5).
- Click "View Results."
- Export to Excel (*see* **Note 20**).
- Repeat for each region of interest.

VI. For counting from stacks, use **steps 1–5**.

Step 1. Set up the subject.

- The user will be asked if they would like to use their already defined serial sections. Click **YES**.
- It is not necessary to do anything else on this screen.

Step 2. Measure mounted thickness.

- Under "Choose your method for measuring," check the box for "measure the mounted thickness while counting."
- Under "Measure mounted thickness while counting," check the box for "measure mounted thickness at sampling sites" and set the evaluation interval to 1.

Step 3. Define disector options.

- If the probe height and guard zones were previously defined in the Acquire SRS Image Stack Workflow (*see* **step 8** in Subheading 3.2.7), the correct values for the top guard zone height and optical disector height are displayed.

Step 4. Count objects.

- In the probe menu, click the plus sign next to the region of interest you would like to count and select ACQUIRED STACKS from the menu.
- Press the blue arrow to begin a new probe run. The stack is loaded, and the user will be directed to the first sampling site.
- The *Focus Top of Section* dialog box is displayed. Using the **PageUp** key, focus to the top of the stack and then click OK in the dialog box.
- The *Focus Bottom of Section* dialog box is displayed. Using the **PageDown** key, focus to the bottom of the stack and then click OK. *This is important to do at each site, even if working from pre-made stacks, to ensure an accurate and unbiased estimate.*.
- Select a marker from the Use Marker drop-down menu in the workflow panel, or from the marker toolbar.
- Count the cells/varicosities following the counting rules (*see* **step 5**: Counting the TH Varicosities).
- Click the "fast forward" button to go to the next site. Repeat this step for all the regions in all the sections.
- Click the *I'm Finished Counting* button when done. Click the *Next Step* button to move to **step 5**.

Step 5.

- Click the *Display Probe Run List* button.
- Highlight all of the entries for **one region of interest at a time** on the probe menu (e.g., PrL Section 1, PrL Section 2, PrL Section 3, PrL Section 4, PrL Section 5).
- Click "View Results."
- Export to Excel (*see* **Note 20**).
- Repeat for each region of interest.

3.2.5 Counting Objects (See *Note 21*)

In **step 10 of the LIVE Optical Fractionator Workflow OR **step 4** of the STACKS Optical Fractionator Workflow, the user will count a random and systematic sample of the TH+ varicosities/cells in the predefined regions of interest.**

I. Place a marker for each TH+ cell or varicosity <u>within the bounds of your probe</u>.

An object is considered to be within your probe when:

- It is located within the X-Y coordinates of the counting probe (i.e., within the box).

- The sides of the probe are green and red. If they are **yellow**, it is outside the Z-plane range of the disector, i.e., in the guard zone. DO NOT place markers to count these objects.
- For objects not clearly completely contained within the box of the probe (e.g., a cell that is halfway outside the box), count everything that touches the green side of the box and DO NOT count ANYTHING touching the red side of the box.

II. When counting is finished, click “I’ve finished counting” and go to **step 11** (live) or **step 5** (stacks) in the optical fractionator workflow (*see* above) to see the results.

III. Always clean oil immersion objectives well with lens paper after counting. Leaving oil on the objective could damage it over time.

3.2.6 Accounting for a Missing Section

I. In practice, this is most accurate if the first and last sections are present, and the one that is missing is in the middle somewhere. This is because the missing section does not change the volume of the region, it **only adjusts the counts**.

II. **BEFORE** counting in the Optical Fractionator workflow OR using the Cavalieri estimator:

Probe → “Account for missing section.”

Choose the analysis and indicate which section is missing.

III. Then, run probe as usual. A column with counts adjusted for missing section will be present; this is the count that should be used.

*3.2.7 OPTIONAL: Taking Image Stacks (*See ***Note 22****).*

I. Set up the serial sections and trace the contours of the predefined regions of interest (the first three steps of the Optical Fractionator Workflow, already completed if the Cavalieri estimate for volume has already been made).

II. Next go to *Acquire* → *SRS Image Stack Work Series* and follow the workflow below:

Step 1. Set up.

- This should be pre-filled (*see* Subheading 3.2.4).
- Choose “Edit Serial Sections” if changes are needed.

Step 2. Set tracing magnification.

- Select the low-magnification objective used to trace the regions of interest (5×).

Step 3. Region(s) of interest.

- This should be automatically filled out if the contours are already made! (*see* **step 3** in Subheading 3.2.4).

Step 4. Region selection.

- Click on the regions for which stacks will be taken (*see* **Note 23**).

Step 5. Set acquire magnification.

- Set magnification to 100× (or to the chosen counting objective). The stage will move in the Z-plane to the correct parfocal position, which will be at or near to the plane necessary for the image to come into focus. The objective will change to 100× automatically if the microscope is equipped to do so; otherwise, change to the 100× objective manually.

Step 6. Counting frame settings.

- The user should set this **EXACTLY** as they plan to count. Photo stacks will **ONLY** be taken where the counting frames are going to be (e.g., each corner of the grid), not necessarily of the entire region of interest.
- Stacks should not be used to pilot an experiment since the settings used here will be automatically applied to the counting workflow.
- For more information on determining counting frame settings, *see* **Note 19**.

Step 7. SRS grid layout settings.

- For more information on determining SRS (systematic random sampling) grid settings, *see* **Note 19**.
- For the settings commonly used in our laboratory for our regions of interest, *see* Table 2.

Step 8. Image stack settings.

- The "Collect Images for:" box should be checked. Set this option to Mounted Thickness.
- Set the guard zone and probe depth for the experiment (*see* **Note 19**).
- Measuring mounted thickness: Notice that the "Mounted Thickness" box is at 0.00 by default. Click the button that says "Set" next to the "Mounted Thickness" box. A window will appear on the screen. In the case of multiple labeling, choose the channel with the most robust signal (usually TH/red channel is good for this) and select *Live Image* so that the section can be seen. Scroll with the fine focus to the top of the section (*see* **Note 24**). Once you are at the top of the section, click the "Set Top of Section" button. Scroll to the bottom of the section and click the "Set Bottom of Section" button. Then click the "Add Measurement to List" button. Repeat for three to five areas in the section, moving to each region using the

Macro View → Go to (*see* **Note 9**). This process will automatically generate the average mounted thickness in the workflow. To change a measurement, highlight that measurement on the list and use the "Revisit Measurement" or "Delete Measurement" buttons below to adjust.

- Additional images can be added to be taken above or below the mounted thickness by checking the appropriate boxes. **Do not exceed ~30 μm total for 35-μm mounted thickness to avoid breaking the slide and/or objective**.
- Step size: The step size is the movement in the Z-plane, and the best step size is a trade-off between file size and the axial resolving power of your objective, which is based on the numerical aperture. With our 100× objective, the axial resolving power is ~770 nm, and we use a minimum step size of **1 μm**.

Step 9. Output location.

- This will default to the folder that the tracing file is in. By clicking the ellipsis button next to the Path setting, a different folder can be chosen to store the image stacks.
- Image base: This will give all of the images a prefix—whatever text is entered in the box (e.g., mouse ID and/or region of interest).
- Image size: Choose "No Camera Binning" to take full resolution pictures.
- Format: MBFJpeg2000.
- Data file: Specify the name of the data file. This will autopopulate to the name of the tracing file, but it can be changed to save the stacks to a new file.
- Remember to save one file with stacks for all of the sections through the region of interest.

Step 10. Prefocus.

- This is the most important step in the stacks workflow! This step will set the top of **all** of the image stacks.
- Click "Set Top of Section." A live view of the first site will appear.
- If nothing can be seen in the live view (e.g., the shutter is closed), go to → *Options* → *Multichannel control* and click on the most robust channel of interest.
- For each site, focus at the top of the section and click "Next." Then hit "Next Step."

Step 11. Save configuration.

- Here the user can save their configuration for future use.

Step 12. Image stack acquire.

- If taking MULTICHANNEL stacks, make sure that the *Multichannel Acquire* box is checked! Select the channels to be included in Multichannel setup.
- Highlight all of the regions of interest and click the "Click to Start Acquiring" button.
- The status menu will either say "Waiting" (*has not started that region yet*), "Acquiring" (*working on that region now*), "Completed" (*that region is done*), or "Cancelled" (*if the workflow is stopped—note that the program* ***will not save stacks for the incomplete region*** *and it will need to be restarted from the beginning*).

After saving the stacks, objects can be counted immediately or at any later time by opening the file containing all of the stacks and continuing from **step 4**, Part VI.

4 Notes

1. Tissue organization: a critical step in stereology protocols
   (i) The organization of tissue while sectioning is the most important preparation step for stereology. The sections must be placed in the correct order when sectioned and in the correct order on the slide in order to know the distance between the sections, which is necessary to meet the stereological assumptions and make accurate calculations.
   (ii) It is for this reason stereology experiments must be designed a priori, and normally tissue sectioned without these constraints in mind will be impossible to use for a stereological experiment.
   (iii) We do all sectioning and immunofluorescence steps in 24-well plates. To reduce single-use plastic use, these plates can be rinsed at the end of the experiment and reused indefinitely. To minimize the number of plates used for a single experiment, the liquid can be removed with a fine-tipped Pasteur or transfer pipet at the end of each step, and the liquid for the next step pipetted into the plate directly, instead of moving the tissue between plates. At the end of each experiment, the plates can be washed and reused instead of replaced.
   (iv) While the exact organization of the counting series can be adjusted (e.g., reducing the space between quantified sections to accommodate a smaller region of interest or adjust

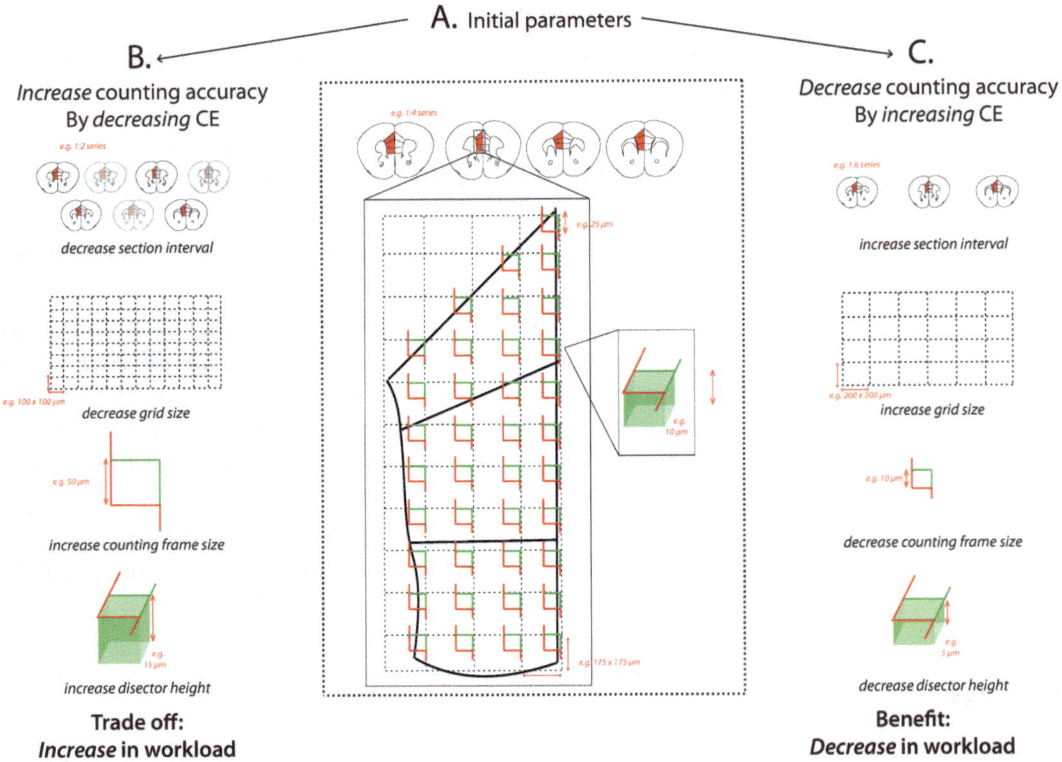

Fig. 3 Selecting parameters for stereological experiments. Designing a stereology experiment requires balancing the accuracy of the outcome with the workload entailed. Accuracy is highest when the maximum number of objects is counted, and thus this balance results from the selection of each counting parameter to determine the number of objects used to make the estimates. (**a**) Representative experimental design with parameters of a section interval of 1:4, a grid size of 175 × 175 μm, a counting frame size of 25 μm, and a disector height of 10 μm in the PFC. By modifying these parameters, the accuracy of the counts obtained can be increased—with a corresponding increase in workload (**b**)—or decreased, with a corresponding decrease in workload (**c**). The **section interval** should span the entire region of interest (ROI). For small ROIs, it is recommended to decrease the section interval in order to count from more sections within the ROI and increase accuracy; however, including more sections will increase the workload. The number of counting sites within experimental contours (tracing of the ROI) is determined according to the **grid size**. Counting site number will increase or decrease with the size of the grid, with fewer counting sites sampled with a larger grid size. The size and density of the objects to count will play an important role in determining the size of the **counting frame**. It is recommended to have a **minimum** of one to five objects per counting site, with accuracy of the estimate increasing as the size of the frame increases. The **disector height** must be within the boundaries of the tissue thickness and large enough to distinguish objects that overlap through the Z-plane. Following the same rules that apply to the counting frame in the X–Y plane, dissector height should be adjusted so that a **minimum** of one to five objects are counted per site, with accuracy of the estimate increasing as the size of the dissector increases, until the limit of measured tissue thickness (less guard zones)

workload/CE parameters; *see* Fig. 3), for the regions of interest discussed in this protocol, we process alternate sections in a 1:2 series for immunofluorescence and then count from every other section on the slide for a final stereology interval of 1:4 (Fig. 1). For example, we process wells 1, 3, and 5 and use wells 2, 4, and 6 for other experiments either immediately or by freezing them for later (*see* **Note 3**).

(v) Keep your slides at 4 °C and shielded from light to maximize the life of their fluorophores.

2. Perfusion tips

(i) Use a pump and maintain the PBS and the PFA on ice for best results. First perfuse with saline until the liver has cleared from dark red to a sallow peach color; the lungs should also be white. This is usually achieved with 20 or 30 mL of PBS. Then switch from saline to PFA and perfuse until the liver is firm when touched with forceps (30 mL or 40 mL of PFA). A sign of a good perfusion with PFA is the limbs and the body becoming rigid and the liver becoming firm.

(ii) Leaving the brains in postfix longer than overnight is not recommended, as it can interfere with the specific labeling of some antibodies.

(iii) A good perfusion is the best start to a stereology experiment; care should be taken to maximize results at this step.

3. Fresh or frozen?

(i) Brains can be sectioned on a vibratome, cryostat, or microtome according to best practices in your laboratory. However, if you decide to freeze the tissue, it must be done slowly as it is essential to avoid flash freezing the brains because it damages DA axons.

(ii) We recommend processing sections for TH immunofluorescence free-floating, so cryostat sections should be collected into a 24-well plate, and not directly mounted onto the slide.

(iii) When freezing tissue at the "pause point," simply move tissue into wells full of cryoprotectant solution (*see* Subheading 2) and seal plate well with Parafilm. Store at −20 °C.

(iv) When starting from frozen sections: wash 3× 10 min in 1× PBS to wash off cryoprotectant solution.

4. Tips for making appropriate stereology sections.

(i) For stereology using the optical fractionator probe, you need a final section thickness on the slide of >20 μm. This will allow you to distinguish between cells/objects that are in the same X-plane (i.e., "on top" of each other) as you adjust the focus and move through the Z-plane of the tissue. We make our vibratome sections at 35 μm. This allows for some shrinkage of the tissue during processing but is also still thin enough that the antibody penetrates well without leaving the center of the tissue unstained in densely innervated regions.

(ii) We always count from the right side of the brain so it is important to mark the left side of the brain with a notch or cut from a razor blade. This way, you will mount all the sections facing the correct direction on the slide and you will know when counting which side of the brain you are on.

5. Primary antibody incubation tips.
   (i) The anti-TH polyclonal antibody raised in rabbit (Millipore AB152) that we typically used can be diluted in block at 1: 1000.
   (ii) However, it is ALWAYS a good idea to try several dilutions in a pilot study before beginning your experiment.
   (iii) If your antibodies are stored diluted in a 50% glycerol solution, you will need to double the amount of antibody solution per well.
   (iv) We recommend to prepare up to 1 mL extra of the block + primary antibody solution to account for liquid loss when pipetting into the 24-well plate.
6. Secondary antibody incubation tips.
   (i) We typically use Alexa Fluor-conjugated secondaries raised in goat or donkey, which we dilute 1:500 in block.
   (ii) If your antibodies are stored diluted in a 50% glycerol solution, you will need to double the amount of antibody solution per well.
   (iii) While we typically use immunofluorescence, it is also possible to perform stereological studies using diaminobenzidine (DAB) staining, as long as your region of interest can be clearly seen while using the 5× objective for tracing. Following primary antibody incubation:
      (a) Wash sections 3× 10 min in 1× PBS and incubate in goat anti-rabbit biotinylated secondary antibody (1: 250 dilution; Vector Laboratories) for 2 h at room temperature.
      (b) Incubate sections in Vectastain Elite ABC reagent (Vector Laboratories) for 2 h at room temperature.
      (c) Wash 3× 10 min in 1× PBS.
      (d) Incubate sections in nickel-intensified DAB to visualize staining (DAB kit; Vector Laboratories). E. Mount sections onto gelatin-coated slides, dehydrate, clear with xylene, and coverslip using Permount (Thermo Fisher Scientific).

7. Coverslipping reagents for immunofluorescence.
   (i) An aqueous (non-hardening or non-curing) medium is necessary for stereological experiments as it minimizes tissue shrinkage. *See* Subheading 2 for our recommended mounting media.
   (ii) Make sure that the slide is sealed perfectly well or the aqueous mounting medium will leak. Clear nail polish works very well for this purpose. Apply to the edges of coverslip and let dry flat overnight in a drawer or light-tight box.
   (iii) Always test your mounting media of choice before starting stereology experiments; if the tissue shrinks below 20 μm, it will not be possible to perform accurate stereological counting.
   (iv) Importantly, we have found that most hard-mount media (with or without DAPI) will significantly shrink tissue to a final thickness of <20 μm from an original thickness of 35 μm and thus should not be used for stereological experiments.
8. Note on *Stereoinvestigator* Software
   (i) The term *stereology* itself refers to the mathematical and statistical concepts used to make estimations of the distribution of objects in three-dimensional space from two-dimensional information. This can be done manually or using software. This protocol covers the use of MicroBrightField (MBF Bioscience) *Stereoinvestigator* and is specifically written with instructions for version 2020.2.3.
   (ii) This chapter is a collection of our stereology protocols and tips for optimal use of *Stereoinvestigator* developed over years by several experimenters. We attempt to include the background information necessary to understand how to design probes and conceptualize experiments, but www.stereology.info is an excellent site curated by MBF which includes further technical reading that may be helpful for advanced users.
   (iii) We recommend contacting MBF online support directly should you have specific questions pertaining to translating this protocol to a different version of *Stereoinvestigator*, or if you encounter software-specific issues.
9. Parcentric and parfocal calibration and moving your slide in *Stereoinvestigator*.
   (i) *Parcentric calibration.*

      This calibration compensates for the deviations from parcentricity that are normally encountered between different microscope objective lenses and ensures that the same X-Y position is found when switching between objectives.

(ii) *Parfocal calibration.*

This calibration compensates for the deviations from parfocality (focal plane) that are normally encountered between different microscope objective lenses and ensures that the same Z depth (i.e., focus) is found when switching between objectives.

(iii) *Moving around your slide in Stereoinvestigator.*

Joy Free will move the stage/slide while *leaving your tracing where it is on the screen.* It gives the impression that the slide is moving underneath the tracing and is useful for aligning the correct area of the section with existing tracings or moving to the first section on the slide to start tracings.

Joy Track will move the *tracing and the slide together.* **Always** use Joy Track to move within or between sections once you have starting making your contours. As you add contours, Stereoinvestigator will automatically create a representation of your slide in the program, so when you move with Joy Track to another contour, you will be in the correct position on the slide.

Macro View → Go to will allow you to jump to any position on the slide once you have made your tracings. The macro view will populate and adjust the space as you add tracings while conserving your position on the slide with *Joy Track*, eventually giving you a miniature "bird's-eye view" of all of the contours on your slide. By clicking on a place in relation to your tracings, the stage will automatically line up the corresponding place on your slide.

10. Blank screen?

If the image window of Stereoinvestigator is empty (no image can be seen):

(i) Click on the LiveImage button.

(ii) Check that your image settings are within the histogram range for the light received by the camera (the black arrowhead should be at the beginning of the histogram and the white at the end of the gray area, which represents the light received by the camera).

11. Grid tune

This will calibrate how well the image transmitted through the camera represents distance in the X-Y plane. The accuracy of this calibration can be confirmed by measuring a line between two of the grid lines. This can be done by going into *Tools → Quick Measure Line* and drawing a line between the vertical and horizontal lines of the gird near the center of the screen and seeing that it is close to 250 or 25 μm depending on the calibration grid used.

12. Not lining up?

    If the calibration grid is crooked in comparison to the white dotted lines, your camera needs to be realigned to the stage. This can usually be achieved by gently rotating the camera until the image on the screen lines up perfectly with the grid (i.e., what is being observed on the stage). When satisfied, carefully retighten camera attachment.

13. Tracing tips

    (i) Inter-experimenter variability in stereological experiments tends to be quite a bit bigger than intra-experimenter variability. For this reason, it is always best to have one experimenter perform the entire stereological experiment to reduce the introduction of variability into the study.

    (ii) It is best to trace all the brains for an experiment together and then go back and count them. This way the tracings will be more consistent across the experiment.

    (iii) To avoid bias, you should blind your experimenter to the condition of each slide.

14. File organization.

    Stereoinvestigator is designed to have all the sections in the same region of interest in one data file. All sections that will be included in the same estimation should be saved into **one file per subject per slide** (e.g., it's okay to have one PFC file and one VTA/midbrain file per animal since those regions will be counted separately and can be mounted on different slides). But all the contours you plan to count together **must be on one slide and in one file** (e.g., all NAc sections need to be in the same file), and contours that will be counted separately but are on the same slide (e.g., PFC subregions and NAc) **can** be in the same or different files. For this latter case, it is generally more efficient to keep them in the same file, as multiple contours can be used in the same file.

15. Tracing contours.

    (i) If you start tracing your contours in the workflow, you need to select which section you are working on starting to trace in order to assign those contours (your tracings) to that specific section (*see* **step 3** in Subheading 3.2.4).

    (ii) If you are tracing contours in a new file (before starting the workflow), simply trace them in order, rostral to caudal, and then assign the contours to sections when prompted to (*see* **step 3** in Subheading 3.2.4). They will appear in the menu in the order that they were traced. You can confirm which section a contour belongs to by going to the "Trace your Region(s) of Interest" menu within the optical fractionator workflow, going to *Options → Display Options* and checking the box for "Display only the current section," and then selecting each section in turn from the optical fractionator workflow.

(iii) If you need to change which contours are assigned to which section, go to *Trace → Serial Section Manager.*

16. Accounting for a missing section.

(i) If a section is missing from your slide, you may need to count two serial sections to stay within the bounds of the atlas plates **OR** skip the section and account for missing sections in your workflow. It is best to account for a missing section when it is in the center of the sections you are counting (*see* Subheading 3.2.6, for more information).

(ii) When the first or last section is missing, it is more accurate to alter the sections counted. For example, instead of counting the sections in rows 1 and 3 of column 2 and row 2 of column 3, if the section in row 1 is missing, you can choose instead the sections in row 2 column 2 and rows 1 and 3 of column 3 if they are within your region of interest.

17. Prefrontal cortex proportions.

(i) The five PFC sections that we count from correspond roughly to plates 14, 15, 16, 17, and 18 of the *Mouse Brain Atlas* (*see* Subheading 2). We recommend making the proportional calculations for each subregion for the atlas edition standardly used in your laboratory.

(ii) For each plate, measure with a ruler the length of the *lateral boundaries* of the three subregions and calculate the total length of the PFC.

(iii) Calculate the proportion of each section within the entire PFC (*see* Fig. 2 and Table 3 for examples). Use these numbers as a reference for your tracings by keeping the

Table 3
Example PFC subregion proportions calculated from atlas

	Plate 15	
	cm	%
Cingulate	0.1	3
Prelimbic	1.5	44.1
Infralimbic	1.8	52.9
Total	3.4	100

contour for each subregion occupying roughly the same proportion of PFC as in the atlas.

(iv) Try to make the proportions as close as you can as the proportions of the atlas. You can choose if 1 or 2% away is fine, but be consistent in all the sections!

(v) Always use the same atlas for reference, and if you want to change to another edition, recalculate these proportions.

18. Cavalieri estimator

(i) We use a grid size of 100 μm and section thickness of 35 μm.

(ii) The Cavalieri estimator is a method of segmentation with point-counting used to estimate the volume of a three-dimensional space from two-dimensional information. It is an essential step not only to determine the volume of an ROI, but also calculating the density of an object within an ROI is of interest. It uses a point grid laid over each 2D plane with a uniform distance between points (e.g., 100 μm as above). It *is* possible to calculate the volume without defining contours by using *Snap to Grid* mode and selecting each point in your region of interest. However, we typically find predefining the contours much easier since it is then sufficient to select the entire contour to fill with markers.

19. Determining the best parameters for your stereological experiment.

(i) **DO NOT change parameters for every section within an animal! Or, for that matter, do not change parameters *between mice* all within an experiment! All counts for the same experiment should use the same parameters for section interval, grid size, probe size, and number of sections to count.**

(ii) It is **ALWAYS** a good idea to run a pilot study to determine if your parameters are well chosen to maximize accuracy of your counts (i.e., minimize error). Stereological parameters are always a trade-off between the amount of work and the accuracy of the counts, which is evaluated by the coefficient of error (CE, *see* **Note 20**). Counting more during the probe runs (e.g., increasing the number of counting sites per section, or increasing the counting frame) will not necessarily increase the estimated total number of objects at the end of your probe run but will affect heavily the *accuracy* of your counts and *decrease* your coefficient of error—albeit *increasing* your workload. The accuracy of your estimate therefore results from a balance between the sampling interval, grid size, counting frame size, and probe depth—increasing or decreasing any of these parameters to alter the work/accuracy trade-off (Fig. 3). It is **ALWAYS** advisable in a stereology experiment to err on the side over oversampling rather than undersampling.

Section Interval.

The interval between sections should be carefully chosen to span the area of interest.

Grid Size (SRS).

Grid size will establish the number of counting sites within each region of interest by laying a grid over the region and placing the counting frame at one corner of each square. The grid incorporates the principle of systematic random sampling, SRS, as the grid is placed *randomly* over the region of interest and the sampling sites are *systematically* placed at the corner of each square formed by the grid. *Increasing* the grid size (i.e., the size of each box formed by the grid) will *decrease* the number of counting sites and *increase* the coefficient of error (CE, *see* **Note 20-II** for a description). Typically we use a square grid, i.e., the length and height of each box formed by the grid are equal; however, a rectangular grid can be used to increase or decrease counting sites in a directional manner (e.g., decreasing the width of the boxes while keeping the height the same will increase the number of sites along each row of the grid while keeping the number of sites along each column the same) (*see* Fig. 3).

Counting Frame Size.

The size of the counting frame should be adjusted so that for each experiment it contains at least between one and five objects per counting site. For this reason, it is generally necessary to take into account the size and density of the object you would like to count when determining the counting frame size—use a larger counting frame for larger objects (e.g., DA cell bodies in the VTA) and a smaller frame for smaller objects (e.g., DA varicosities in the PFC), and for densely distributed objects (e.g., DA varicosities in the NAc), the counting frame size can be further decreased. *See* Table 2 for examples.

Disector Depth.

The optical disector allows you to move through the Z-plane of your tissue and distinguish objects that overlap in this plane. This must be sufficient to distinguish objects, but not too large as to exceed the boundaries of some tissue sections or not accommodate the use of guard zones. In regions densely innervated with small objects (e.g., the NAc, *see* Table 2), disector size can be reduced.

Guard Zone.

The guard zone ensures you do not have an under or overestimation of objects due to issues with objects at the surface of the tissue, for example, having an uneven cut surface or having objects that are cut or damaged close to the surface. Normally the guard zone will be between 3 and 5 μm at the surface of the tissue. A bottom guard zone is not necessary to explicitly define in the software but should be implicitly taken into account when designing your parameters (e.g., for tissue cut at 35 μm, with a minimum measured depth of 20 μm, guard zone of 5 μm + disector probe depth of 10 μm = implicit bottom guard zone of at least 5 μm). You should choose a total depth for the disector and guard zones that can be accommodated in your thinnest piece of tissue.

20. Interpreting your results sheet.

    I. You will get several estimated totals for your object of interest:

    - **Estimated Population Using User-Defined Section Thickness**

      This is the least accurate measure, and if you did not manually enter a section thickness, it will be 0.

    - **Estimated Population Using Mean Section Thickness.**

      This estimate is calculated based on the average thickness of the tissue that you measured within the workflow at every site, whether cells were counted or not.

    - **Estimated Population Using Mean Section Thickness with Counts.**

      This estimate is calculated based on the average thickness of the tissue only for the sites where you marked cells.

    - **Estimated Population Using Number-Weighted Section Thickness**

      The estimate takes into account wavy tissue.

    We use the **Estimated Population Using Mean Section Thickness with Counts** measure.

    II. You will also get several error measurements. We use the **Coefficient of Error (Gundersen), *m* = 1** which typically needs to be ≤0.1 to ensure that your sampling is accurate. The coefficient of error (CE) has no meaning in a biological sense (i.e., it is not measuring inter- or intra-subject variability), and it is used solely for evaluating the precision of stereological estimates [28, 36]. It can be defined mathematically as the standard error of the mean of repeated estimates divided by the mean [37, 38] and is a *prediction* of the difference between the estimate and the true value [39], which is impossible to know definitively in biological samples.

21. Counting objects.

    Rules for Including Cells.

    When counting cells, each experimenter will want to define a priori the point of the cell (such as the top of the cell, the widest perimeter of the cell, the nucleus, the widest perimeter, etc.) that they will count. Then you will place a marker in your counting frame only when that point of the cell comes into focus; you will place only **as long as it meets the rules of the counting frame** (*see* Subheading 3.2.5). The subjectivity in this step is one reason why it is always preferable to have one experimenter do all of the counting for an experiment: while inter-subject variability is generally low in experiments done by a single experimenter, it can vary much more significantly between different experimenters.

    What Is a Varicosity?

    TH+ varicosities are defined as *circular enlargements along a TH+ fiber* [16, 31, 35, 40, 41]. TH+ varicosities should only be counted if they are clearly associated with an axon when visualized with a 100× objective; do not count spots that are not clearly part of an axon! The easiest way to count is to follow an axon while moving through the Z-plane with the fine focus (page up/down in an image stack), marking all the varicosities along the axon. Then repeat for all of the axons within your probe.

22. Why take image stacks? (Or not?)

    (i) Image stacks allow you to continue your counting from any computer with *Stereoinvestigator* installed, not only the one which is attached to the microscope.

    (ii) Counting double-labeled cells or varicosities is easier, since you can count directly from the merged image, whereas in live counting, you will need to toggle between the two emission filters on the microscope.

    (iii) You can store and keep images for later reference.

    (iv) The only disadvantage of taking image stacks is that it adds time to the experiment; an experienced investigator can count varicosities live in less than the time it takes to make the image stacks.

23. How many regions should I take stacks for at the same time?

    While you can take stacks for multiple sections at once, it's better to take them one section at a time so that the focus of the microscope does not drift. In this case, you can set up the stacks for all regions of interest on the same section at the same time (e.g., Cg1, PrL, IL cortices). Taking image stacks for the PFC with our parameters takes about an hour per section.

24. Section top or bottom?

    If it is difficult to determine the top of the section from the bottom, the Z-meter will tell you which direction you are moving in. For live counting, this is less important, but for stacks, it is essential to set the top of your section to correctly take the images.

Acknowledgments

CF is supported by the NIDA (R01DA037911), the Canadian Institutes of Health Research (MOP-74709; MOP-119543), and the Natural Sciences and Engineering Research Council of Canada (2982226).

LMR was supported by a NIDA—Inserm Postdoctoral Drug Abuse Research Fellowship from the National Institute on Drug Abuse (NIDA, USA) and Institut national de la santé et de la recherche médicale (Inserm, France).

References

1. Wise RA (2004) Dopamine, learning and motivation. Nat Rev Neurosci 5:483–494
2. Schultz W (2007) Behavioral dopamine signals. Trends Neurosci 30:203–210
3. Bromberg-Martin ES, Matsumoto M, Hikosaka O (2010) Dopamine in motivational control: rewarding, aversive, and alerting. Neuron 68:815–834
4. Cools R, D'Esposito M (2011) Inverted-U-shaped dopamine actions on human working memory and cognitive control. Biol Psychiatry 69:e113–e125
5. Orsini CA, Moorman DE, Young JW, Setlow B, Floresco SB (2015) Neural mechanisms regulating different forms of risk-related decision-making: insights from animal models. Neurosci Biobehav Rev 58:147–167
6. Coddington LT, Dudman JT (2019) Learning from action: reconsidering movement signaling in midbrain dopamine neuron activity. Neuron 104:63–77
7. Reynolds LM, Flores C (2021) Mesocorticolimbic dopamine pathways across adolescence: diversity in development. Front Neural Circuits 15:735625
8. Walker DM, Bell MR, Flores C, Gulley JM, Willing J, Paul MJ (2017) Adolescence and reward: making sense of neural and behavioral changes amid the chaos. J Neurosci 37: 10855–10866
9. Paus T, Keshavan M, Giedd JN (2008) Why do many psychiatric disorders emerge during adolescence? Nat Rev Neurosci 9:947–957
10. Lee FS, Heimer H, Giedd JN, Lein ES, Šestan N, Weinberger DR, Casey BJ (2014) Mental health. Adolescent mental health--opportunity and obligation. Science 346: 547–549
11. Barth B, Portella AK, Dubé L, Meaney MJ, Silveira PP (2019) The interplay between dopamine and environment as the biological basis for the early origins of mental health. In: Vaiserman A (ed) Early life origins of ageing and longevity. Springer International Publishing, New York, pp 121–140
12. Reynolds LM, Flores C (2021) Adolescent dopamine development: connecting experience with vulnerability or resilience to psychiatric disease. In: Martin CR, Preedy VR, Rajendram R (eds) Diagnosis, management and modeling of neurodevelopmental disorders. Academic Press. https://doi.org/10.1016/b978-0-12-817988-8.00026-9
13. Reynolds LM, Flores C (2019) Guidance cues: linking drug use in adolescence with psychiatric

disorders. Neuropsychopharmacology 44(1): 225–226

14. Vassilev P, Pantoja-Urban AH, Giroux M, Nouel D, Hernandez G, Orsini T, Flores C (2021) Unique effects of social defeat stress in adolescent male mice on the Netrin-1/DCC pathway, prefrontal cortex dopamine and cognition (social stress in adolescent vs. adult male mice). eNeuro 8(2):ENEURO.0045-21.2021
15. Kopec AM, Smith CJ, Ayre NR, Sweat SC, Bilbo SD (2018) Microglial dopamine receptor elimination defines sex-specific nucleus accumbens development and social behavior in adolescent rats. Nat Commun 9:3769
16. Manitt C, Eng C, Pokinko M, Ryan RT, Torres-Berrío A, Lopez JP et al (2013) dcc orchestrates the development of the prefrontal cortex during adolescence and is altered in psychiatric patients. Transl Psychiatry 3:e338
17. Reynolds LM, Pokinko M, Torres-Berrío A, Cuesta S, Lambert LC, Pellitero EDC et al (2018) DCC receptors drive prefrontal cortex maturation by determining dopamine axon targeting in adolescence. Biol Psychiatry 83: 181–192
18. Naneix F, Marchand AR, Scala GD, Pape J-R, Coutureau E (2012) Parallel maturation of goal-directed behavior and dopaminergic systems during adolescence. J Neurosci 32: 16223–16232
19. Xu B, Goldman JS, Rymar VV, Forget C, Lo PS, Bull SJ et al (2010) Critical roles for the netrin receptor deleted in colorectal cancer in dopaminergic neuronal precursor migration, axon guidance, and axon arborization. Neuroscience 169:932–949
20. Peretti-Renucci R, Feuerstein C, Manier M, Lorimier P, Savasta M, Thibault J et al (1991) Quantitative image analysis with densitometry for immunohistochemistry and autoradiography of receptor binding sites—methodological considerations. J Neurosci Res 28:583–600
21. Zhou FC, Zhang JK, Lumeng L, Li T-K (1995) Mesolimbic dopamine system in alcohol-preferring rats. Alcohol 12:403–412
22. Phillips GD, Setzu E, Vugler A, Hitchcott PK (2003) An immunohistochemical examination of the effects of sensitisation on the mesotelencephalic dopaminergic response to d-amphetamine. Neuroscience 117:741–753
23. Vos PE, Steinbusch HWM, Ree JMV (1996) Reinnervation after destruction of the dopaminergic system in the rat nucleus accumbens: a quantitative immunohistochemical analysis. Neurosci Lett 207:21–24
24. Brown DL (2017) Bias in image analysis and its solution: unbiased stereology. J Toxicol Pathol 30:183–191
25. Bratu VA, Erpenbeck VJ, Fehrenbach A, Rausch T, Rittinghausen S, Krug N et al (2014) Cell counting in human endobronchial biopsies – disagreement of 2D versus 3D morphometry. PLoS One 9:e92510
26. Geinisman Y, Gundersen HJG, Zee EVD, West MJ (1996) Unbiased stereological estimation of the total number of synapses in a brain region. J Neurocytol 25:805–819
27. West MJ, Slomianka L, Gundersen HJG (1991) Unbiased stereological estimation of the total number of neurons in the subdivisions of the rat hippocampus using the optical fractionator. Anat Rec 231:482–497
28. Schmitz C, Hof PR (2005) Design-based stereology in neuroscience. Neuroscience 130: 813–831
29. Deniz ÖG, Altun G, Kaplan AA, Yurt KK, von Bartheld CS, Kaplan S (2018) A concise review of optical, physical and isotropic fractionator techniques in neuroscience studies, including recent developments. J Neurosci Methods 310:45–53
30. West MJ (2002) Design-based stereological methods for counting neurons. Prog Brain Res 135:43–51
31. Reynolds LM, Makowski CS, Yogendran SV, Kiessling S, Cermakian N, Flores C (2015) Amphetamine in adolescence disrupts the development of medial prefrontal cortex dopamine connectivity in a dcc-dependent manner. Neuropsychopharmacology 40:1101–1112
32. Cuesta S, Nouel D, Reynolds LM, Morgunova A, Torres-Berrío A, White A et al (2020) Dopamine axon targeting in the nucleus accumbens in adolescence requires netrin-1. Front Cell Dev Biol 8:487
33. Hoops D, Reynolds LM, Restrepo-Lozano JM, Flores C (2018) Dopamine development in the mouse orbital prefrontal cortex is protracted and sensitive to amphetamine in adolescence. eNeuro 5:ENEURO.0372-17.2017
34. Cuesta S, Restrepo-Lozano JM, Popescu C, He S, Reynolds LM, Israel S et al (2019) DCC-related developmental effects of abused- versus therapeutic-like amphetamine doses in adolescence. Addict Biol 25:e12791
35. Manitt C, Mimee A, Eng C, Pokinko M, Stroh T, Cooper HM et al (2011) The netrin receptor DCC is required in the pubertal organization of mesocortical dopamine circuitry. J Neurosci 31:8381–8394
36. Slomianka L, West MJ (2005) Estimators of the precision of stereological estimates: an

example based on the CA1 pyramidal cell layer of rats. Neuroscience 136:757–767

37. Gundersen HJG, Jensen EB (1987) The efficiency of systematic sampling in stereology and its prediction*. J Microsc 147:229–263
38. Gundersen HJG, Jensen EBV, Kiêu K, Nielsen J (1999) The efficiency of systematic sampling in stereology—reconsidered. J Microsc 193: 199–211
39. Glaser EM, Wilson PD (1998) The coefficient of error of optical fractionator population size estimates: a computer simulation comparing three estimators. J Microsc 192:163–171
40. Parish CL, Finkelstein DI, Drago J, Borrelli E, Horne MK (2001) The role of dopamine receptors in regulating the size of axonal arbors. J Neurosci 21:5147–5157
41. Sesack SR, Hawrylak VA, Matus C, Guido MA, Levey AI (1998) Dopamine axon varicosities in the prelimbic division of the rat prefrontal cortex exhibit sparse immunoreactivity for the dopamine transporter. J Neurosci 18: 2697–2708

Chapter 3

Dendrites Crossing Borders: Quantifying the Distribution of Individual Substantia Nigra Dopaminergic Neurons' Dendritic Tree in Pars Compacta and Pars Reticulata

Rafael Ignacio Gatica, Guillermo Lazcano, and Pablo Henny

Abstract

In the central nervous system, the dendritic tree of individual neurons normally extends beyond the borders of the nucleus or nucleus' subdivision where its cell body locates. We refer to this as heterotopic dendritic distribution (Montero et al., Front Neural Circuits. 15:769342, 2021). This phenomenon has important implications for individual neurons afferent synaptic integration; insofar dendritic trees extending across different nuclei or subdivisions can receive differentiated input. A typical example of neurons exhibiting heterotopic dendrites is the substantia nigra pars compacta (SNc) neuron, whose dendritic domain can extend into the underlying substantia nigra pars reticulata (SNr). In this chapter, we describe three basic approaches to quantify dendritic extensions into adjacent nucleus using 3D reconstructions of previously labeled individual rodent SNc dopaminergic neurons. We used the software Neurolucida, which allows vector-based 3D reconstruction of both individual neurons and nuclei and offers various analytical tools. The first method involves the use of cyto- or chemo-architectonics to identify boundaries between subdivisions and marking, onto the dendrite being traced, the exact point where it crosses the border. The second method utilizes instead rendered 3D reconstructions of the relevant subdivisions (e.g., SNc or SNr) to identify dendritic border crossing, followed by computation of total dendritic length inside or outside the subdivision. We also propose a method based on MATLAB scripts to make the process of quantification more efficient. We finish discussing and listing the advantages and disadvantages of each approach.

Key words Substantia nigra, Dendrites, Neurolucida, MATLAB, Reconstructed neurons

1 Introduction

Dendritic trees play a critical role in defining the neuron's firing pattern of activity due to the varied and complex electric properties they exhibit [1] as well as being a recipient of most synaptic inputs [2]. The architecture of the dendritic domain is complex and characteristic of a given neuronal population. Yet it also exhibits large neuron-to-neuron variation. Because dendritic architecture itself affects both intrinsic properties and synaptic integration

José Antonio Fuentealba-Evans and Pablo Henny (eds.), *Dopaminergic System Function and Dysfunction: Experimental Approaches*, Neuromethods, vol. 193, https://doi.org/10.1007/978-1-0716-2799-0_3,

[1, 2], defining individual neurons dendritic organization in detail is critical to our understanding of single-cell firing mechanisms.

Dendritic trees are diverse and exhibit high branching complexity and size. And, although always depending on the type of neuron and species, the dendritic domain can extend for longer distances, up to over a millimeter [2]. Complexity is also reflected in that, for many neuronal types, the individual neuron's dendritic domain extends across different nuclei, subdivisions, or layers. A typical example comes from neocortical pyramidal cells, with dendritic trees that extend through several cortical layers [3], being each cortical layer differentiated, among other things, by the type of afferent they receive. For instance, layer 2/3 somatosensory cortex pyramidal cells receive somatostatin-positive (SOM) interneuron inputs exclusively in the dendritic tuft located in layer 1 and parvalbumin (PV) interneuron inputs exclusively in dendrites located in layer 2/3 [4] (Fig. 1a). Similarly, layer 5 visual cortex pyramidal cells receive excitatory thalamic inputs in dendrites that are located in layer 4 [5], overall emphasizing that the dendritic tree of individual pyramidal cell is under differential, region-specific (i.e., layer specific), afferent control [4, 5]. Another example of neurons exhibiting heterotopic dendritic distribution are ventral horn motoneurons of the spinal cord which, apart from dendrites extending in the grey matter neuropil, can exhibit dendrites that bridge into the white matter [6, 7]. Such dendrites are innervated by axon terminals, likely granting descending and ascending spinal systems direct control of motor neurons activity, as proposed in ref. 6 (Fig. 1b). Also, dopaminergic neurons of the parabrachial pigmented nucleus (PBP) subdivision of the ventral tegmental area (VTA) extend their dendritic domain to adjacent VTA subdivisions and even the substantia nigra [8] (Fig. 1c).

A further, and well-reported, case of heterotopic dendritic distribution is that of substantia nigra pars compacta (SNc) dopaminergic neurons. Many SNc neurons have dendrites that descend into the substantia nigra pars reticulata (SNr) [8–11]. Afferents to SNc originate from regions or neuronal types that are different to those that provide afferents to SNr, which in turn results in SNc and SNr dendrites exhibiting contrasting proportions of inhibitory and excitatory inputs (discussed in ref. 11). Critically, dendrites located in the SNr appear to conform a distinctive compartment that endows neurons with distinctive physiological properties. For example, SNc dopaminergic neurons with a higher proportion of their dendrites in the SNr exhibit more regular firing *in vivo* [8] (Fig. 1d), are more likely to be inhibited following nociceptive stimulation [11], and are responsible for SNc neuron's inhibition after local current injection [12].

In this chapter, we describe three approaches we have used to quantify dendritic extensions of individual SNc neurons into the SNr using 3D reconstructions of previously labeled SNc

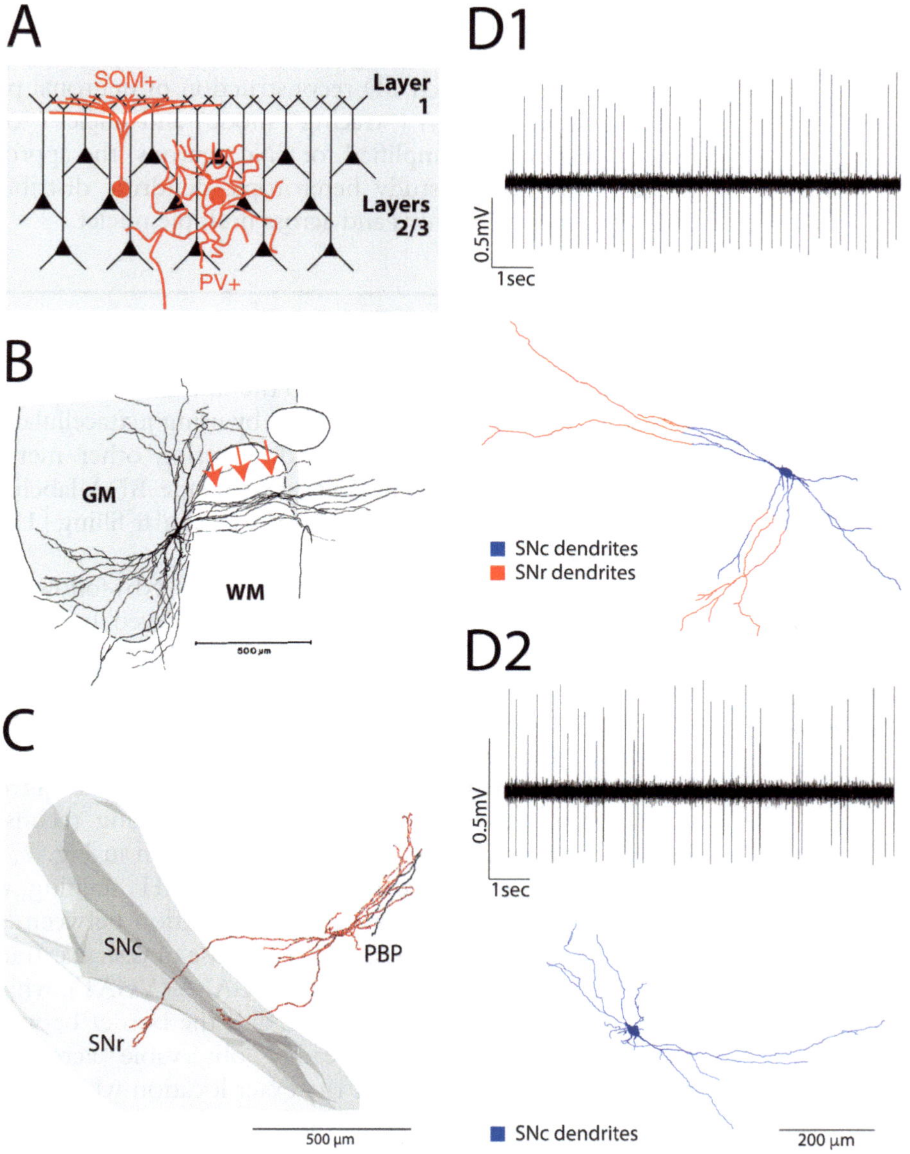

Fig. 1 Examples of heterotopic dendritic domain distribution. (**a**) The dendritic domain of mouse layer 2/3 pyramidal cells extends across layers 1, 2, and 3. Dendrites in layer 1 receive axonal innervation of local somatostatin (SOM) interneurons, whereas dendrites in layers 2 and 3 receive axonal innervation of local parvalbumin (PV) interneurons. (Modified from Packer et al. [4], https:/doi.org/10.1093/cercor/bhs210, with permission under fair use licensing). (**b**) The dendritic domain of a motor neuron is distributed beyond ventral horn gray matter into adjacent white matter anterior columns where there may receive distinctive, yet sparser, synaptic innervation [6]. (Modified from Rose and Richmond [6] with permission). (**c**) A PBP neuron from the VTA with dendrites extending beyond the PBP and VTA borders and into the substantia nigra. (Modified from Montero et al. [8] with permission under Creative Commons use licensing). (**d**) A differential extension of their dendrites in the SNr is found in SNc dopaminergic neurons (red dendrites are inside the SNr; blue dendrites are in SNc). (**d1**) As measured during previous juxtacellular recording, neurons with a high proportion of their dendrites in SNr show a much more regular firing pattern. (**d2**) On the other hand, irregular firing was found in neurons with a low proportion of their dendrites in the SNr. (Modified from Montero et al. [8])

dopaminergic neurons [8, 13]. We used the reconstruction software Neurolucida (MicroBrightField Biosciences, Vermont, USA), which uses vector-based 3D reconstruction of neuronal processes and individual neurons ("tracing" mode) and nuclei ("contour" mode). Although exemplified for SNc neurons, the approach can be easily applied to study heterotopic dendritic distribution in different types of neurons and across multiple nuclei.

2 Methods

All three methods require access to the somatodendritic domain of individual neurons. We achieved this by using juxtacellular labeling [8, 11, 13], but this could be done using other methods for individual cell labeling, including retrograde BDA labeling [14], sparse viral labeling [15], or *ex vivo* intracellular filling [15, 16].

2.1 Method 1: Manual Measurement of Dendrites in a Subdivision During Neuron Reconstruction

Single neurons are labeled with neurobiotin (Nb) following *in vivo* juxtacellular labeling. Revealing Nb is achieved by using either nickel-enhanced 3,3′-diaminobenzidine (Ni-DAB) avidin-biotin reaction (ABC) or Cy3-conjugated streptavidin for light and fluorescent microscopy, respectively. Reconstruction is carried out as explained elsewhere [17–19]. Critical for this approach is that borders between nuclei must be recognizable by using cyto- or chemo-architectonics, while the neuron is being reconstructed. For SNc dopaminergic neurons, and as shown in Fig. 2, this can be achieved by using tyrosine hydroxylase (TH) staining, which in the substantia nigra allows a clear delimitation between SNc and SNr. Other alternatives exist such as the use of vesicular transporter proteins for glutamate (VGluT2) or GABA (VGAT), which show reciprocal density in SNc and SNr, and the border between these two subdivisions can be readily observable across sections [11]. After determination of the exact location where the labeled dendrite is crossing across the two structures, a marker is added to the tracing to indicate the exact point at which the two subdivisions border each other (Fig. 2). This is repeated for all dendrites that penetrate the SNr, as observed across all sections in which the dendritic tree is present, until the entire dendritic tree is traced. For the sake of simplicity, we are omitting the presence of dendritic fragments that locate elsewhere than SNc or SNr, as described in detail in ref. 8. Then, the full dendritic length of the neuron is measured using Neurolucida Explorer (MicroBrightField Biosciences, Vermont, USA) (Fig. 2).

Next, markers are revisited in Neurolucida, and a tracing point is manually added at the exact marker's location. This latter step allows splitting the dendrite at that point. This is repeated for all the dendrites which presented markers to isolate all fragments present in the SNr. In some cases, dendrites located in the SNr can reenter

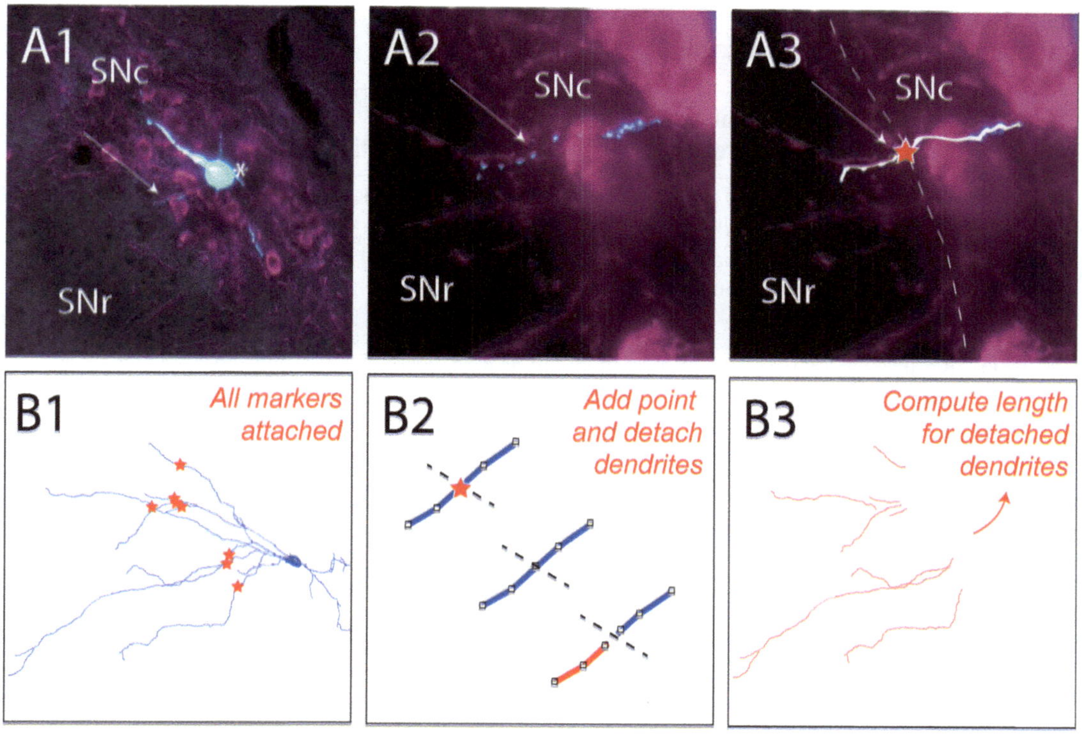

Fig. 2 *Method 1.* Estimating the proportion of dendrites crossing into an adjacent nucleus by marking borders on the tissue while reconstructing. As part of the reconstruction process, dendritic fragments are observed at high magnification (streptavidin-Cy3 revealed Nb+ dendritic fragment, in magenta, **a1–2**) and vector-based reconstructed in Neurolucida (in white, **a3**). Using double co-staining for TH in the same tissue section (revealed with secondary Alexa 488 antibodies, in cyan) the border between SNc (top, with TH+ cells) and SNr (bottom, no TH+ cells) is readily identified (**a1–3**). A marker is attached to the traced dendritic fragment to indicate the border (arrow in **a2**, **a3**; red star in **a3**). Once all dendritic and somatic compartments have been traced, all fragments are spliced up into a full reconstructed neuron, which maintains the SNc-SNr markers attached to the dendrites (**b1**). In a separate file, all dendritic fragments distal to the markers are detached from the parent dendritic tree by adding a point at the marker's location and manually detaching it (**b2**). The detaching process is repeated for all dendritic fragments. Then, dendritic fragments outside the SNr are deleted, and length computation for SNr dendrites is done using Neurolucida Explorer (**b3**)

the SNc, in which case a special marker was added to indicate reentry of the dendritic fragment into the SNc. Then, all dendritic portions inside the SNr are detached from the parent dendrites. Afterward, all dendritic fragments located outside the SNr are deleted, and the file is saved apart. Finally, the file containing the dendrites located in the SNr is opened in Neurolucida Explorer, and the dendritic length is obtained.

The advantage of this method is that it allows an accurate representation of the borders between two structures. On the other hand, it is time-consuming as it requires visiting dendrite by dendrite and then adding markers and tracing points and then detaching each dendrite from its parent tree (Table 1). In the case

Table 1
Summary of advantages and disadvantages of each method

	3D map	Identification of dendrite entering adjacent regions	Length estimation	Advantages	Disadvantages
Method 1	Not required	Marker on dendritic fragment as it enters adjacent region, as assessed by cytoarchitectonic or neurochemical identification of regions	Manual	Accuracy of limits Does not require 3D map	Time investment
Method 2	Required	Marker on dendritic fragment as it enters adjacent region, as assessed by examination of the semitransparent, rendered 3D map	Manual	Common standard map	Time investment. Less accuracy in limits
Method 3	Required	Automatic	Automatic	Time	Less accuracy in limits

of neurons with denser dendritic domains, this method may not be practical, such as with motoneurons or pyramidal cells which may present much more complex dendritic trees (Fig. 1b).

2.1.1 Method 1: Step by Step

1. *Labeling, reconstruction, and marking borders*
   - (a) Label an individual neuron dendritic domain. *In our case, we record neurons* in vivo *and label them with Nb using juxtacellular labeling. Alternative means include Golgi impregnation or genetically engineered labeling of individual neurons.*
   - (b) Process the tissue to reveal the neuron's somatodendritic domain. *In our case, we perfuse the animal, then cryoprotect the brain, cut it in serial sections, and then reveal Nb with fluorescently tagged streptavidin or using the ABC–DAB reaction to visualize the Nb in a light microscope.*
   - (c) Co-label tissue to provide a mean to distinguish subdivisions: use Nissl staining or immunohistochemistry for cyto- or chemo-architectonics identification of nucleus boundaries. *In our case, we carry out immunohistochemistry for TH* (Fig. 2) *or other types of staining that can distinguish borders between SNc and SNr.*
   - (d) Reconstruct the somatodendritic domain of individual neurons using 3D vector-based tracing. *In our case, we used serial section reconstruction of neurons with Neurolucida coupled with a 100× microscope, or using confocally acquired stacks of images containing dendritic fragments, based on reconstruction protocols described elsewhere* [17, 19].

(e) As tracing is being conducted, identify the border between one subdivision to another, for all the dendritic fragments being traced (Fig. 2a2).

(f) Add a marker to the tracing to indicate the location of the subdivision border. Attach the marker to the traced dendritic fragment to the dendrite so it keeps attached to the tracing, irrespective of tracing displacement, rotation, or orientation (Fig. 2a3). *In our case, we use the "add marker" command → "attach marker to branch" option. First, click to add the marker, and then using right click over the marker, select "attach marker to branch" and click on the dendritic fragment. Repeat for all dendritic fragments crossing over a border.*

(g) Splice each serial section's dendritic fragments with the correspondent dendritic fragment of adjacent sections to finish reconstruction.

2. *Quantification of dendritic length in a given subdivision*

(a) Open the reconstruction file and save a copy. Identify the markers attached to the dendrites. *In our case, we used the software Neurolucida.*

(b) Add a tracing point to the dendritic fragment in the exact location the marker locates. *In our case, we did that by right-clicking next to the dendrite (not over the dendrite) and selecting "add point." Then, click on the exact location on the dendrite to add that point.*

(c) Detach the dendrite at the point added in **step 2(b)** and repeat for all dendrites with markers. Note that the detached dendritic fragment now has a new color. Save another copy of the file. *In our case, we place the cursor over the new point added previously and right-click menu. Then select "detach branch." The dendritic fragment will be detached.*

(d) Select all proximal dendrites, as well as the cell body, and delete them. Save a new file. *In our case, they can be easily identified because they keep the original tracing color. They are also proximal to the cell body. Once selected, delete them from that file, using the "delete" command from right-click menu.*

(e) Open the new file in an analytic software and extract the length of all dendrites in the file. Compare that length with the length of the original neuron. *In our case we opened the file in Neurolucida Explorer, and from the top bar menu, select Analysis → branch → structure analysis.*

2.2 Method 2: Manual Measurement of Dendrites in Subdivision Using a Rendered 3D Map

The second approach combines single-cell reconstruction with a 3D reconstruction of the nucleus or subdivision where dendritic length needs to be determined [8]. Thus, the proportion of dendrites in one nucleus is determined based on whether they locate inside or outside a rendered 3D reconstruction of the subdivision. Reconstruction of the neuron is achieved as described elsewhere [17] (Fig. 3a1). The 3D reconstruction of the subdivision is carried out by tracing the contours of that subdivision across serial sections using the "Serial Section Manager" module on Neurolucida. Contours can be traced from images taken from a standard brain map [20] or a more detailed serial map of the region of interest [21] or from brain series from own's material. Once all the contours are traced, the file is saved. Then, the neuronal reconstruction is copied from its original file and pasted onto the file containing the contour's serial reconstructions (Fig. 3a2). In this case, it is critical that the reconstructed neuron is placed as close as possible to its original

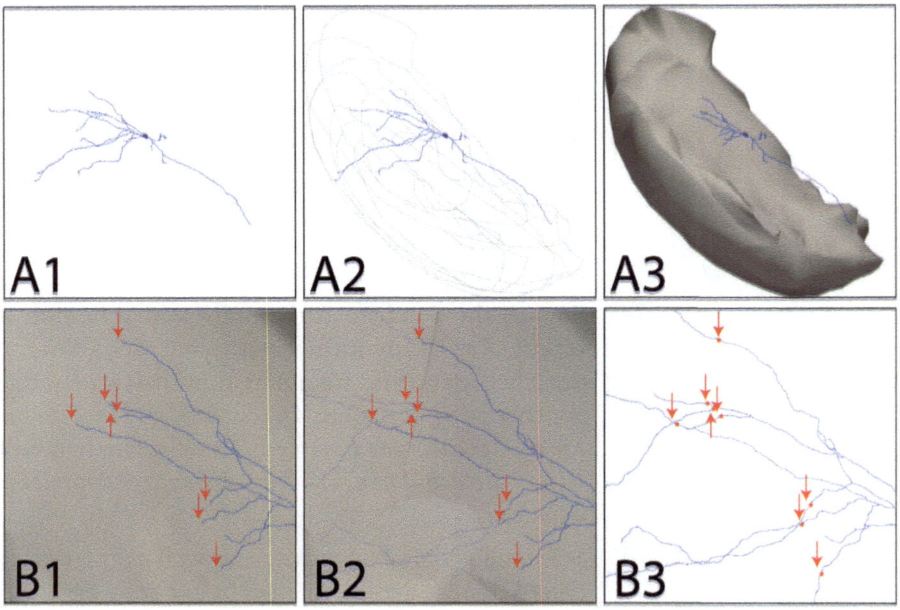

Fig. 3 *Method 2.* Measuring the proportion of dendrites crossing into an adjacent nucleus by marking the borders on a rendered 3D map after single-cell reconstruction. (**a**) This method requires, in addition to the reconstruction of the neuron (**a1**), the tracing of the contours of the nucleus or subdivision that dendrites innervate. In this case, the SNr can be traced based on standard or region-specific maps [24] (**a2**). Alternatively, contours can be traced using neurochemical staining (TH), which allows defining the overlying upper border of the SNr. Using the "3D visualization function" on Neurolucida, a rendered view of the SNr and the reconstructed neuron (that is localized in the SNc) can be obtained (**a3**). Using the "zoom in" function, it can be observed that some dendrites enter the SNr, as some dendritic fragments are hidden from view inside the SNr (**b1**). If necessary, the SNr rendered view is made semitransparent, which enables to see the dendrites that are inside the SNr with diminished intensity (**b2**). Then, a marker is attached to the dendritic fragments to indicate the border (**b3**). As in Method 2, all dendritic points in which there are markers are detached for subsequent length estimation (*see* Fig. 2b1–3)

location, not only in terms of cell body localization within its parent nucleus but also in the proper orientation, which can be achieved by tilting correction, as explained below and in ref. 8.

The file containing both the neuron and nucleus contour reconstruction is opened using the "3D visualize" function within Neurolucida, to see the rendered version of the nucleus (Fig. 3a3). Both the neuronal reconstruction and the rendered 3D reconstruction of the nucleus are fully opaque; thus, the portion of the dendritic tree that is inside the nucleus is hidden (Fig. 3b1). This signals the exact place one dendrite is entering one nucleus (Fig. 3b2). If necessary, transparency of the nucleus can be adjusted so that dendrites inside and outside are easily observed, and then a screenshot is taken. The "3D visualize" window is closed, and, in the reconstruction file in Neurolucida, a marker to the dendritic fragment point in the border between SNc and SNr is attached (Fig. 3b3), in the same way it was described for Method 1. Repeat this process for all dendrites that are crossing the border and continue as in Method 1: add points at the marker border sites, detach dendrites, duplicate files, erase dendrites, and measure the length (*see* Fig. 2b1–3).

The advantage of this approach is that it allows multiple neurons to be mapped on the same rendered 3D reconstruction. Disadvantages, on the other hand, are that it is less accurate in determining the borders and therefore the proportion of dendrites inside or outside a given nucleus (*see* Table 1).

2.2.1 Methods 2: Step by Step

1. *Generating a rendered 3D reconstruction of the nucleus of interest*
   (a) Reconstruct a 3D contour map from brain serial sections containing the nucleus or subdivision of interest with enough contour-to-contour resolution. *In the case of the mouse SNr, it is recommended to sample between six and eight sections/mm of tissue. For example, using 40-μm serial sections, you can sample one of every four sections to achieve a good depiction of the SNr.*
      (i) Use a microscope coupled with a motor stage, and a reconstruction software. *In in our case it was Neurolucida.*
      (ii) At low magnification, trace the relevant contour across all sections under the microscope, as indicated elsewhere [8, 17]. *Use Nissl staining or immunohistochemistry for TH, or other staining to define the boundaries of SNr.*
   (b) Alternatively, use a series of images acquired previously in a different microscope, or a previously published map that contains the subdivision of interest. Open these images in Neurolucida, sort them using the serial section manager with the proper section evaluation interval, calibrate the

lens to the acquired image scale, and trace the relevant contour. *We have used this late option* [8], *in which the 3D map was reconstructed from a previously published characterization of the substantia nigra from mice brain series* [21].

(c) Open the contour on Neurolucida, click on the "3D visualization" function, and familiarize with the rendering display. *To view a rendered display of the subdivision contour, we recommend using the shell option: click in Contour → Select the contour → Show as: Shell.*

2. *Placing the neuron on the 3D map and marking the borders on the neuron tracing*

(a) Copy the neuron from the original reconstructed file of Neurolucida and paste it onto the file containing the 3D map. *You must first open the neuron file in Neurolucida, press select all, and then copy. Close Neurolucida and then open the file with the 3D map and press paste.*

(b) Place the neuronal cell body on the corresponding position in the map, as it was on the brain from where the neuron was reconstructed. That implies not only displacing the cell body and reconstruction on mediolateral, dorsoventral, and anteroposterior directions to match the original location but also tilting the entire reconstruction if necessary. *To assure that the tilting is correct, find the distal dendrites in the stained brain sections, look where they locate, and then correct the position in the 3D map to match what is found in the stained brain section using the rotation function for each axis* [8].

(c) Observe the neuron and 3D map on the "3D visualization" window of Neurolucida, displaying the rendered view of the contours (*shell view, as explained in* ***step 1 (c)***). Display the contour rendering with no transparency, to detect the exact location where dendrites enter the rendered contour. Dendrites inside the subdivision of interest should not be visible. Take a screenshot.

(d) Make the contour 3D view rendering semitransparent, such as to observe dendrites crossing from one subdivision to the other. Take a screenshot. Compare the screenshots of **steps 2(c)** and **2(d)** and determine the exact location where dendrites cross (Fig. 3b1–2). Close the "3D visualization window." *To decrease the transparency of the 3D rendering of the subdivision, move the horizontal lever of the transparency bar (under the "Show as:" option) to the right until the dendrites inside the subdivision can be observed.*

(e) Add a marker to that exact location where the dendrite crosses to the analyzed subdivision, using the normal Neurolucida view file (similar to Subheading 2.1, **step 1 (f)**) (Fig. 3b3).

(f) Repeat **steps 2(c)–2(e)** till all the dendritic fragments entering the subdivision of interest are marked.

(g) Quantify the proportion of dendrites in the subdivision of interest following Subheading 2.1, **step 2**.

2.3 Method 3: Automatic Measurement of Dendrites Length in a Subdivision

The third method is a novel procedure that we developed using MATLAB (MathWorks, Massachusetts, USA) (Fig. 4). The main aim here was the create a faster method to quantify dendritic extension in a nucleus or subdivision. As in Method 2, it also uses the rendered 3D map of the subdivision of interest. First, the neuron is placed in the appropriate location within the rendered 3D map in Neurolucida and saved it as a containing the reconstructed neuron and the 3D map. Then, this file is used to extract the following two data files that are required as inputs in the MATLAB script: segment points-dendrites data (Fig. 4a2) and an object file (.obj) from the rendered 3D map (Fig. 4b2). This object file contains all the necessary information to create the triangulations required to create the shell polyhedron of the subdivision in MATLAB. Following, the segment points-dendrites data file and the rendered 3D map object file are read by the MATLAB script (Fig. 4c1–2). With these data, the in_polyhedron function [22] is used to obtain the dendritic fragments that are inside (and outside) the rendered 3D map. In the example shown in Fig. 4, the SNc neuron had 41% of its dendrites inside the SNr, and 59% outside the SNr (Fig. 4c3). To validate the MATLAB method, Method 2 and Method 3 were compared, using a sample of 15 SNc dopaminergic neurons from ref. 8 (all freely available in NeuroMorpho.Org database) and found that both methods showed basically the same results (Fig. 4d).

Compared to Methods 1 and 2, Method 3 is a faster procedure, considering that no manual sampling is required and it is only needed to extract a pair of data files from Neurolucida and Neurolucida Explorer. This could be relevant also if sampling of multiple nuclei or subdivisions is required (as in ref. 8), a task that is more time-consuming in Methods 1 and 2. As in Method 2, and in contrast to Method 1, Method 3 has the disadvantage that borders between subdivisions are less accurate.

2.3.1 Method 3: Step by Step

1. *Preparation of the required files for the automatic analysis*

(a) Place the neuron in the appropriate location within the rendered 3D map in Neurolucida (as in Subheading 2.2, **steps 2(a)** and **2(b)**) and save it (ASC file, now named "neuron + 3D map file").

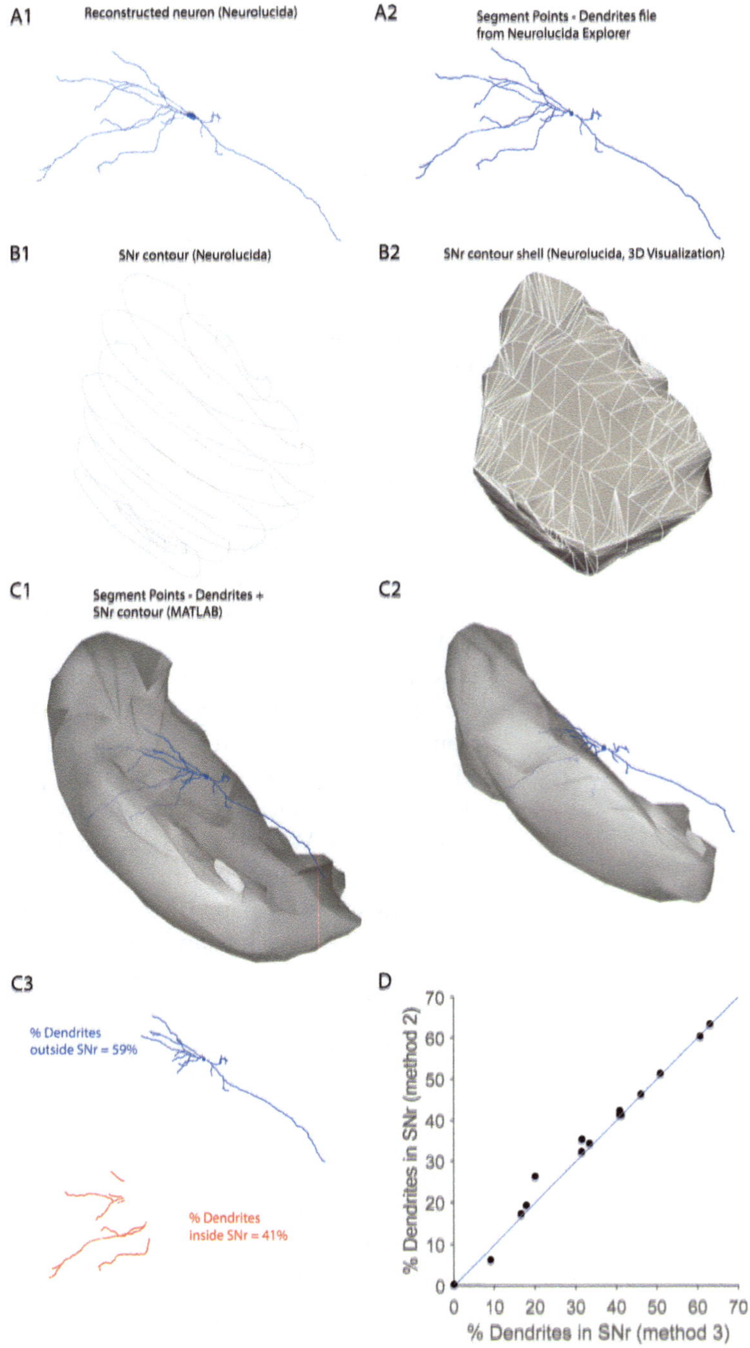

Fig. 4 *Method 3.* Automatic measurement of the proportion of dendrites inside a subdivision. (**a1**) Reconstructed dopaminergic neuron from ref. 8. Image was obtained from the "3D visualization" function on Neurolucida. Frontal view. (**a2**) Scatter representation of the segment point-dendrite data from the neuron in **a1**. Segment point-dendrite data is obtained from Neurolucida Explorer. Figure was generated in MATLAB. Cell body is depicted with a solid point. Small points represent the start of each dendritic fragment. (**b1**) SNr contours reconstructed

(b) Open the "neuron + 3D map file" in Neurolucida Explorer (Fig. 4a1), select all the dendrites, and obtain the segment points-dendrites data (in Structure → Branch Structure → Segment Points-Dendrites) and save it as an excel file (.xlsx) (Fig. 4a2). *This file contains the coordinates where each dendritic fragment begins, as well as the length of each fragment, among other data.*

(c) Open the rendered 3D map in Neurolucida (Fig. 4b1), open the 3D visualizer, select all the contours from the subdivision where dendrites will be measured, and create a shell object of the contours by clicking on "show as" and selecting the shell option (Fig. 4b2). Then, click on "save 3D model" on the top left menu and save it as a Wavefront file (.obj). *This object file contains all the necessary information to create the triangulations required to create the shell polyhedron in MATLAB. If the neuron + 3D map file for this step is opened, all the neuron-related traces (cell body, dendrites, axon) must be hidden. If contours from other subdivisions are present in the rendered 3D map, they must be hidden (the method only works with one subdivision at a time). For hiding extra contours, click on Trace → View (gear symbol, bottom left corner) → Tracing → Contours → Customize. Select as visible only the contours needed. Contours can also be hidden in the 3D visualization window: Contour → Select contour → Hide Selection.*

2. *Measurement of dendrites in a subdivision using MATLAB*

(a) Read the .obj file in MATLAB and assign it to variable named "obj." *To read the .obj file, we use the readObj function* [23]. *The output from readObj corresponds to a structure array. In this array, "obj.v" corresponds to the*

Fig. 4 (continued) using Neurolucida, as in Fig. 3. (**b2**) SNr contour shell representation, using the "3D visualizer" function on Neurolucida. White lines represent the binding points between contours, which form triangulations. (**c1**) Three-dimensional representation of the neuron in **a2** together with the SNr contours in **b2**. Neuron was in its position on the SNc (as in ref. 8, SNc contours not shown). SNr shell was created in MATLAB, using the triangulation data from the SNr shell object file from Neurolucida. (**c2**) Another angle of view from the SNc neuron and SNr shell contour. Dendrites inside SNr have a lighter blue color. (**c3**) Measurement of dendrites inside and outside SNr. Neuron is shown as segment-point dendrites (as in **a2**). A MATLAB script was used to measure the percentages inside and outside SNr. (**d**) Validation of the manual method (Method 2) vs automatic method (Method 3). The percentage of dendrites inside the SNr was calculated for 15 SNc dopaminergic neurons from ref. 8 and the obtained values were compared to those obtained using Method 2. Methods were comparable, as all points located on or near the 1:1 function line *vertices of the contours, and "obj.f.v" corresponds to the facet data.*

(b) Extract from "obj" the vertices and facets data and assign them a structure named "PK": assign "PK.vertices" to the data from the vertices of the contours ("obj.v," three-column array, corresponding to the *x*, *y*, and *z* axis) and "PK.faces" to the facet data ("obj.f.v," three-column array). The facets in .obj file correspond to triangulation values. Each row of "PK.facets" has three values, which are the row values from "PK.vertices" required to create a triangulation (Fig. 4b2). *The 3D contour can be visualized in MATLAB (as in* Fig. 4c*) using the "trisurf" function. We recommend adding light to the figure (light properties), which allows a clearer view of the 3D contour surface. It might happen that in some axis of the exported contours, the signs could be changed from + to - (in our case, it happens with the values from column 2 and 3 from "PK.vertices"). Multiply for* -1 *to correct this issue.*

(c) Read the segment points-dendrites data file in MATLAB and assign it to variable named "den_data": the "read-table" function can be used to read the .xlsx and create a table array, followed by the "table 2array" function to transform the table variable into an array variable. Create a variable named "xyz_den" and assign data from columns 3 to 5 (now column 1 to 3 in xyz_den) of "den_data," which correspond to the coordinates (*x*, *y*, and *z* axis) from each dendrite fragment origin, and column 10 (now column 4 in xyz_den) of "den_data," which correspond to the length of each dendritic fragment. *We plotted all the neurons in* Fig. 4 *(except* Fig. 4a1*) using the scatter function. If both 3D contour and the neuron are plotted together (*Fig. 4c*), we recommend using "FaceAlpha" values lower than 1 in the "trisurf" function to visualize the dendrites inside the contour.*

(d) Use the in_polyhedron function [14] as follows:

$$\mathrm{inSNr} = \mathrm{in}_{\mathrm{polyhedron}}(\mathrm{PK}, \mathrm{xyzden}(:\,,\ 1:3));$$

The first input of the function must be the "PK" structure created in **step 2(b)**. The second input must be "xyz_den," but only reading the first three columns, which correspond to the coordinates (*x*, *y*, *z* axis) of the dendritic fragment origins. The output from the function is named "inSNr" and contains a logical array. Each row of "inSNr" is equivalent to the same row of "xyz_den." Logical 1 (true) indicate that a given dendritic fragment is inside SNr, whereas logical 0 (false) correspond to a given fragment that is outside SNr.

(e) Perform the following indexing, using "inSNr" and "xyz_den":

xyzden(inSNr, 4)

Here, the length of the dendritic fragments that are inside SNr will be extracted. Assign the result of this indexing to a variable named "den_len_snr." Use the "sum" function with "den_len_snr" as input to obtain the dendritic length that is inside the SNr. *To obtain the percentage of dendritic length inside the SNr, first obtain the total dendritic length by performing the sum of column 4 of "xyz_den," assign to a variable named "den_len_tot," and then divide "sum(den_len_snr)" to "den_len_tot" and multiply for 100.*

3 Conclusions and Future Perspectives

Here we showed the development of three methods to analyze the heterotopic distribution of dendrites, using the case of SNc dopaminergic neurons that extend their dendrites into SNr. Although the methods shown here were applied only to analyze dendritic extension in one nucleus or subdivision, these methods can be extended to analyze multiple subdivisions. For example, we have previously used Method 2 to quantify the heterotopic distribution of VTA dopaminergic neurons across neighboring subdivisions including the PBP, paranigral nucleus (PN), and parainterfascicular nucleus (PIF), among others [8]. Now, with the development of Method 3, quantification of dendritic extension in different nuclei could be performed much faster than with Method 2. Even though it is time-consuming, Method 1 could still be an alternative when only a few subdivisions are needed to be analyzed and a 3D model of the subdivisions of interest is not available.

We used this method with neurons obtained from *in vivo* juxtacellular recordings, but it could be applied also to any labeling technique that can allow the 3D reconstruction of individual neurons, such as *ex vivo* techniques or genetically labeled neurons. We have previously highlighted the functional correlates related to the heterotopic distribution of dendrites [8, 11]. We expect that the methods presented in this chapter help other researchers to continue exploring the functional consequences of dendritic extension into different subdivisions in the brain.

Acknowledgments

We thank our funding agency, Fondecyt Regular grant 1191497, Agencia Nacional de Investigación y Desarrollo (ANID) to P.H.

References

1. Stuart GJ, Spruston N (2015) Dendritic integration: 60 years of progress. Nat Neurosci 18: 1713–1721
2. Stuart G, Spruston N, Häusser M (2016) Dendrites. Oxford University Press, London
3. Nieuwenhuys R, Voogd J, van Huijzen C (2008) The human central nervous system, 4th edn. Springer, Heidelberg
4. Packer AM, McConnell DJ, Fino E, Yuste R (2013) Axo-dendritic overlap and laminar projection can explain interneuron connectivity to pyramidal cells. Cereb Cortex 23:2790–2802
5. Peters A, Feldman ML (1976) The projection of the lateral geniculate nucleus to area 17 of the rat cerebral cortex. I. General description. J Neurocytol 5:63–84
6. Rose PK, Richmond FJR (1981) White-matter dendrites in the upper cervical spinal cord of the adult cat: a light and electron microscopic study. J Comp Neurol 199:191–203
7. Fukuda S, Maeda H, Sakurai M (2020) Reevaluation of motoneuron morphology: diversity and regularity among motoneurons innervating different arm muscles along a proximal--distal axis. Sci Rep 10:13089
8. Montero T, Gatica RI, Farassat N et al (2021) Dendritic architecture predicts in vivo firing pattern in mouse ventral tegmental area and substantia nigra dopaminergic neurons. Front Neural Circuits 15:1–16
9. Juraska JM, Wilson CJ, Groves PM (1977) The substantia nigra of the rat. J Comp Neurol 172: 585–600
10. Bolam JP, Smith Y (1990) The GABA and substance P input to dopaminergic neurones in the substantia nigra of the rat. Brain Res 529:57–78
11. Henny P, Brown MTC, Northrop A et al (2012) Structural correlates of heterogeneous in vivo activity of midbrain dopaminergic neurons. Nat Neurosci 15:613–619
12. Hajós M, Greenfield SA (1994) Synaptic connections between pars compacta and pars reticulata neurones: electrophysiological evidence for functional modules within the substantia nigra. Brain Res 660:216–224
13. Meza RC, López-Jury L, Canavier CC, Henny P (2018) Role of the axon initial segment in the control of spontaneous frequency of nigral dopaminergic neurons in vivo. J Neurosci 38: 733–744
14. Megías M, Emri Z, Freund TF, Gulyás AI (2001) Total number and distribution of inhibitory and excitatory synapses on hippocampal CA1 pyramidal cells. Neuroscience 102: 527–540
15. Vrieler N, Loyola S, Yarden-Rabinowitz Y et al (2019) Variability and directionality of inferior olive neuron dendrites revealed by detailed 3D characterization of an extensive morphological library. Brain Struct Funct 224:1677–1695
16. Turrini P, Casu MA, Wong TP et al (2001) Cholinergic nerve terminals establish classical synapses in the rat cerebral cortex: synaptic pattern and age-related atrophy. Neuroscience 105:277–285
17. Glaser JR, Glaser EM (1990) Neuron imaging with neurolucida – a PC-based system for image combining microscopy. Comput Med Imaging Graph 14:307–317
18. Ascoli GA (2006) Mobilizing the base of neuroscience data: the case of neuronal morphologies. Nat Rev Neurosci 7:318–324
19. Henny P, Brown MTC, Micklem BR et al (2014) Stereological and ultrastructural quantification of the afferent synaptome of individual neurons. Brain Struct Funct 219:631–640
20. Franklin KBJ, Paxinos G (2007) The mouse brain in stereotaxic coordinates, 3rd edit edn. Academic Press, New York
21. Fu YH, Yuan Y, Halliday G et al (2012) A cytoarchitectonic and chemoarchitectonic analysis of the dopamine cell groups in the substantia nigra, ventral tegmental area, and retrorubral field in the mouse. Brain Struct Funct 217:591–612
22. Tuszynski J (2022) in_polyhedron. https://www.mathworks.com/matlabcentral/fileexchange/48041-in_polyhedron, MATLAB Central File Exchange. Retrieved May 23, 2022
23. Abayowa B (2022) readObj. https://www.mathworks.com/matlabcentral/fileexchange/18957-readobj, MATLAB Central File Exchange. Retrieved May 23, 2022
24. Franklin KBJ, Chudasama Y (2012) Prefrontal cortex. In: Watson C, Paxinos G, Puelles L (eds) The mouse nervous system. Elsevier Inc, pp 727–735

Part II

Cell Physiology

Chapter 4

Characterizing Dopamine Transporter Mediated Uptake and Efflux in Brain Slices: A New Approach

Nora Awadallah, Jose A. Pino, Dimitri N. Koutzoumis, and Gonzalo E. Torres

Abstract

The dopamine transporter (DAT) plays a fundamental role in dopamine (DA) neurotransmission by retrieving the transmitter from the extracellular space back into the DA nerve terminal. Several lines of evidence suggest that in addition to uptake or forward transport, DAT can also function to release DA. This process, which is referred to as DAT-mediated release or efflux, is the mechanism used by potent and highly addictive psychostimulants, such as amphetamine (AMPH) and its analogues, to increase extracellular DA levels in motivational and reward areas of the brain. Conventionally, dopaminergic neurons release DA through a calcium-dependent exocytotic vesicular process. In the past several years, our lab and others have identified signaling pathways that promote DA release through DAT. In this review chapter and step-by-step protocol, we will describe assays to study endogenous DA efflux and DAT dynamics using acute brain slices combined with high-performance liquid chromatography (HPLC) for DA detection along with a modification of a classically used radioactive uptake assay to measure levels of DA uptake. Further examining DAT functions and DA dynamics is key for the development of novel therapeutics in DA-related disorders including substance abuse. Here, we described the use of acute brain slices further dissected into precise regions of interest to examine the dual function of DAT.

Key words Dopamine, Dopamine transporter, Release, Uptake, Efflux, Brain slices, Dorsal striatum, Nucleus accumbens

1 Introduction

Dopamine (DA) is a critical and multidimensional monoamine neurotransmitter found throughout the body. Within the central nervous system (CNS), dopaminergic neurons originate primarily in the midbrain, sending projections predominantly to the dorsal striatum, nucleus accumbens, and prefrontal cortex [1]. DA neurotransmission is highly regulated by a specialized transporter protein, the dopamine transporter (DAT), involved in the reuptake of extracellular DA back into the cell (Fig. 1). DA and DAT play a heightened role in several physiological processes [2], as well as the

José Antonio Fuentealba-Evans and Pablo Henny (eds.), *Dopaminergic System Function and Dysfunction: Experimental Approaches*, Neuromethods, vol. 193, https://doi.org/10.1007/978-1-0716-2799-0_4,

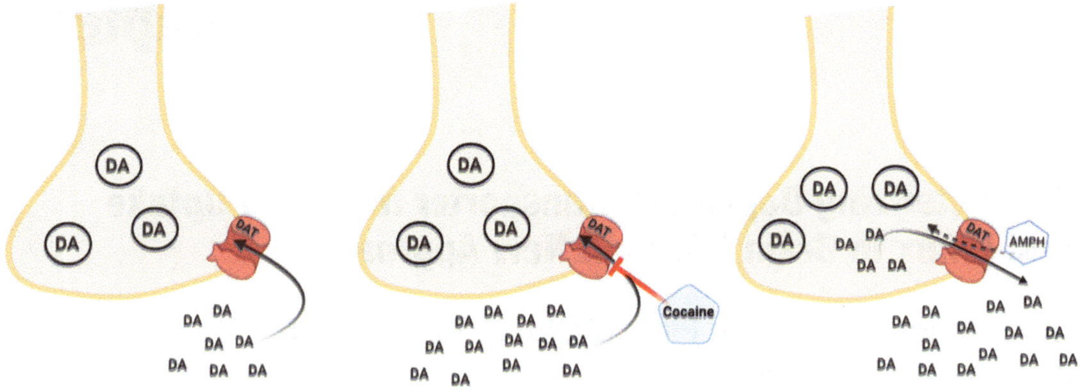

Fig. 1 Diagram depicting the two functions of the dopamine transporter and the pharmacological actions of psychostimulants. The methods described here are used to examine uptake and efflux. (**a**) The dopamine transporter mediates dopamine neurotransmission through the process of uptake. (**b**) Cocaine is a common psychostimulant that blocks uptake. (**c**) Amphetamine and its analogues produce the effect of DAT-mediated DA efflux or release into the extracellular space

pathophysiology of diseases such as substance use disorder [3–5], attention-deficit/hyperactivity disorder (ADHD) [6], and Parkinson's Disease [7, 8].

As a member of the solute carrier family (SCLA6A3), DAT takes up one molecule of DA while co-transporting two Na^+ and one Cl^- [9]. In terms of anatomical distribution, DAT is largely found in the presynaptic nerve terminal of DA neurons [10, 11] although the transporter is also localized to dendrites [12]. DA is conventionally released from nerve terminals via Ca^{2+}-dependent vesicular release and taken back up through DAT [13]. In addition, DAT is the target for potent and highly addictive psychostimulants such as cocaine and amphetamine, which dramatically increase extracellular DA via an interaction with DAT. However, these psychostimulants have opposing mechanisms of action. Cocaine is a known DAT inhibitor and attenuates uptake [14]. Amphetamines, on the other hand, produce a significant non-physiological increase in DA efflux into reward and motivational areas of the brain (Fig. 1), such as the nucleus accumbens [15–18]. This process is referred to as DAT-mediated release or efflux. Further, amphetamine including its analogues such as methylphenidate is used in the treatment for ADHD [19, 20]. These medications work by increasing the extracellular levels of DA (also norepinephrine, NE) in key brain areas involved in attention and motivation. Overall, DAT is a dynamic protein that primarily uptakes DA but is also known to release or "efflux" DA [21]. Interestingly, our group has recently discovered a novel interaction between DAT and the βγ subunits of G proteins (Gβγ) [22]. According to the classic paradigm, upon binding of an agonist to a G protein-coupled receptor (GPCR), G proteins dissociate into active Gα and Gβγ subunits, resulting in the activation of multiple effector systems [23]. Our results indicate that Gβγ binding to DAT is an endogenous modulator, important to promote DA efflux not only under physiological conditions but also in the actions of AMPH [17, 21]. Thus, our results suggest that the transporter releases DA under physiological (Gβγ activation) as well as pharmacological (AMPH) conditions. The purpose of this paper is to describe conditions to measure DAT-mediated DA efflux in brain slices.

There are several in vitro and in vivo techniques available to characterize DA dynamics and DAT function such as radioactivity-based assays, microdialysis, fast-scan cyclic voltammetry (FSCV), amperometry, optogenetics, and genetic sensors. In vivo microdialysis is an excellent tool for investigating extracellular DA dynamics under varying pharmacological conditions with a significant advantage of using awake and moving animals [17, 24]. FSCV in vivo or in brain slices measures stimulated exocytotic release and clearance with both a greater temporal and spatial resolution [25–27]. More recent advancements in optogenetics permit cell-type specificity to control dopaminergic activity in awake and moving animals [28, 29]. Further, genetically encoded fluorescent DA sensors in vivo and in brain slices have been implemented to determine the role of specific signaling pathways in DA activity [30]. Radioactive uptake assays used in heterologous cells or brain synaptosomes have been useful for the initial characterizations of the DAT uptake kinetics [21, 31, 32]. High-speed chronoamperometry is another powerful technique that examines in vivo DA clearance [33].

While the described methods above are useful for examining DA uptake and release mechanisms, it is challenging to distinguish between classic exocytotic release and DAT-mediated efflux. In most studies, the measurement of DAT uptake and DAT-mediated efflux largely relies on radioactive assays. Here, we describe a simplified method to measure DA efflux from brain slices. This method takes advantage of pharmacological agents to distinguish between vesicular release and transporter-mediated DA efflux. The released DA can then be detected through high-performance liquid chromatography (HPLC) analysis. We discuss the use of acute brain slices to measure endogenous DA efflux using both a superfusion system and a 24- or 48-well plate. The optimizations regarding DA efflux are a necessary progression for they permit the study of endogenous DA efflux while simultaneously allowing the experimenter to distinguish between DAT-mediated efflux and vesicular release. Given the dual function of the DAT and for comparison purposes, we also describe DAT-mediated uptake using brain slices and radiolabeled DA ([^{3}H]-DA).

In the efflux assay, rat brain slices are dissected and placed in an oxygenated superfusion chamber or a multiwell plate. Slices are then incubated with pharmacological agents to induce DA efflux, which is then recovered from the supernatant and measured using a

HPLC. In the radioactive uptake assay, slices are incubated with [^{3}H]-DA, and incorporated radioactivity is measured via a scintillation counter. Therefore, without [^{3}H]-DA, the experimenter is unable to distinguish between endogenous DA found in vesicles versus DA taken up through the transporter. Overall, radioactivity is necessary for studying DAT uptake but is no longer necessary for studying efflux [34]. As radioactive assays typically use heterologous cells or brain synaptosomes, we will describe an alternative approach by using acute brain slices to measure [^{3}H]-DA uptake and endogenous DA efflux. The use of acute brain slices allows for the measurement of endogenous DA and maintains both the neuroarchitecture and endogenous neurochemistry in each region of interest.

In general, all described techniques involve incubating acute brain slices further dissected into discrete regions of interest in varying pharmacological conditions and measuring DA uptake via a scintillation counter or DA efflux using a HPLC.

2 Materials

2.1 Obtaining Acute Rat Brain Slices

1. Adult Sprague-Dawley rats (250–300 g body weight).
2. Dissection Buffer: For one rat, make 1 L. Krebs-Henseleit (KH) buffer (116-mM NaCl; 3-mM KCl, 1-mM $MgSO_4$, 1.2-mM KH_2PO_4, 25-mM $NaHCO_3$, 11-mM D-glucose; pH 7.4) supplemented with 0.4 mM of ascorbic acid.
3. Vibrating microtome, vibratome, or compresstome (used here: Precisionary Compresstome® VF 310-0Z Vibrating Microtome).
4. Rodent guillotine.
5. 1 and 2 mm sample corers.
6. Carbogen (95% O_2/5% CO_2) tank.

2.2 Rat Brain Superfusion Experiments for Measuring DA Efflux Using HPLC

1. 6-chamber superfusion apparatus (Brandel Suprafusion 1000 System, Gaithersburg, MD).
2. Whatman GF/B filter paper for superfusion systems (Brandel, USA).
3. Carbogen (95% O_2/5% CO_2) tank.
4. 7-mL copolymer plastic vials (Thermo Fisher Scientific, USA).
5. Microcentrifuge nanosep filter tubes (0.22 μm).
6. *Dissection buffer* (Subheading 2.1, **item 2**).
7. *Efflux buffer:* For one rat, make 200 mL. Krebs-Henseleit (KH) buffer supplemented with 1.8-mM $CaCl_2$, 10-μM pargyline, and 400-μM ascorbic acid.

8. *Compounds for experiment*: 10-μM amphetamine (AMPH, Sigma-Aldrich); 40-mM potassium chloride (KCl, Sigma-Aldrich); 10-μM GBR12935 (DAT blocker, Sigma-Aldrich); 20-μM gallein (a Gβγ inhibitor, Tocris) dissolved in oxygenated KH efflux buffer.

2.3 High-Performance Liquid Chromatography HPLC to Measure DA Release

1. HPLC and appropriate software to acquire and analyze the chromatographic data (*see* **Notes** in Subheading 4).
2. Dopamine and, if required, dopamine metabolite standards (3,4-dihydroxyphenylacetic acid [DOPAC] and homovanillic acid [HVA]), prepared at 0.5–100 nM in 0.1-N $HClO_4$.
3. DC protein assay (Bio-Rad Laboratories, Hercules, CA).
4. Microcentrifuge nanosep filter tubes (0.22 μm).

2.4 Multiwell-Based Assay for the Measurement of Endogenous DA Release Using HPLC

1. Slide warmer.
2. Petri dishes.
3. Ice bucket(s).
4. 24- or 48-well plate(s).
5. Tissue holder custom made of six microcentrifuge filter units without the inset filters (Ultrafree-MC, Millipore, USA).
   (a) *To make the holder*, use a sturdy plastic rod (e.g., cell scrapper) and super glue the microcentrifuge filter units without the inset filters to it. Let dry for 1–2 days.
6. Carbogen (95% O_2/5% CO_2) tank with proper tubing to oxygenated buffers and the 24- or 48-well plate.
7. HPLC and appropriate software to acquire and analyze the chromatographic data (*see* **Notes** in Subheading 4).
8. Dopamine and, if required, dopamine metabolite standards (3,4-dihydroxyphenylacetic acid [DOPAC] and homovanillic acid [HVA]), prepared at 0.5–100 nM in 0.1-N HClO4.
9. DC protein assay (Bio-Rad Laboratories, Hercules, CA).
10. *Dissection buffer* (Subheading 2.1, **item 2**) for one rat, make 1 L.
11. *Efflux buffer:* For one rat, make 200 mL. Krebs-Henseleit (KH) buffer supplemented with 1.8-mM $CaCl_2$, 10-μM pargyline, and-400 μM ascorbic acid.
12. *Compounds for experiment*: 10-μM amphetamine (AMPH, Sigma-Aldrich); 40-mM potassium chloride (KCl, Sigma-Aldrich); 10-μM GBR12935 (DAT blocker, Sigma-Aldrich); 20-μM gallein (a Gβγ inhibitor, Tocris) all dissolved in oxygenated KH efflux buffer.

2.5 Use of Acute Brain Slices for Ex Vivo [3H]-DA Uptake Assay

1. Adult Sprague-Dawley rats (250–300 g body weight).
2. Tritiated dopamine (PerkinElmer, 27.8 Ci/mmol).

   Uptake buffer: 116-mM NaCl, 3-mM KCl, 1-mM $MgSO_4$, 1.2-mM KH_2PO_4, 25-mM $NaHCO_3$, 0.4-mM ascorbic acid, 11-mM D-glucose, 1.8-mM $CaCl_2$, 0.01-mM pargyline, pH 7.4.

   Wash buffer: 116-mM LiCl, 3-mM KCl, 1-mM $MgSO_4$, 1.2-mM KH_2PO_4, 25-mM $NaHCO_3$, 0.4-mM ascorbic acid, 11-mM D-glucose, 1.8-mM $CaCl_2$, 0.01-mM pargyline, pH 7.4.

   Lysis buffer: 0.1% SDS.
3. Drug compounds for experiment: GBR12935 (DAT blocker, Sigma-Aldrich).

3 Methods

3.1 Acute Rat Brain Slices for Both Uptake and Efflux Assays

1. All animal procedures were used in accordance with the City College of New York Institutional Animal Care and Use Committee (IACUC #1071).
2. Prepare 1 L of the fresh dissection buffer (KH buffer) and saturate with bubbling 95%/5% (O_2/CO_2) for at least 20 min on ice before beginning the experiment.
   (a) Remove 50 mL of dissection buffer and chill on ice in a small beaker/petri dish.
3. Anesthetize one to two adult rats (250–350 g body weight) with 1–2% isoflurane.
   (a) Decapitate the animal(s) using a guillotine, and immediately remove the brain(s).
   (b) Place the brain(s) in the beaker/petri dish from Subheading 3.1, **step 2** (Fig. 2).
4. Using a vibrating microtome or compresstome, cut 300–400 μm coronal sections until each region of interest is obtained (Fig. 2). A brain atlas should be used when dissecting regions of interest.
   (a) For example, the prefrontal cortex is obtained in the initial slices, but several cuts must first be made to obtain the dorsal striatum. Note that the first slice should be discarded as it is not always the correct thickness.
   (b) Carefully transfer brain slices into a *new* petri dish filled with ice-cold oxygenated dissection buffer.
   (c) Further dissect brain slices (e.g., punches, cut out) to study specific brain regions using a brain atlas as reference. Use tissue corers to create individual punches at desired

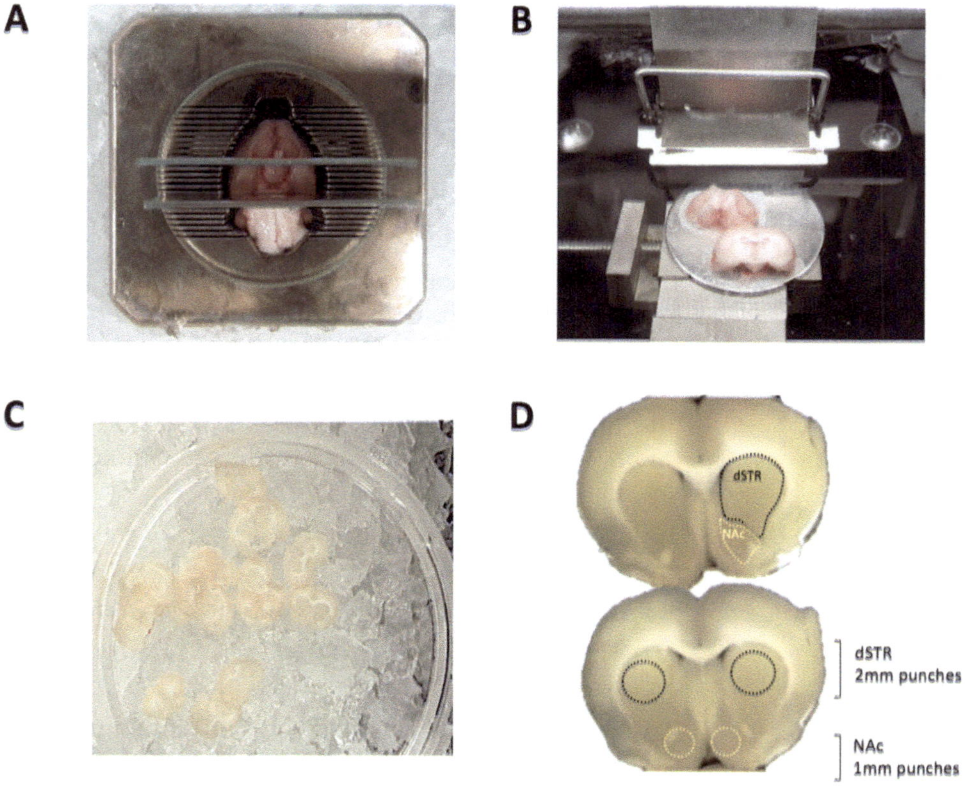

Fig. 2 Producing acute brain slices using a vibratome. (**a**) To obtain the dorsal striatum (dSTR) and nucleus accumbens (NAc), generate two gross cuts, placing one razor blade at the base of the posterior limit of the hypothalamus and the other at the optic chiasm. (**b**) Once secured to the vibratome, cut 300–400 μm slices, ensuring slices are released into dissection buffer with bubbling O_2/CO_2 (95%/5%). (**c**) Carefully transfer brain slices into a petri dish on ice with constant bubbling O_2/CO_2 (95%/5%). (**d**) Transfer brain slices to a glass slide and cut micropunches of the dorsal striatum (2 mm) and nucleus accumbens (1 mm)

sizes. Here, the dorsal striatum and nucleus accumbens were dissected into 2-mm and 1-mm punches, respectively (Fig. 2d). The dorsal striatum is identified based on its dark, striated structure. The nucleus accumbens is identified as immediately below the dorsal striatum.

(d) Using a plastic transfer pipette with the tip cut off or a stainless steel spatula, transfer brain slices or punches into a petri dish filled with ice-cold dissection buffer with 95%/5% O_2/CO_2 bubbling on ice. The tissue samples are now ready for experimentation.

(e) Note: when tissue samples are not being used for experimentation, they must be on ice with consistent oxygenation (95%/5%, O_2/CO_2).

3.2 Use of Acute Brain Slices for Ex Vivo [3H]-DA Uptake Assay

1. Turn on slide warmer and set to 37 °C.
2. Obtain 300–400 μm acute brain slices and micropunches (Subheading 3.1).
   (a) Ensure there are enough slices for each condition (at least three slices/condition).
3. *Tissue recovery:* Incubate tissue for 20–40 min at 37 °C in oxygenated uptake buffer (Fig. 3).
4. *Drug condition:* Incubate the tissue in each well of the tissue holder with or without a DAT blocker (e.g., GBR12935 10 μM) for 5 min at 37 °C in a 48-well tissue culture plate.
5. *Uptake:* Move the tissue holder to a new set of wells containing 50-nM [^{3}H]-DA in 500 μL of pre-oxygenated uptake buffer (Subheading 2.2) with or without the DAT blocker still present, and incubate for 10 min at 37 °C.
6. *Wash:* Move the tissue holder to a new set of wells containing pre-oxygenated wash buffer.
7. *Radioactive count measurements using a scintillation counter:*
   (a) Homogenize tissue in 300 μL of 1% SDS for 1 h at room temperature.
   (b) Add 250 μL of each sample to 2 mL of scintillation fluid (Bio-Safe II, RPI, Mt. Prospect, IL, USA).
   (c) Measure counts per minute (CPM) using radioactive measurement by scintillation counter.
8. *Protein quantification:*
   (a) Use the leftover 50 μL of homogenized samples for protein quantification using the DC™ Protein Assay Kit from Bio-Rad at 750 nm.
9. Normalize [^{3}H]-DA uptake by protein quantification and express in fmol*min/μg protein (Fig. 3c).

3.3 Rat Brain Superfusion for Measuring Endogenous Dopamine Efflux

1. Prepare superfusion system.
   (a) Pre-oxygenate buffer and reagents with carbogen using gassing probe (Fig. 4a1).
   (b) Pass reagents from the test tubes (Fig. 4a2) through the Reagent Probe (Fig. 4a3) via the peristaltic pump (Fig. 4a4), through the superfusion chambers (Fig. 4a5).
2. *Tissue activation:* Load the tissue (2-mm dSTR or NAc brain punches) onto GF/B Whatman filters, and place into a superfusion chamber (Fig. 4b).
   (a) Perfuse tissue at a flow rate of 1 mL/min at 37 °C for 20 min with oxygenated efflux buffer (supplemented with 1.8-mM $CaCl_2$ and 0.01-mM pargyline).

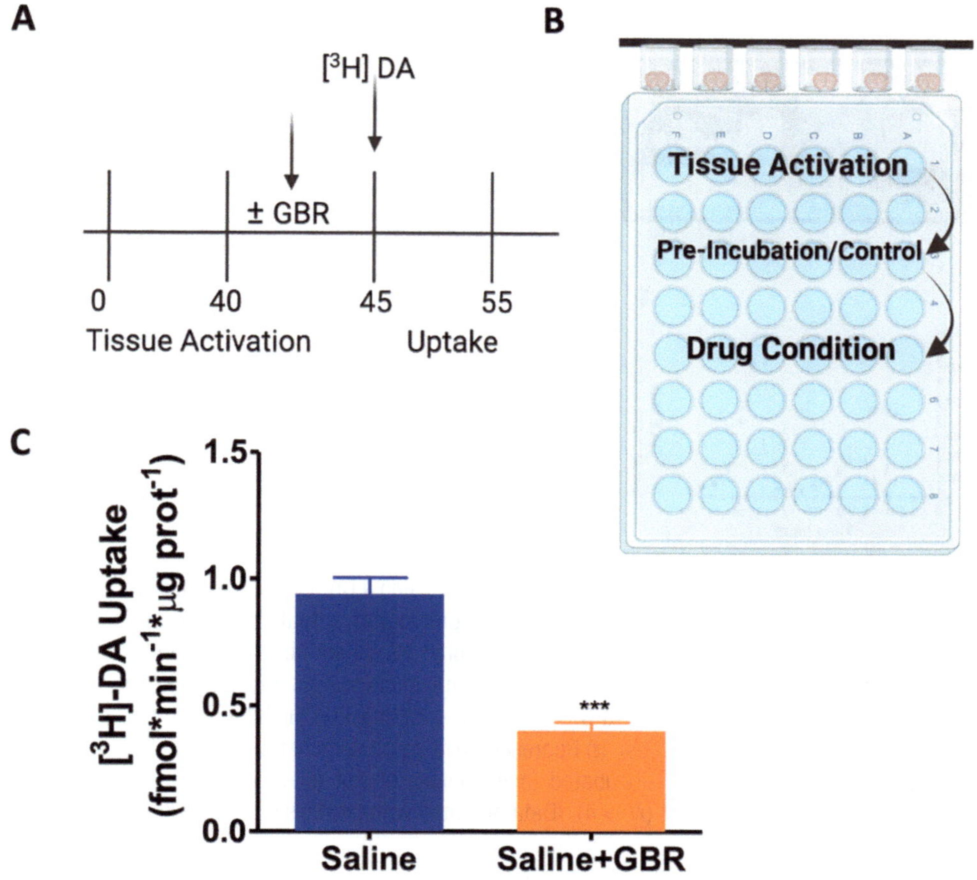

Fig. 3 Experimental schematics of the radioactive dopamine ([^{3}H]-DA) uptake assay using acute brain slices dissected into micropunches of the dorsal striatum. (**a**, **b**) Optimization to both the superfusion and radioactive DA uptake assays using a 48-well plate. Nanosep filter tubes are glued to a cell scraper rod to fit in each well of a 48-well plate. Together this is used as a tissue holder to move the tissue between each row. When the tissue holder is moved to a new row, the tissue samples are incubated in a different drug condition. Tissue activation occurs when the tissue slices or punches are moved from the ice-cold buffer to incubation in KH buffer at 37 °C on a slide warmer. (**c**) Once complete, the amount of striatal [^{3}H]-DA uptake into the micropunches is measured using a scintillation counter. Representative data indicating a decrease in DA uptake when tissue was preincubated in GBR12395, a DAT inhibitor. Data is presented as fmol*min per μg protein ± SD; ***$p < 0.001$ (unpaired t-test). (Data obtained with permission from authors [35])

3. *Drug preincubation:* Perfuse the tissue at a flow rate of 1 mL/min at 37 °C for 20 min while incubating with 20-μM gallein or oxygenated efflux buffer.
   (a) Collect four superfusate samples with 1 min collection times (Fig. 4a6).
4. *Drug conditions:* Incubate tissue with 1 mL of 10-μM amphetamine through a 2 min bolus injection into the superfusion chamber (Fig. 4a7), followed by 1 min superfusate collection every 20 min.

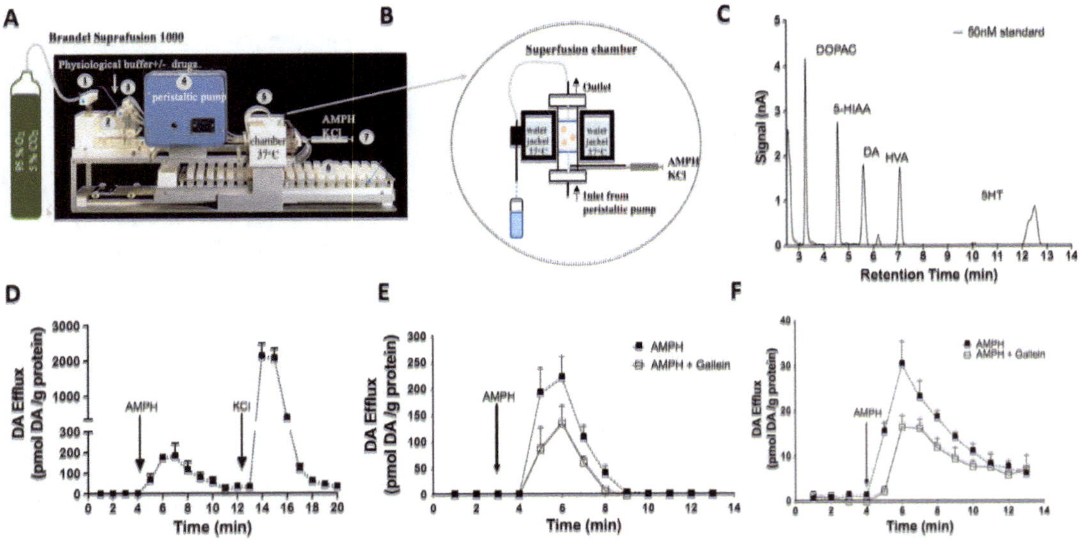

Fig. 4 Superfusion experimental setup to examine DA efflux using a Brandel Suprafusion System 1000 and an HPLC detection system. (**a**, **b**) Depiction of system. (**c**) Chromatogram output from HPLC depicting metabolites, dopamine, and serotonin. (**d**) Representative graph demonstrating data obtained from striatal micropunches from acute rat brain slices when incubated with 10-μM amphetamine (first arrow) and KCl (second arrow) and (**e**) when preincubated in 20-μM gallein, a G protein beta gamma (Gβγ) inhibitor (open boxes) prior to AMPH exposure, or AMPH alone (filled in boxes). (**f**) Representative graph demonstrating data obtained from nucleus accumbens micropunches either preincubated (open boxes) or not (filled in boxes) with gallein. Graphs represent pmol DA/g protein ± SEM ($N = 4$). (Data obtained with permission from authors [17])

(a) Optional incubation: perfuse with a 2 min bolus of 40-mM KCl to induce exocytotic DA release.

5. Sample preparation for HPLC:
   (a) Add $HClO_4$ (or the appropriate preservative depending on the HPLC) to all of the samples (final concentration, 0.1 N).
   (b) Filter the samples using microcentrifuge Nanosep® microcentrifuge filter tubes (0.22 μm). Samples are now ready for HPLC analysis or can be stored at −80 °C before HPLC analysis.
   (c) Load the desired volume of samples and standards into the HPLC (dependent upon type of HPLC and column used).
      (i) Standards should be made between 0.1 and 100 nM in mqH_2O (dependent upon type of HPLC column used, be sure to check manufacturer's recommendation for preparation of standards and samples).
6. Protein quantification from brain punches (Subheading 3.2, **step 8**):

(a) Place tissue samples in labeled Eppendorf tubes.

(b) Homogenize using lysis buffer supplemented with 1% Triton X-100 and protease inhibitor cocktail.

(c) Centrifuge samples at 16,000 g × 15 min at 4 °C.

(d) Determine protein quantification using DC protein assay (Bio-Rad Laboratories, Hercules, CA).

(e) After analysis using HPLC-specific software and protein quantification is obtained, DA content is expressed as pico-moles DA/g protein.

(i)

$$\text{pmol DA/gprotein} = \frac{\text{pM DA} \times \text{HPLC injection volume } (\mu\text{L}) \times \text{dilution factor} \left(\frac{1000\ \mu\text{L}}{\text{HPLC injection volume in } \mu\text{L}}\right)}{\mu\text{g protein (brain punches)}}$$

3.4 Multiwell-Based Assay for the Measurement of Endogenous Dopamine Release

1. Turn on the slide warmer and set to 37 °C.
2. *Tissue activation:* Transfer brain sections or punches (Subheading 3.1) from the petri dish on ice to each well of the premade tissue holder on the slide warmer.

   (a) Incubate for at least 30 min at 37 °C in 0.5–1 mL of oxygenated KH buffer (supplemented with 1.8-mM CaCl2 and 0.01 mM pargyline) with constant, gentle bubbling 95%/5% O_2/CO_2 (Fig. 5).

3. *Preincubation/control:* Transfer the tissue holder to new wells containing 0.5 mL of oxygenated and supplemented KH buffer (supplemented with 1.8-mM CaCl2 and 0.01-mM pargyline) and incubate the tissue for 20 min at 37 °C with constant, gentle bubbling of 95%/5% O_2/CO_2.

   (a) *Collection:* Transfer buffer from each well to microcentrifuge tubes containing the appropriate preservative (e.g., $HClO_4$: 0.1 N final concentration). Maintain microcentrifuge tubes on ice and label appropriately.

   (b) *Optional*: Incubate tissue samples with drugs dissolved in an oxygenated and supplemented KH buffer (e.g., 40-μM cocaine) to test different pharmacological conditions.

4. *Drug condition*: Transfer the tissue holder to a new set of wells including 0.5 mL of desired compounds and incubate for 20 min at 37 °C with constant, gentle bubbling of 95%/5% O_2/CO_2.

   (a) *Compounds may include:*

      (i) KH buffer

      (ii) 10–30-μM AMPH

      (iii) 10–30-μM AMPH and 40-μM cocaine

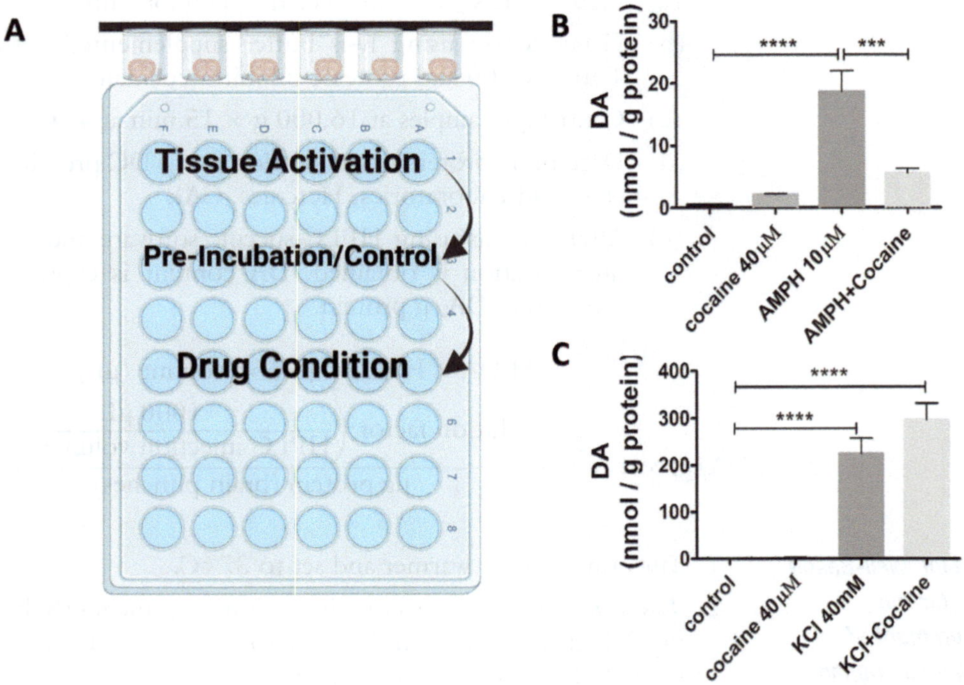

Fig. 5 Optimization to the superfusion assay to measure endogenous DA efflux using a 48-well plate. (**a**) Schematic of the experimental setup, nanosep filters are glued to a cell scraper rod and used as a tissue holder to move tissue between each column. When they are moved to a new column, they are incubated in a different drug condition. Representative images from the (**b**) dorsal striatum incubated in, from left to right, control dissection buffer, 40-μM cocaine, 10-μM amphetamine, and both amphetamine and cocaine, and (**c**) dorsal striatum incubated in, from left to right, control dissection buffer, 40-μM cocaine, 40-μM KCl, and both KCl and cocaine. Graphs represent nmol DA/g protein ± SD; $^{***}p < 0.001$, $^{****}p < 0.0001$ (one-way ANOVA and Tukey's post-test)

(iv) 10-μM GBR12935

(v) 10-μM fluoxetine

(vi) High K^+ KH buffer

(b) *Collection:* Transfer buffer from each well to microcentrifuge tubes containing the appropriate preservative (e.g., $HClO_4$). Maintain microcentrifuge tubes on ice and label appropriately.

5. *Tissue collection:* Carefully remove the tissue using small tweezers and transfer to microcentrifuge tubes.

6. *Filtration and HPLC analysis:* Filter the supernatant from Subheading 3.4, **step 4(b)** using Nanosep® microcentrifuge filter tubes (0.22 μm) at 2500 g × 2 min at 4 °C.

   (a) Load the desired volume of samples and standards into the HPLC (dependent upon type of HPLC and column used).

   (i) Standards should be made between 0.1 and 100 nM in mqH_2O (dependent upon type of HPLC column used, be sure to check manufacturer's recommendation for preparation of standards and samples).

(b) Use the filtrate to determine DA content using HPLC with electrochemical detection. DA content is expressed as nmol DA/g protein in brain tissue.

(c) nmol DA /g protein =

$$\frac{\text{nM DA} \times \text{HPLC injection volume } (\mu\text{L}) \times \text{dilution factor} \left(\frac{500\ \mu\text{L}}{\text{HPLC injection volume in } \mu\text{L}}\right)}{\mu\text{g protein (brain punches)}}$$

4 Notes

Throughout all three experiments described, there are certain aspects that the experimenter must pay attention to or may need troubleshooting.

(a) When studying DAT, ensure correct dissect from neuroanatomical locations where DAT is known to be expressed (e.g., primarily dorsal striatum and nucleus accumbens, while there is limited expression in the prefrontal cortex).

(b) Although all the data shown here is based off experiments done on adult male rats, all experiments described are flexible regarding type of animal, developmental time point, sex, and perhaps organ.

  (i) Keep in mind the amount of tissue required if using a smaller animal such as a mouse or in juvenile/adolescent animals.

(c) Be consistent with how many pieces of tissue are used in each condition to maintain equality between conditions.

(d) The efflux experiments described require gentle but consistent oxygenation (95%/5% O_2/CO_2). A lack of oxygenation may significantly interfere with tissue viability.

  (i) An MTT assay should be performed to determine tissue viability at the end of experimentation in a separate set of tissue.

(e) The efflux assay performed in the 48-well plate uses a buffer containing Ca^{2+}. If concern arises regarding the influence of Ca^{2+} synaptic release, remove Ca^{2+} from the efflux buffer.

(f) The efflux assay performed in the 48-well plate may also be performed in a 24-well plate. However, the experimenter will need to create a new tissue holder and will have to use up to 1 mL in buffer per well.

(g) Check with the HPLC manufacturer's recommendations prior to loading samples and analyzing data.

(i) The experiments listed here used the following settings for HPLC analysis: Decade amperometric HPLC system (Antec LC-EC system, Antec, Leyden BV, Netherlands) with a reverse phase column (Acquity UPLC BEH C-18 1.7 μm, 1 × 100 mm, Waters Co., Milford, MA) using an electrochemical detector (2-mm glassy carbon electrode, SenCell, Antec) set at a potential of 0.8 V. The *mobile phase* used included 100-mM phosphoric acid, 100-mM citric acid, 0.1-mM EDTA-Na_2, 600-mg/L octanesulfonic acid (pH 3.0), 8% acetonitrile.

5 Discussion

In this chapter we sought out two goals: to review the current literature on DAT dynamics and to provide updated and optimized protocols to study DA release and uptake via DAT in acute brain slices using both HPLC and radioactive assays. Thus far, we have detailed optimizations to traditional methods used to study the dynamic nature of DAT. DAT has a dual function: uptake and efflux. The transporter uptakes DA from the extracellular space after release, thus regulating the intensity and duration of DA transmission [36]. However, the transporter is also capable of releasing DA, which is an effect that has been widely characterized for amphetamines [16–18, 37]. Our lab and others have proposed that DA released through the transporter occurs physiologically through activation of endogenous signaling pathways [17, 21, 38–40]. Thus, the interplay between DA uptake and efflux is an area under intense investigation.

To measure uptake in brain slices, we took advantage of the use of radioactive DA ([^{3}H]-DA). We are unable to use non-radioactive DA as we would not be able to distinguish it from endogenous tissue DA in regular uptake assays. Further, studying uptake allows us to compare different experimental conditions as well as examine kinetic parameters (e.g., K_m and V_{max}). In terms of DA efflux assays, we no longer need to rely on [^{3}H]-DA as endogenously released DA into the media can be directly measured via HPLC. This is a preferred approach as the goal is to measure endogenous DA. Our group has successfully used all methods described in this chapter to study uptake and efflux in acute brain slices [21, 34, 35]. By using acute brain slices further dissected into micropunches, we are able to characterize DA uptake and endogenous DA efflux throughout the brain. A major advantage of the techniques reviewed in this chapter is the ability to distinguish vesicular vs DAT-mediated DA release. In addition, the use of acute brain slices offers a more appealing model in comparison to heterologous cells or synaptosomes. Such reasons include the following: (1) the microarchitecture and neuroanatomy remain intact, although connections are severed; (2) this does not require the onerous task of obtaining synaptosomes or growing and transfecting cells; and (3) the slices are viable throughout the experiment when properly maintained.

Our group has specifically shown that the dynamics of DAT can be studied using modifications of these well-known techniques. Specifically, we have shown that preincubating dorsal striatum micropunches with GBR12935, a DAT uptake inhibitor, leads to a significant decrease in [^{3}H] DA uptake (Fig. 3) [35]. We have further shown that exposure to a Gβγ inhibitor, gallein, prior to amphetamine exposure decreases the amount of DA released by the transporter in both the dorsal striatum and nucleus accumbens (Fig. 4) [17]. In conjunction with previous work, we show that amphetamine leads to a significant increase in DA efflux in the dorsal striatum using the 48-well plate system (Fig. 5) [34]. Dissecting out the intricate details of DAT-mediated DA efflux is fundamental and crucial for developing novel and highly targeted therapeutics for DA-/DAT-related disorders.

Acknowledgments

We thank the members of the Torres laboratory for their helpful discussions. We also acknowledge the support of the following agencies: Fondecyt Initiation Grant N11191049 (JAP), NIH R01 Grant DA038598 (GET).

References

1. Zahm DS, Trimble M (2008) The dopaminergic projection system, basal forebrain macrosystems, and conditioned stimuli. CNS Spectr 13:32–40
2. Giros B, Caron MG (1993) Molecular characterization of the dopamine transporter. Trends Pharmacol Sci 14:43–49. https://doi.org/10.1016/0165-6147(93)90029-J
3. Volkow ND, Fowler JS, Wang GJ, Goldstein RZ (2002) Role of dopamine, the frontal cortex and memory circuits in drug addiction: insight from imaging studies. Neurobiol Learn Mem 78:610–624
4. Di Chiara G (2002) Nucleus accumbens shell and core dopamine: differential role in behavior and addiction. Behav Brain Res 137:75–114. https://doi.org/10.1016/S0166-4328(02)00286-3
5. Wanat MJ, Willuhn I, Clark JJ, Phillips PEM (2009) Phasic dopamine release in appetitive behaviors and drug addiction. Curr Drug Abuse Rev 2(2):195–213. https://doi.org/10.2174/1874473710902020195
6. Volkow ND, Wang GJ, Newcorn J et al (2007) Brain dopamine transporter levels in treatment and drug naïve adults with ADHD. NeuroImage 34:1182–1190. https://doi.org/10.1016/j.neuroimage.2006.10.014
7. Leenders KL, Palmer AJ, Quinn N et al (1986) Brain dopamine metabolism in patients with Parkinson's disease measured with positron emission tomography. J Neurol Neurosurg Psychiatry 49:853–860. https://doi.org/10.1136/jnnp.49.8.853
8. Scherman D, Desnos C, Darchen F et al (1989) Striatal dopamine deficiency in Parkinson's disease: role of aging. Ann Neurol 26:551–557. https://doi.org/10.1002/ana.410260409
9. Aggarwal S, Mortensen O (2017) Overview of monoamine transporters. Curr Protoc

Pharmacol 79:12.16.1–12.16.17. https://doi.org/10.1002/CPPH.32

10. Nirenberg MJ, Vaughan RA, Uhl GR et al (1996) The dopamine transporter is localized to dendritic and axonal plasma membranes of nigrostriatal dopaminergic neurons. J Neurosci 16:436–447. https://doi.org/10.1523/jneurosci.16-02-00436.1996
11. Blanchard V, Raisman-Vozari R, Vyas S et al (1994) Differential expression of tyrosine hydroxylase and membrane dopamine transporter genes in subpopulations of dopaminergic neurons of the rat mesencephalon. Mol Brain Res 22:29–38. https://doi.org/10.1016/0169-328X(94)90029-9
12. Hersch SM, Yi H, Heilman CJ, Edwards RH, Levey AI (1997) Subcellular localization and molecular topology of the dopamine transporter in the striatum and substantia nigra. J Comp Neurol 388(2):211–227
13. Patel JC, Witkovsky P, Avshalumov MV, Rice ME (2009) Mobilization of calcium from intracellular stores facilitates somatodendritic dopamine release. J Neurosci 29:6568–6579. https://doi.org/10.1523/JNEUROSCI.0181-09.2009
14. Ritz MC, Lamb RJ, Goldberg SR, Kuhar MJ (1987) Cocaine receptors on dopamine transporters are related to self-administration of cocaine. Science (80-) 237:1219–1223. https://doi.org/10.1126/science.2820058
15. Fischer JF, Cho AK (1979) Chemical release of dopamine from striatal homogenates: evidence for an exchange diffusion model. J Pharmacol Exp Ther 208:203–209
16. Kahlig KM, Binda F, Khoshbouei H et al (2005) Amphetamine induces dopamine efflux through a dopamine transporter channel. Proc Natl Acad Sci U S A 102:3495–3500. https://doi.org/10.1073/pnas.0407737102
17. Mauna JC, Harris SS, Pino JA et al (2019) G protein βγ subunits play a critical role in the actions of amphetamine. Transl Psychiatry 9:1–11. https://doi.org/10.1038/s41398-019-0387-8
18. Sulzer D, Maidment NT, Rayport S (1993) Amphetamine and other weak bases act to promote reverse transport of dopamine in ventral midbrain neurons. J Neurochem 60:527–535. https://doi.org/10.1111/j.1471-4159.1993.tb03181.x
19. Punja S, Shamseer L, Hartling L et al (2016) Amphetamines for attention deficit hyperactivity disorder (ADHD) in children and adolescents. Cochrane Database Syst Rev 2:CD009996
20. Castells X, Blanco-Silvente L, Cunill R (2018) Amphetamines for attention deficit hyperactivity disorder (ADHD) in adults. Cochrane Database Syst Rev 8(8):CD007813
21. Garcia-Olivares J, Baust T, Harris S et al (2017) Gβγ subunit activation promotes dopamine efflux through the dopamine transporter. Mol Psychiatry 22:1673–1679. https://doi.org/10.1038/mp.2017.176
22. Garcia-Olivares J, Torres-Salazar D et al (2013) Inhibition of dopamine transporter activity by G protein βγ subunits. PLoS One 8(3):e59788. https://doi.org/10.1371/journal.pone.0059788
23. Gurevich VV, Gurevich EV (2019) GPCR signaling regulation: the role of GRKs and arrestins. Front Pharmacol 10:125. https://doi.org/10.3389/FPHAR.2019.00125
24. Bungay PM, Newton-Vinson P, Isele W et al (2003) Microdialysis of dopamine interpreted with quantitative model incorporating probe implantation trauma. J Neurochem 86:932–946. https://doi.org/10.1046/j.1471-4159.2003.01904.x
25. Yorgason JT, España RA, Jones SR (2011) Demon voltammetry and analysis software: analysis of cocaine-induced alterations in dopamine signaling using multiple kinetic measures. J Neurosci Methods 202(2):158–164. https://doi.org/10.1016/j.jneumeth.2011.03.001
26. Alonso IP, Pino JA, Kortagere S et al (2021) Dopamine transporter function fluctuates across sleep/wake state: potential impact for addiction. Neuropsychopharmacology 46:699–708. https://doi.org/10.1038/s41386-020-00879-2
27. Schelp SA, Brodnik ZD, Rakowski DR et al (2018) Diazepam concurrently increases the frequency and decreases the amplitude of transient dopamine release events in the nucleus accumbens. J Pharmacol Exp Ther 364:145–155. https://doi.org/10.1124/jpet.117.241802
28. Bass CE, Grinevich VP, Vance ZB et al (2010) Optogenetic control of striatal dopamine release in rats. J Neurochem 114(5):1344–1352. https://doi.org/10.1111/j.1471-4159.2010.06850.x
29. Xiao N, Privman E, Venton BJ (2014) Optogenetic control of serotonin and dopamine release in drosophila larvae. ACS Chem Neurosci 5:666–673. https://doi.org/10.1021/cn500044b
30. Sun F, Zeng J, Jing M et al (2018) A genetically encoded fluorescent sensor enables rapid and specific detection of dopamine in flies, fish, and mice. Cell 174:481–496.e19. https://doi.org/10.1016/j.cell.2018.06.042
31. Raiteri L, Raiteri M (2000) Synaptosomes still viable after 25 years of superfusion.

Neurochem Res 25:1265–1274. https://doi.org/10.1023/a:1007648229795

32. Broening HW, William S (1998) Ontogeny of neurotransmitters: monoamines. In: Handbook of developmental neurotoxicology. Academic Press, pp 245–256. https://doi.org/10.1016/B978-012648860-9.50017-0
33. Owens WA, Sevak RJ, Galici R et al (2005) Deficits in dopamine clearance and locomotion in hypoinsulinemic rats unmask novel modulation of dopamine transporters by amphetamine. J Neurochem 94:1402–1410. https://doi.org/10.1111/J.1471-4159.2005.03289.x
34. Pino JA, Awadallah N, Norris AM, Torres GE (2021) A plate-based assay for the measurement of endogenous monoamine release in acute brain slices. J Vis Exp (174). https://doi.org/10.3791/62127. PMID: 34459801
35. Koutzoumis DN, Vergara M, Pino J et al (2020) Alterations of the gut microbiota with antibiotics protects dopamine neuron loss and improve motor deficits in a pharmacological rodent model of Parkinson's disease. Exp Neurol 325:113159. https://doi.org/10.1016/j.expneurol.2019.113159
36. Torres G, Gainetdinov R, Caron M (2003) Plasma membrane monoamine transporters: structure, regulation and function. Nat Rev Neurosci 4:13–25. https://doi.org/10.1038/nrn1008
37. Sulzer D, Chen T, Lau Y et al (1995) Amphetamine redistributes dopamine from synaptic vesicles to the cytosol and promotes reverse transport. J Neurosci 15:4102–4108. https://doi.org/10.1523/JNEUROSCI.15-05-04102.1995
38. Fog JU, Khoshbouei H et al (2006) Calmodulin kinase II interacts with the dopamine transporter C terminus to regulate amphetamine-induced reverse transport. Neuron 51(4):417–429. https://doi.org/10.1016/j.neuron.2006.06.028. PMID: 16908408
39. Johnson LA, Guptaroy B, Lund D, Shamban S, Gnegy ME (2005) Regulation of amphetamine-stimulated dopamine efflux by protein kinase C beta. J Biol Chem 280(12):10914–10919. https://doi.org/10.1074/jbc.M413887200. Epub 2005 Jan 12. PMID: 15647254
40. Leviel V (2011) Dopamine release mediated by the dopamine transporter, facts and consequences. J Neurochem 118:475–489

Chapter 5

Modeling Pacemaking, Bursting, and Depolarization Block in Midbrain Dopamine Neurons

Christopher Knowlton and Carmen C. Canavier

Abstract

Here we present methodology to simulate the electrical activity of dopamine neurons by using a freely available software package to numerically integrate a set of coupled nonlinear equations that describe the equivalent circuit that generates the membrane potential, the nonlinear dynamics of channel gating, and a material balance on Ca^{2+} ions. The general methodology is conductance-based single-neuron computational models. We begin with Hodgkin-Huxley (H-H)-type conductance-based single-compartment models of pacemaking in vitro, which is described mathematically as a limit cycle. We illustrate phase plane methods to gain insight into this activity. Next, we address modeling rhythmic bursting activity. To illustrate that the methodology can be extended beyond the H-H formalism in which activation and inactivation gates operate independently of each other, we include a Markov model of a K+ channel in which they are dependent on each other. We then add random synaptic input to the model to illustrate the hypothesized balanced state in dopamine neurons, which is quite distinct from the balanced state of neocortical pyramidal neurons. We use this model to explain transient bursts and pauses. A major advantage of models is that parameter sweeps can be conducted quickly to determine the robustness of predicted activity. Finally, we move to a multi-compartmental model that captures the full morphology of a dopamine neuron to illustrate the role of the axon initial segment (AIS) in action potential initiation.

Key words Computational model, Nonlinear dynamics, Pacemaking, Bursting, Depolarization block, Oscillations, Substantia nigra, Ventral tegmental area, Dopaminergic neuron

1 Introduction

Midbrain dopamine neurons are thought to be involved in reward-based learning, motivation, action, and cognition [1–4], as well as in major brain disorders such as Parkinson disease [5], schizophrenia [6], and drug abuse [7]. Dopamine neurons in a slice preparation exhibit spontaneous pacemaking at low frequencies [8–10]. Blocking the Na^+ channel can often uncover smaller amplitude slow oscillatory potentials [11, 12] that are thought to bring dopamine neurons to their unusually depolarized spike threshold [13], enabling pacemaking. Both pacemaking and the slow oscillatory

José Antonio Fuentealba-Evans and Pablo Henny (eds.), *Dopaminergic System Function and Dysfunction: Experimental Approaches*, Neuromethods, vol. 193, https://doi.org/10.1007/978-1-0716-2799-0_5,

potentials are accompanied by an oscillation in cytosolic Ca^{2+} concentration [14, 15]. Only low frequencies can be sustained because at higher frequencies these neurons enter a state of depolarization block [16–18]. Certain manipulations can induce rhythmic burst firing in vitro [11, 19–21]. There is some evidence for bouts of rhythmic bursting in vivo [22, 23]. In general, bursting requires a non- or slowly inactivating regenerative inward current to support a plateau and a slowly activating outward current to terminate the plateaus. Pacemaking can also be observed in vivo, along with less regular discharge patterns, including transient high-frequency bursts and long pauses of electrical silence [24]. Pacemaking can be maintained in a balanced state in which inhibition and excitation are balanced [25–28], whereas a burst occurs when excitation dominates and a pause can occur when inhibition dominates.

In this chapter we describe techniques for computational modeling of the electrical activity of single dopamine neurons, including all of the abovementioned patterns, using the methodology for which Hodgkin and Huxley received a Nobel prize [29]. We have chosen to describe these techniques using the NEURON software package [30–33] because it is freely available, highly portable, and user-friendly. The package is designed to model neurons and has built-in mechanisms to perform efficient numerical integration of the differential equations. Modern NEURON is a python toolbox. It has a modular design in which the user can simply insert mechanisms, such as a currents or material balances on ions, into any given compartment of the neuron. We will start with a single-compartment model of an isopotential "point" neuron and finish with a morphologically realistic multi-compartmental model in which action potentials are initiated at the axon initial segment. For a review of models of single dopaminergic neurons, *see* ref. 25.

2 Materials

The NEURON software package is freely downloadable from https://neuron.yale.edu/neuron/. The NEURON code utilized in this chapter can be downloaded from https://modeldb.yale.edu/267357.

3 Methods

3.1 Conductance-Based Models for Pacemaking

The basic computational model of a neuron starts with an equivalent electrical circuit (Fig. 1a). The equivalent circuit shown in Fig. 1a corresponds to a particular, simplified conductance-based model adapted from a single dendritic compartment of a model of a substantia nigra dopamine neuron [34] that illustrates spontaneous pacemaking. This model consists the following currents: a voltage-

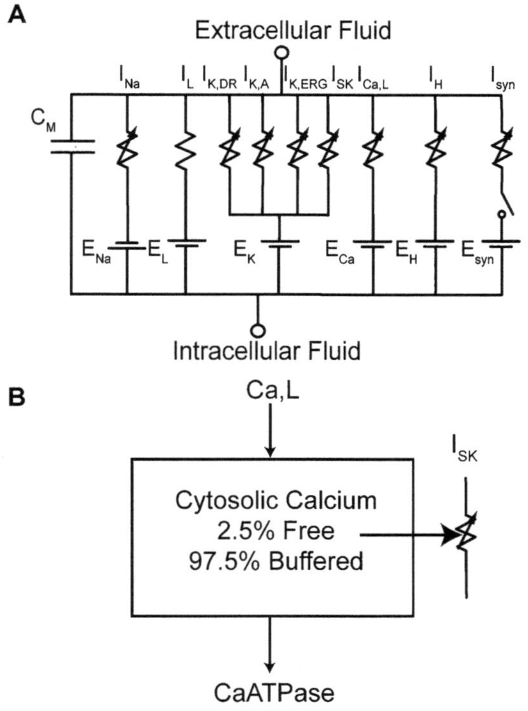

Fig. 1 Conductance-based neuron model. (**a**) Equivalent circuit diagram. Channels are equivalent to a variable resistor in series with an electromotive force given by the Nernst potential of the ion(s). The ERG channel is only present in models for Figs. 3, 4, and 5, while the synaptic currents are present only for Fig. 5. All channels are in parallel with the capacitor formed by the lipid bilayer. (**b**) Ca^{2+} material balance. Ca^{2+} enters via a low-threshold voltage-gated Ca^{2+} channel and is removed via Ca-ATPase pump. Internal calcium is assumed be buffered instantaneously with only 2.5% of the calcium contributing to the available pool. Free Ca^{2+} activates the SK K^+ current

gated Na^+ current (Na), a leak current (L), a delayed rectifier K^+ current (K,DR), an A-type K^+ current (K,A), a Ca^{2+}-activated K^+ current (SK), a single low-threshold Ca^{2+} channel (Ca,L), and a nonspecific hyperpolarization-activated cation current (H). The simulated quantity of interest is the electrical potential between the inside and outside of the cell, with the extracellular potential in this case assumed to be grounded. The charge accumulation on the lipid bilayer of the membrane is modeled as a capacitance C_M, and the capacitive current is $C_m dV/dt$ (by taking the time derivative of both sides of the definition of capacitance $CV = Q$), where Q is charge, V is the membrane potential, and dQ/dt is current I. Ionic channels are approximated as conductances following Ohm's law in which the current flow is given by $I = g(V - E_{REV})$ where g is the conductance, E_{REV} is the Nernst potential of the ion (s) conducted by that channel, and $(V - E_{REV})$ is the driving force. At the Nernst potential, the electrical and chemical driving forces on the channel are equal, and this potential is indicated as a battery in the diagram to represent the chemical potential for each ionic species (Na^+, K^+, Ca^{2+}) created by the different ionic concentrations inside and outside the cell. The L and H currents are permeable to more than one ion. All conductances are in parallel with the capacitance, the potential drop across this capacitor is the desired output of the simulation. Kirchhoff's current law that the total current entering and exiting any node be zero mandates that the ionic current is equal and opposite the current flowing through the capacitor, which gives the differential equation for the time evolution of the membrane potential $C_M = \frac{dV}{dt} = -\sum I_{\text{ionic}}$ which can be solved by numerical integration. In simplest terms, the time derivative gives the slope of the waveform for the variable, and for small time steps, one can assume the slope is constant and predicts the value at the next time point (*see* **Note 1**). The other important point to note is that all the conductances are nonlinear, with the exception of the passive leak conductance. In general, they depend on membrane potential and time and require differential equations to describe their time evolution. Hodgkin and Huxley approximated these dependencies by assuming activation and/or inactivation gates that are state variables that take values between 0 and 1 and are represented by a in the equations below. The effective conductance then becomes the product of these gating variables and a maximum conductance $g_{\max}$. Generally, these will be of the form:

$$I = g_{\max} a^n (V - E)$$

$$\frac{da}{dt} = \frac{a_\infty(V) - a}{\tau(V)}$$

$$a_\infty(V) = 1/(1 + \exp[-(V - a_v)/a_s])$$

Parameter values and kinetics are largely adapted from those of the dendritic compartments of the model in reference [34]. Figure 1b shows the material balance on the calcium pool that activates the SK channel. Calcium influx occurs through voltage-gated Ca^{2+} current and is extruded by a nonelectrogenic Ca-ATPase Ca^{2+} pump. Ca^{2+} buffering is approximated by assuming a small, fixed fraction of the total Ca^{2+} remains free in the cytosol-free Ca^{2+} in the cytosol [35]. We designate this one compartment model dendritic, meaning it has a large surface area to volume ratio, to capture the larger dendritic Ca^{2+} oscillations that accompany pacemaker firing [14, 15]. Although this model is illustrative, the diversity among populations of midbrain dopamine neurons has been established over the last decade [36], and therefore the model is most applicable to neurons with conventional electrophysiology and low maximal firing rates [18].

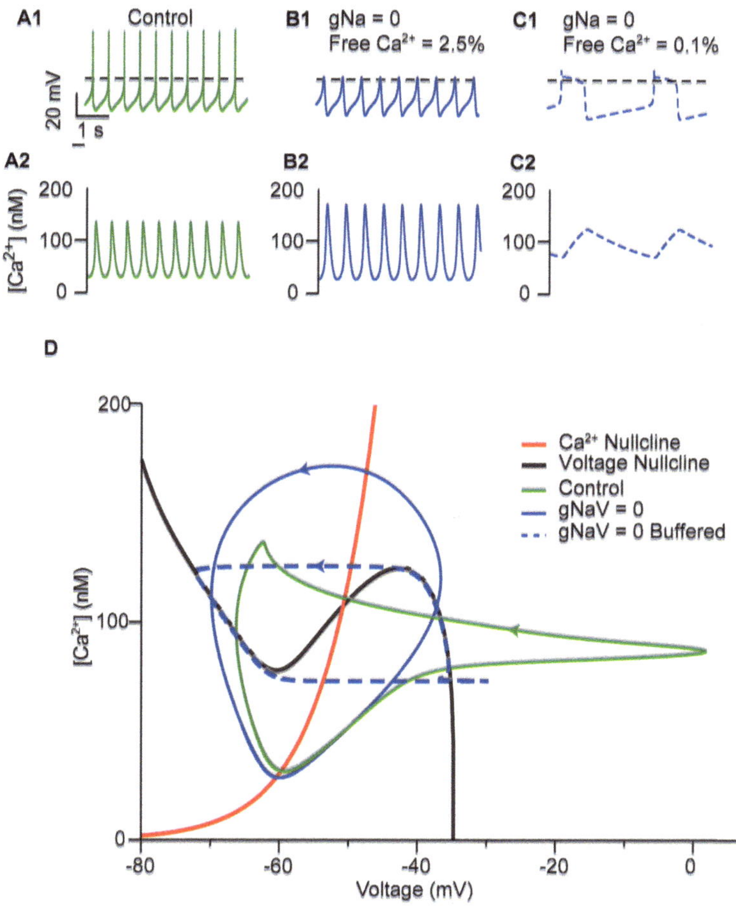

Fig. 2 Fast/slow analysis explains how one compartment model reproduces pacemaking and Ca^{2+} oscillations. (**a**) Pacemaking model neuron with control parameter settings. ERG and synaptic conductances are set to zero. (**a1**) Spontaneous pacemaking. (**a2**) Oscillation in free [Ca^{2+}]. (**b**) Slow oscillation persists with gNa = 0. (**b1**) Slow oscillatory potentials. (**b2**) Oscillation in free [Ca^{2+}]. (**c**) Slow oscillatory potentials are converted to a square wave oscillation when Ca^{2+} dynamics are rendered extremely slow by simulation a powerful exogenous Ca^{2+} buffer. (**c1**) Square wave oscillation with plateau potentials. (**c2**) Oscillation in free [Ca^{2+}]. (**d**) Phase plane (free Ca^{2+} concentration vs membrane potential) for one compartment dopamine model. Ca^{2+} nullcline (red) is set of points where $d[Ca^{2+}]/dt = 0$ for fixed voltage values. Voltage nullcline (black) indicates set of points where $dV/dt = 0$ given fixed Ca^{2+} values. The fixed point at the intersection of the two nullclines is unstable and repels trajectories. The trajectories for a single cycle of pacemaking from panel (**a**) (green), slow oscillatory potentials from panel (**b**) (solid blues), and square wave oscillation in panel (**c**) (blue dashed) all oscillate around the unstable fixed point

Initially, we consider the model without the synaptic or ERG channels. The model can exhibit spontaneous pacemaking in the absence of any applied current (Fig. 2a), because like a real dopamine neuron, the model has no stable resting potential. In addition, an oscillation in free cytosolic Ca^{2+} accompanies pacemaking (Fig. 2a2). We have modeled a single Ca^{2+} current as an L-type current for simplicity, which drives the slow depolarization in membrane potential and the rising phase of the Ca^{2+} waveform in the model. Consistent with experimental data, slow subthreshold oscillation (Fig. 2b1) can persist with the Na^+ conductance set to zero, which abolishes action potentials. The oscillation is due to an interplay between the voltage-gated Ca^{2+} current and the Ca^{2+}-activated SK current [15, 37]. The oscillation in $[Ca^{2+}]$ increases in amplitude (Fig. 2b2) since it is no longer being reset by the repolarizing spike [25]. If the free Ca^{2+} fraction is reduced to a very small number to represent the addition of a fast exogenous Ca^{2+} buffer, a square wave oscillation (Fig. 2c1) takes the place of slow oscillatory potential due the much slower time course of Ca^{2+} (Fig. 2c2) as explained below. Square oscillations with their plateau potentials often underlie rhythmic bursting activity [38, 39] as we will demonstrate later in the chapter.

3.2 Phase Plane Analyses

Phase plane analysis [40] is a common tool in the branch of mathematics called nonlinear dynamics, which is strictly only applicable to two-dimensional systems. The dimension is the number of differential equations, one for each state variable like V, $[Ca^{2+}]$, and the gating variables. In the phase plane (Fig. 2d), time is implicit with one state variable plotted against the other, in this case free Ca^{2+} concentration $[Ca^{2+}]$ versus membrane potential V (*see* **Note 2**). The fixed point (or points) of a nonlinear dynamic system is a very useful concept and one that requires the concept of a nullcline. A nullcline is a set of points, in this case the pair (V, $[Ca^{2+}]$), that cause the differential equation for a given variable to be zero. Hence, on the voltage nullcline (black), the steady-state net ionic currents at each point on the voltage nullcline is zero. In this model, the only calcium-sensitive channel is the SK channel; therefore, at any fixed V, all other currents are set to their steady-state values, and the $[Ca^{2+}]$ that makes the net current zero by adjusting the SK current can be found algebraically or by trial and error (*see* **Note 3**). On the $[Ca^{2+}]$ nullcline (red), Ca^{2+} influx must balance Ca^{2+} efflux. At a given membrane potential, the steady $[Ca^{2+}]$ influx is determined by the voltage-dependent Ca^{2+} current, and the $[Ca^{2+}]$ value required to activate sufficient pump current to balance the influx can be calculated. Alternatively it can be found by using the voltage clamp (SEClamp) built in feature of NEURON to clamp the membrane potential to any desired voltage until the concentration equilibrates, analogous to the experimentally determined Ca^{2+} nullcline in these neurons [15].

Since the time derivatives of all variables are zero at the intersection of the nullclines, this fixed point is the only possible resting membrane potential for the model. The Ca^{2+} nullcline is monotonic, but the V nullcline is cubic, and this cubic shape is the key to the oscillation. The bend that forms the middle branch of the V nullcline is due to the Na^+ channels opening regeneratively; that is, depolarization opens Ca^{2+} channels which causes more depolarization in a positive feedback loop. Conversely, closing these channels causes hyperpolarization which closes more channels. If we can separate the time scales for the two variables, by assuming that that $[Ca^{2+}]$ varies slowly compared to V, we can use fast/slow analysis [41] to show that the middle branch is unstable because of this positive feedback loop. On the other hand, the left and right branches of the V nullcline are stable; trajectories near these curves will converge toward the nullcline. Under this assumption, one can predict the flow of trajectories. Since the resting potential is unstable but the trajectories are attracted to the stable branches by restorative rather than regenerative current, the dynamics are compelled to cycle around the fixed point indefinitely. This point is illustrated most clearly by the blue dashed trajectory corresponding to Fig. 2c, with action potentials blocked by setting the Na+ conductance to zero. Free Ca^{2+} varies so slowly because 99.9% is presumed to be immediately buffered; thus, the fast-slow analysis of the trajectory following the nullclines and jumping at the bending point from one stable branch to the other is fairly accurate. With a substantially faster time course for $[Ca^{2+}]$ as in Fig. 2b, the separation of time scales is not as complete, and the trajectory (solid blue) is not as predictable, but the trajectory still oscillates in a roughly predictable manner. For the pacemaking (green trace) trajectory, the voltage-dependent variable is far from steady state, causing large deviations from the nullcline calculated based on steady state, but the cyclic motion that causes spontaneous pacemaking is intact. Thus, the simple prediction that the resting membrane is unstable and the stable branches constrain the trajectory to remain nearby is sufficient to predict and explain pacemaking. This picture is not unique, as many combinations of currents and parameter values can produce similar activity [42].

3.3 Markov Models

Before progressing to models of bursting activity, we note that conductance-based models are not restricted to Hodgkin-Huxley models with independent and multiplicative gates. The proteins that form the pores through the membrane for current flow are complex 3D structures, and independent gates are not always sufficient to capture some phenomena of interest, so we can increase the complexity of the model using interdependent states of the channel. The ether-a-go-go-related K^+ channel (ERG) has been implicated in bursting in dopamine neurons in vitro [20, 43]. This current is well suited to terminate bursts and prolong the interburst interval because it accumulates slowly during depolarization but inactivates quickly. In a Markov model such as the one in Fig. 3a, the transition rates between two states are generally voltage-dependent and depend linearly upon the relative occupancy in

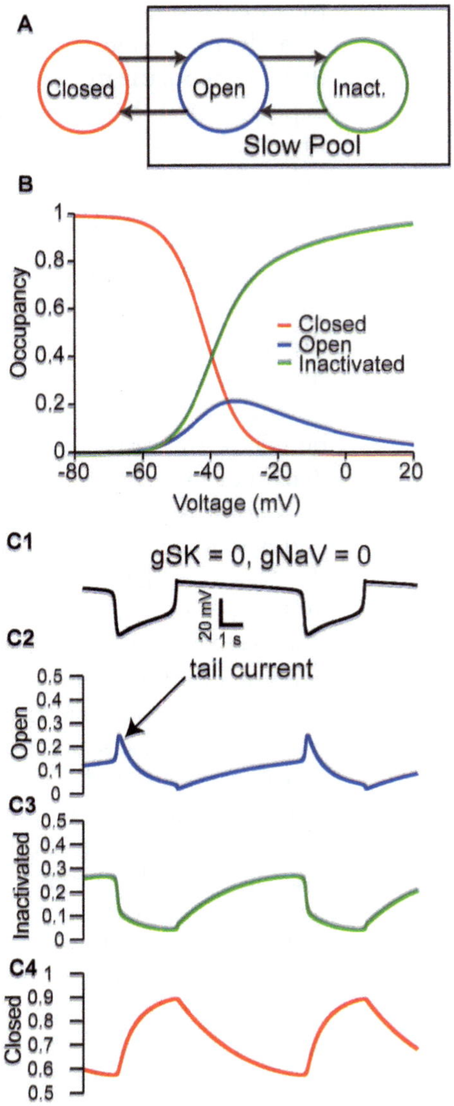

Fig. 3 Square wave oscillations repolarized by ERG K^+ current instead of SK K^+ current. (**a**) Schematic of Markov channel model for ERG with the slow pool indicated in the black box. (**b**) Steady-state values of closed, open, and inactivated states vs voltage. (**c1**) Square wave oscillations with both gNa and gSK set to zero. (**c2**) Fractional occupancy of ERG channels in the open state. (**c3**) Fractional occupancy of ERG channels in the inactivated state. (**c4**) Fractional occupancy of ERG channels in the closed state

each state. Since occupancy is defined as the fraction of all channels that occupy a particular state, the total occupancy among all states remains constant at one. The equations for the ERG channel included here are adapted from Ji et al. [43]. Figure 3b gives the steady-state occupancy in the three states. However, in this case we can also find a slow variable for the fast/slow analysis, namely, the combined slow pool governed by the slow transition between the open and closed states and the fast transition between the open and inactivated states (*see* Fig. 3a). To illustrate mechanisms of burst firing while using a Markov model for the slow K^+ channel, we added this channel to the model from Figs. 1 and 2 (*see* **Note 4**).

The SK channel is known to play a role in regularizing pacemaking [11, 44–46], and blocking this channel leads to bursting both in vivo [42, 47] and in vitro [11, 20, 46]. Bursting results largely because blocking this channel produces plateau potentials [20, 37] because the after-hyperpolarizing potential (AHP) is not sufficient to repolarize the membrane after a single spike; instead, multiple spikes occur during each depolarizing event [25–27, 48]. As a prelude to burst firing, in Fig. 2c1 we show that this current could also serve to repolarize the plateau observed during square wave oscillation as in Fig. 2c, in which the weakened SK channel repolarized the plateaus. We set gNa = 0 and gSK = 0 to reveal this oscillation. As in Fig. 2c2 and the dashed blue trace in Fig. 2d, a relaxation oscillation (in which the trajectory quickly relaxes to the nullclines) is formed by the interaction of the regenerative L-type calcium channel with a slow restorative K^+ channel. During the depolarized plateau, occupancy in the open state accumulates slowly (Fig. 3c2); the occupancy in the inactivated state also increases (Fig. 3c3). Once sufficient K^+ current has accumulated, the Ca^{2+} current turns off regeneratively and the plateau terminates. There is no direct path from the inactivated state to the open state, so upon hyperpolarization, the inactivated channels must open before they can close (Fig. 3c4), providing a large additional transient hyperpolarizing current (arrow in Fig. 3c2 indicating large tail currents). This transient characteristic of the ERG channel cannot be captured by Hodgkin-Huxley-type models. When a sufficient number of K^+ channels have closed, the Ca^{2+} current is able to regeneratively active, leading to another plateau. Again, this picture is not unique, and multiple currents likely contribute to bursting in dopamine neurons. A persistent Na^+ current or an NMDA receptor current could also function as the regenerative current underlying a plateau, and several slowly developing outward currents, such as the electrogenic Na^+/K^+ pump [49], the K^+ ATP current [50], and the M-type K^+ [51], have also been suggested to contribute to plateau repolarization and burst termination in these neurons.

3.4 Intrinsic Bursting

As stated in the previous sections, one common way to generate bursting activity is for multiple spikes to occur during a single depolarizing event. Often, when spiking is blocked, the underlying depolarizing envelope can be observed. We will use the square wave oscillation, with depolarized plateaus supported by a low-threshold Ca^{2+} current, from the previous section to build two bursting models. The addition of spiking currents complicates the simple two-dimensional square wave concept, as we shall see in the following examples. Figure 4a shows pacemaking activity qualitatively

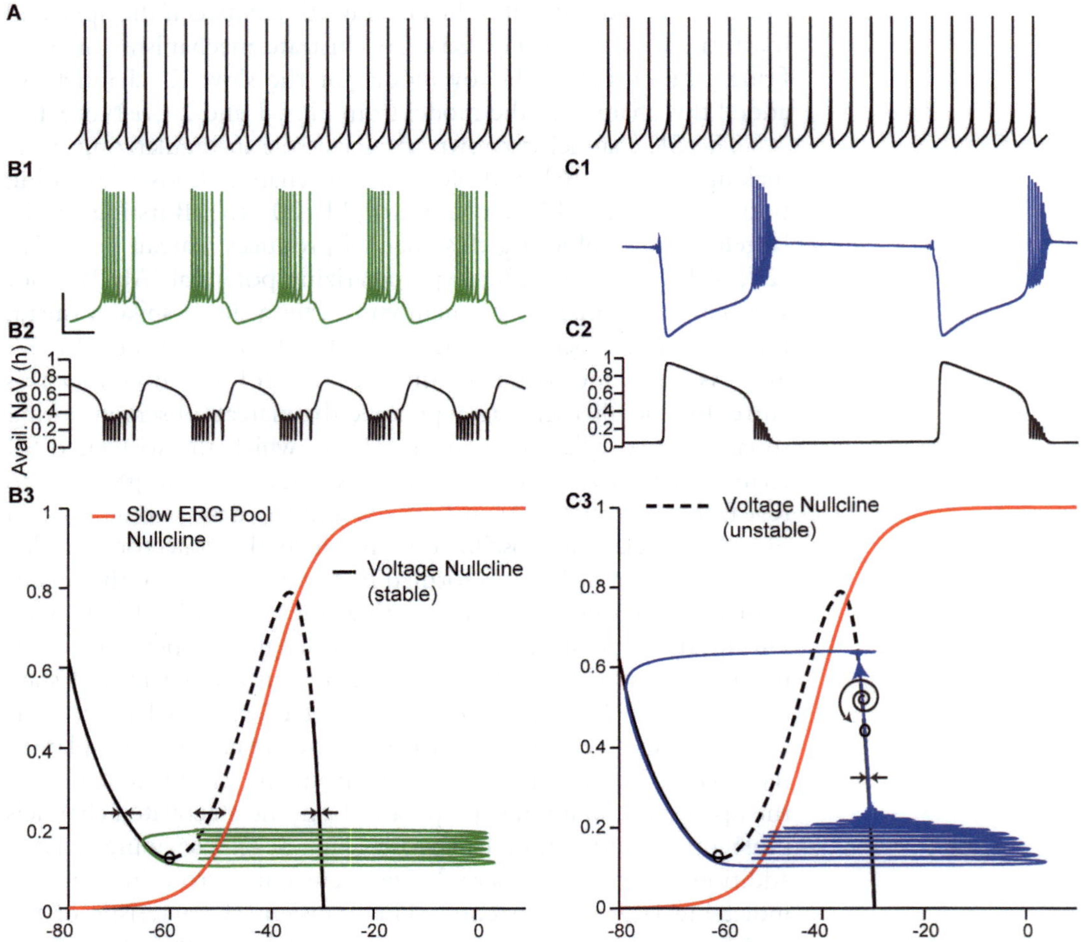

Fig. 4 Bursting induced by setting gSK to zero without and with depolarization block. (**a**) Pacemaking as in Fig. 2a1 but with gERG nonzero. (**b**) Rhythmic bursting with gSK = 0. (**b2**) Non-inactivated fraction (*h*) of Na^+ channels. (**b3**) Phase plane analysis with of slow pool of ERG channel occupancy (*O* + *I*) vs membrane potential of trajectory from **b1** (green). (**c1**) Rhythmic bursting terminating in depolarization block with the same parameters as **b1** except 1% less gKdr. (**c2**) Non-inactivated fraction (*h*) of Na^+ channels. (**c3**) Phase plane analysis with of slow pool of ERG channel occupancy (*O* + *I*) vs membrane potential of trajectory from **c1** (blue)

identical to that observed in Fig. 2a except the ERG current is nonzero. When the SK conductance is set to zero in Fig. 4b1, rhythmic burst results because the AHP does not have sufficient depth or duration to turn off the low-threshold Ca^{2+} current. Qualitatively similar bursting activity has been observed experimentally in dopamine neurons in vitro when an SK blocker was applied concurrently with NMDA [19] or an L-type Ca^{2+} channel agonist [21]. Both manipulations strengthen the available pool of non-inactivating, regenerative inward currents that support the plateau.

The phase plane analysis in Fig. 4b3 is qualitatively similar to that of Fig. 3d, with a different slow variable. Since the SK conductance is set to zero, Ca^{2+} dynamics are no longer relevant to the model. Instead, the slow variable for the fast-slow analysis is the slow pool from Fig. 3a consisting of ERG channels in the open or inactivated state; they form a single pool because the transition rates between those states are fast relative to the transition rates between the closed and inactivated states. The V-nullcline (black) was calculated as before except the quantity to adjust is fractional occupancy in the slow pool that determines the level of the ERG conductance. It retained the cubic shape that enables oscillations. The slow pool nullcline (red) is simply the sum of the blue and green steady-state occupancy curves from Fig. 3b. However, the nullcline analysis is further complicated by the presence of the spiking currents. Whereas the depolarized plateaus in Figs. 2c and 3c1 initiate and terminate at the bends in the nullclines, spiking initiates at a bend but terminates when the trough of the spike touches the unstable middle branch and causes a switch to the leftmost stable branch.

Since the critical aspect of burst termination is the depth of the AHP, in Fig. 4c, we reduced by 1% the magnitude of the delayed rectifier conductance that (in the absence of SK in the model) is the major contributor to the depth of the AHP. Whereas pacemaking is indistinguishable from Fig. 4a, the effect of setting the SK conductance to zero was quite different (compare Fig. 4c1 to b1). Instead of rhythmic bursting, spiking terminates into a depolarized plateau potential, reminiscent of burst activity evoked by SK alone [11, 43, 52], or of the mode of failure of a subpopulation of electrophysiologically atypical dopamine neurons [18]. The stable part of the rightmost branch of the nullcline corresponds to a state of depolarization block that is often observed in dopamine neurons [16, 43, 53, 54]. Depolarization block occurs when the pool of available, non-inactivated sodium channels is too small to support a regenerative action potential. In Fig. 4b2, the available pool recovers sufficiently during the interspike intervals for spiking to be sustained during the depolarized phase. However, the depolarized phase in Fig. 4c is too prolonged and intense and depletes the available pool (Fig. 4c2) so that spiking ceases during the depolarized plateau. Thus, the depth of the after-hyperpolarization relative to the middle branch of the nullcline determines which of these two types of bursting, that on the surface appear very different, will be exhibited (*see* **Note 5**).

3.5 The Balanced State of Dopamine Neurons In Vivo

In neocortical neurons, the balanced state between excitation and inhibition [55] causes pyramidal neurons to spend most of their time below spike threshold such that spikes are driven by noisy fluctuations in their synaptic input [56, 57] causing the spike trains to resemble the output of a noisy Poisson process. We describe a very different balanced state for dopamine neurons. It is well known that, as described above, in the absence of synaptic input, dopamine neurons spontaneously fire repetitive single spikes in a pacemaker-like fashion. This oscillatory regime is referred to as the mean driven as opposed to fluctuation driven because the firing rate reflects the average input current. A fairly regular single-spiking mode can be preserved in the presence of balanced excitation and inhibition, as confirmed by in vitro dynamic clamp experiments [28] and simulations [25–27] and also supported by in vivo patch clamp recordings of single spike firing with low CV [24].

To demonstrate the in vivo balanced state, we used the Neuron NetStim class (*see* **Note 6**) to apply two simultaneous spike trains (1000 Hz $GABA_A$, 100 Hz NMDA) from a Poisson process to our single-compartment model, one to activate inhibitory $GABA_A$ synapses and the other to activate excitatory synapses with NMDA receptor kinetics (AMPA kinetics were omitted for simplicity). Figure 5a shows the results of a parameter sweep, in which the parameter space of the model is systematically characterized into three main regions in the plane representing the conductance strengths associated with inhibitory GABAergic inputs and excitatory NMDA inputs. The excitatory/inhibitory balanced state was observed in a diagonal band in this plane, which includes the origin representing no synaptic input. In this band, spiking was regular, with approximately one spike per depolarization (representative trace in Fig. 5b1). Above and to the left of that band, the model neuron was largely quiescent, because inhibition was dominant. Below and to the right of that band, excitation was dominant, and the firing became bursty (representative trace in Fig. 5b2). If on average 1.5 or more spikes were observed per depolarizing wave, the firing was characterized as bursting.

3.6 Biophysically Realistic Morphologies

Real neurons are not isopotential point neurons, but rather have spatial extent including specialized regions such as the axon and its initial segment, the soma, and various dendrites. A database of detailed 3D neural morphologies is available at http://www.neuromorpho.org/ (*see* **Notes 7** and **8**). Some features of the electrical activity of neurons cannot be captured in a single-compartment model. One example of such a feature is the "kink" or sudden increase in slope at action potential onset that results from distant action potential initiation at the axon initial segment in most mammalian central nervous system neurons [58], including dopamine neurons. In dopamine neurons microns, the initial segment can be located up to 200 microns from the soma in rats [59] and up to 70 microns away in mice [60]. In this section, we switch to a published model [61] that is freely downloadable from https://senselab.med.yale.edu/ModelDB/ShowModel?model=244599#tabs-1 to illustrate the effect of morphology on the electrical activity. Figure 6a shows the NEURON multi-compartmental implementation of the 3D morphology of a reconstructed

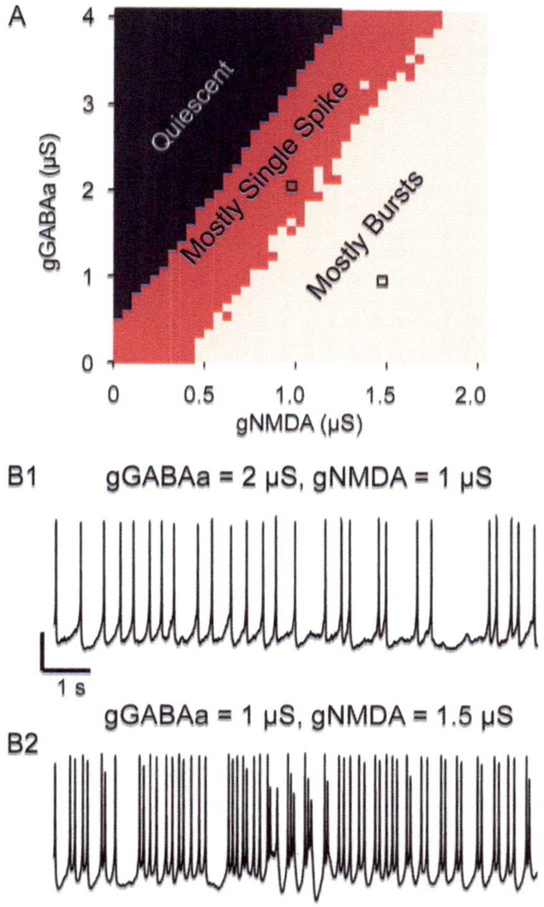

Fig. 5 The single spike firing balanced state in dopamine neurons. (**a**) Results of a parameter sweep for inhibitory and excitatory synaptic conductance strengths. Quiescent (black), balanced (red), and bursting (beige) regions emerged from this analysis. Activity type was characterized by the number of spikes per depolarizing wave, which is defined as the interval between crossings of −60 mV. <0.5 spikes per wave was considered quiescent, 0.5–1.5 was considered a balanced state, and greater considered bursting. The number of spikes per wave generally increases toward higher NMDA and lower GABAa under thresholds between −60 and −50 mV. (**b1**) Example trace of single spike firing balanced state. (**b2**) Example trace of bursting mode

dopamine neuron from the substantia nigra pars compacta, with a blowup of the region containing the soma and axon-bearing dendrite (ABD) and the axon initial segment (AIS) shown in red. The actual length of the AIS was previously determined using ankyrin staining [60]. The model has a very similar equivalent circuit in each compartment as in Fig. 1 and exhibits spontaneous pacemaking activity shown at the top, but the conductances are spatially heterogeneous. For example, the spiking conductances are far

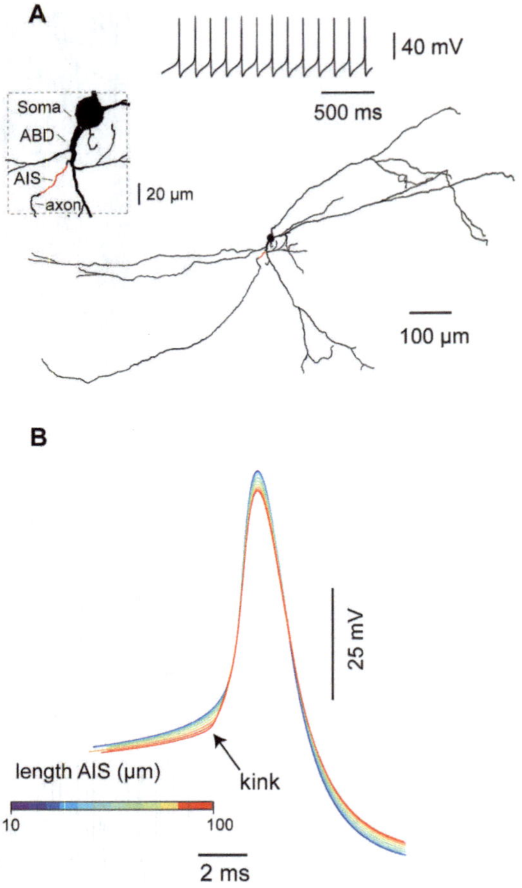

Fig. 6 Morphologically realistic model of dopamine reproduces the kink in the somatic action potential waveform due to distal spike initiation. (**a**) Clockwise from top: Somatic voltage trace during spontaneous pacemaking in the morphologically realistic model. Morphology of model with axon initial segment (AIS) highlighted in red. Expanded view of soma, axon-bearing dendrite (ABD), and axon initial segment (AIS). (**b**) "Kink" in voltage trace is more pronounced with larger AIS compartment

larger in the AIS than elsewhere, so the action potential is always initiated in the AIS and propagates to the soma. Dopamine neurons have an axonal and a somato-dendritic component to their action potential [62], sometimes separated by a kink as shown in Fig. 6b, or peak in the second derivative of the membrane potential. The change in slope occurs when the somato-dendritic membrane potential becomes regenerative rather than passively reflection axonal depolarization. In this model, the somatic Na^+ conductance density was set relatively low to capture some features of the extracellular action potential (*see* **Note 9**) not illustrated here, and the ABD became the major contributor to the somato-dendritic component of the action potential. In a model, one can change parameters that cannot be manipulated experimental. For example, here we kept the location of the proximal end of the AIS constant but extended the position of the distal end of the AIS to manipulate its total length. The resultant simulated intracellular action potentials are color coded with hotter colors corresponding to longer AIS. The larger currents evoked in a larger AIS cause a more pronounced kink, starting the second, regenerative phase of the action potential more abruptly at lower membrane potential. This sharp bend is impossible to achieve in single-compartment model [63, 64].

4 Notes

1. NEURON by default uses a staggered fixed time step (set with h.dt) which is reasonably efficient and accurate provided the dynamics are much slower than your time step. However, as problems become larger and/or require increased accuracy at faster timescales, the multi-order ODE solver CVODE can be called (see lines 201-202 in SimpleModel_Fig12.py). As CVODE allows for significantly longer steps when the dynamics allow for it, e.g., during hyperpolarizations when fast spiking currents are inactivated, cv.solve (TSTOP) should be used over h.run() or iterative calls to h.fadvance(). However, this integration routine requires more in-depth procedure for recording and printing state variables (*see* **Note 2**). For models with large numbers of neurons such as with a network or parameter sweep on single morphology, ParallelContext and pc.psolve() (see lines 488-489 and 527-529 in SimpleModel_Fig12.py) can be used over CVODE and cv.solve(). ParallelContext automatically distributes cells to CPU threads allowing for multithreaded runs. Note that the integration steps for ParallelContext and thread-gather condition are based on the event queue of any netcon objects in the system. To get sensible results, each neuron created must be paired with a netcon object reading a voltage state of that neuron—even if that netcon is not broadcasting to any synapse. This netcon should be created as part of the python class for the neuron model.
2. To get the phase trajectories, you need to use vector record (see lines 242-246 in SimpleModel_Fig12.py) to save the values of membrane potential and calcium at the exact same times and then plot them against each other rather than time. This was implemented here by creating a python array of NEURON vector objects with a vector for each variable to record, and then for each vector, call the record function calling the *pointer* to that variable and either a fixed recording time step or a preset

vector of time points at which to record. Note that to be recordable, a variable must either be a default process (e.g., Ca^{2+} concentration, voltage) or a STATE or RANGE variable from an imported mechanism from a hoc file. Note that time is a recordable variable as well (see line 412 in SimpleModel_-Fig12.py).

3. A steady-state voltage is only possible over a narrow range of voltages as the calcium-sensitive current is limited by the maximal (and minimal) conductance of the SK channel.
4. Markov models should use the KINETIC block rather than DERIVATIVE block as it allows for more efficient "sparse" integration and automatically handles conservation statements. See IERG.mod for an example implementation.
5. The nullcline analysis is further complicated by the presence of the spiking currents; the rightmost branch is only partially stable. The dashed portion is not destabilized by the Ca^{2+} current that destabilizes the middle branch, but rather by the regenerative effects of the Na^{+} current that mediate spiking. This is only relevant to the dynamics because the system is not truly 2D in the presence of the spiking currents. Instead of simply repelling trajectories like the middle branch does, there is a circular flow around the nullcline in a direction perpendicular to the phase plane that causes spike amplitude to decrease as it spirals into the rightmost branch [65]. The stability was calculated by numerically calculating the Jacobian matrix in the vicinity of the nullcline. The largest nonzero eigenvalue of the resulting matrix determines the stability. If the real part of the largest eigenvalue is positive, the point is unstable. This method can also be used to identify the types of bifurcation based on the creation and destruction of complex pairs. Dynamics follows right branch upward until supercritical Hopf (circle at right) destabilizes nullcline. Stability transitions do not map neatly onto knees of nullclines due to the high dimensional space.
6. Netcon objects in NEURON translate spiking events in presynaptic neurons into summated or stereotyped synaptic currents in a post-synaptic cell. Outside of a fixed network, netcons can also receive event times via either a Netstim object or "manually" via repeated invoking of the event() function. NetStim allows for both regular and noisy inputs with a toggle (netstim.noise) that ranges between perfectly regular (0) and negative exponential distributed ISIs (1).
7. Importing morphologies will vary with the program that was used to do the z-stack imaging but will generally use the cellbuilder tool in NEURON—from which data from common neuron imaging software can be imported. A set of scripts to convert neurolucida3 data with labeled compartment types is

included. First, the morphology must be loaded by using nrngui import_morphology.py <your_nl3_filename> Confirm morphology by clicking topology, subsets in cellbuild window that appear. If no errors, use "Management" and then "export to file" to save as hoc. Template_Transform.py can be used to change the syntax of the resulting morphologies into a template per **Note 10**.

8. In models with multiple compartments, axial current flow is handled automatically with axial resistivity (ohm-cm) set on a per section basis with sec.Ra. The default value is 34.5, but 100–200 is consistent with two site recordings of dendritic propagation [59].
9. Extracellular compartments can be added via section.insert ("extracellular").
10. NEURON in python allows for the use of class objects that are ideal for creating a template for creating neurons, vectors to store measurements, associated synapses, and current clamps.

References

1. Athalye VR, Carmena JM, Costa RM (2020) Neural reinforcement: re-entering and refining neural dynamics leading to desirable outcomes. Curr Opin Neurobiol 60:145–154
2. Gershman SJ, Uchida N (2019) Believing in dopamine. Nat Rev Neurosci 20:703–714
3. Langdon AJ, Sharpe MJ, Schoenbaum G et al (2018) Model-based predictions for dopamine. Curr Opin Neurobiol 49:1–7
4. Schultz W (2015) Neuronal reward and decision signals: from theories to data. Physiol Rev 95:853–951
5. Obeso JA, Stamelou M, Goetz CG et al (2017) Past, present, and future of Parkinson's disease: a special essay on the 200th Anniversary of the Shaking Palsy. Mov Disord 32:1264–1310
6. McCutcheon RA, Abi-Dargham A, Howes OD (2019) Schizophrenia, dopamine and the striatum: from biology to symptoms. Trends Neurosci 42:205–220
7. Lüscher C (2016) The emergence of a circuit model for addiction. Annu Rev Neurosci 39: 257–276
8. Kang Y, Kitai ST (1993) A whole cell patch-clamp study on the pacemaker potential in dopaminergic neurons of rat substantia nigra compacta. Neurosci Res 18:209–221
9. Yung WH, Häusser MA, Jack JJ (1991) Electrophysiology of dopaminergic and non-dopaminergic neurones of the guinea-pig substantia nigra pars compacta in vitro. J Physiol 436:643–667
10. Nedergaard S, Flatman JA, Engberg I (1993) Nifedipine- and omega-conotoxin-sensitive Ca^{2+} conductances in guinea-pig substantia nigra pars compacta neurones. J Physiol 466: 727–747
11. Ping HX, Shepard PD (1996) Apamin-sensitive Ca^{2+}-activated K+ channels regulate pacemaker activity in nigral dopamine neurons. Neuroreport 7:809–814
12. Kang Y, Kitai ST (1993) Calcium spike underlying rhythmic firing in dopaminergic neurons of the rat substantia nigra. Neurosci Res 18: 195–207
13. Grace A, Onn S (1989) Morphology and electrophysiological properties of immunocytochemically identified rat dopamine neurons recorded in vitro. J Neurosci 9:3463–3481
14. Chan CS, Guzman J, Ilijic E et al (2007) "Rejuvenation" protects neurons in mouse models of Parkinson's disease. Nature 447: 1081–1086
15. Wilson CJ, Callaway JC (2000) Coupled oscillator model of the dopaminergic neuron of the substantia nigra. J Neurophysiol 83: 3084–3100
16. Richards CD, Shiroyama T, Kitai ST (1997) Electrophysiological and immunocytochemical characterization of GABA and dopamine neurons in the substantia nigra of the rat. Neuroscience 80:545–557
17. Tucker KR, Huertas MA, Horn JP et al (2012) Pacemaker rate and depolarization block in

nigral dopamine neurons: a somatic sodium channel balancing act. J Neurosci 32: 14519–14531

18. Lammel S, Hetzel A, Häckel O et al (2008) Unique properties of mesoprefrontal neurons within a dual mesocorticolimbic dopamine system. Neuron 57:760–773
19. Johnson SW, Seutin V, North RA (1992) Burst firing in dopamine neurons induced by N-methyl-D-aspartate: role of electrogenic sodium pump. Science 258:665–667
20. Canavier CC, Oprisan SA, Callaway JC et al (2007) Computational model predicts a role for ERG current in repolarizing plateau potentials in dopamine neurons: implications for modulation of neuronal activity. J Neurophysiol 98:3006–3022
21. Johnson SW, Wu Y-N (2004) Multiple mechanisms underlie burst firing in rat midbrain dopamine neurons in vitro. Brain Res 1019:293–296
22. Freeman AS, Meltzer LT, Bunney BS (1985) Firing properties of substantia nigra dopaminergic neurons in freely moving rats. Life Sci 36:1983–1994
23. Schiemann J, Schlaudraff F, Klose V et al (2012) K-ATP channels in dopamine substantia nigra neurons control bursting and novelty-induced exploration. Nat Neurosci 15: 1272–1280
24. Otomo K, Perkins J, Kulkarni A et al (2020) In vivo patch-clamp recordings reveal distinct subthreshold signatures and threshold dynamics of midbrain dopamine neurons. Nat Commun 11:6286
25. Canavier CC, Evans RC, Oster AM et al (2016) Implications of cellular models of dopamine neurons for disease. J Neurophysiol 116: 2815–2830
26. Canavier CC, Landry RS (2006) An increase in AMPA and a decrease in SK conductance increase burst firing by different mechanisms in a model of a dopamine neuron in vivo. J Neurophysiol 96:2549–2563
27. Komendantov AO, Komendantova OG, Johnson SW et al (2004) A modeling study suggests complementary roles for GABAA and NMDA receptors and the SK channel in regulating the firing pattern in midbrain dopamine neurons. J Neurophysiol 91:346–357
28. Lobb CJ, Wilson CJ, Paladini CA (2010) A dynamic role for GABA receptors on the firing pattern of midbrain dopaminergic neurons. J Neurophysiol 104:403–413
29. Hodgkin AL, Huxley AF (1952) A quantitative description of membrane current and its application to conduction and excitation in nerve. J Physiol Lond 117:500–544
30. Hines ML, Carnevale NT (1997) The NEURON simulation environment. Neural Comput 9:1179–1209
31. Hines ML, Davison AP, Muller E (2009) NEURON and python. Front Neuroinform 3:1
32. Carnevale NT, Hines ML (2006) The neuron book. Cambridge University Press
33. Hines ML, Carnevale NT (2001) NEURON: a tool for neuroscientists. Neuroscientist 7: 123–135
34. Kuznetsova A, Huertas M, Kuznetsov A et al (2010) Regulation of firing frequency in a computational model of a midbrain dopaminergic neuron. J Comput Neurosci 28: 389–403
35. Foehring RC, Zhang XF, Lee JCF et al (2009) Endogenous calcium buffering capacity of substantia nigral dopamine neurons. J Neurophysiol 102:2326–2333
36. Roeper J (2013) Dissecting the diversity of midbrain dopamine neurons. Trends Neurosci 36:336–342
37. Amini B, Clark JW Jr, Canavier CC (1999) Calcium dynamics underlying pacemaker-like and burst firing oscillations in midbrain dopaminergic neurons: a computational study. J Neurophysiol 82:2249–2261
38. Bertram R, Butte MJ, Kiemel T et al (1995) Topological and phenomenological classification of bursting oscillations. Bull Math Biol 57:413–439
39. Rinzel J (1987) A formal classification of bursting mechanisms in excitable systems. In: Mathematical topics in population biology, morphogenesis and neurosciences. Springer, Berlin, pp 267–281
40. Izhikevich EM (2007) Dynamical systems in neuroscience. MIT Press
41. Bertram R, Rubin JE (2017) Multi-timescale systems and fast-slow analysis. Math Biosci 287:105–121
42. Drion G, Massotte L, Sepulchre R et al (2011) How modeling can reconcile apparently discrepant experimental results: the case of pacemaking in dopaminergic neurons. PLoS Comput Biol 7:e1002050
43. Ji H, Tucker KR, Putzier I et al (2012) Functional characterization of ether-a-go-go-related gene potassium channels in midbrain dopamine neurons: implications for a role in depolarization block. Eur J Neurosci 36:2906–2916

44. Lovejoy LP, Shepard PD, Canavier CC (2001) Apamin-induced irregular firing in vitro and irregular single-spike firing observed in vivo in dopamine neurons is chaotic. Neuroscience 104:829–840
45. de Vrind V, Scuvée-Moreau J, Drion G et al (2016) Interactions between calcium channels and SK channels in midbrain dopamine neurons and their impact on pacemaker regularity: contrasting roles of N- and L-type channels. Eur J Pharmacol 788:274–279
46. Shepard PD, Bunney BS (1991) Repetitive firing properties of putative dopamine-containing neurons in vitro: regulation by an apamin-sensitive Ca^{2+}-activated K+ conductance. Exp Brain Res 86:141–150
47. Waroux O, Massotte L, Alleva L et al (2005) SK channels control the firing pattern of midbrain dopaminergic neurons in vivo. Eur J Neurosci 22:3111–3121
48. Oster A, Faure P, Gutkin BS (2015) Mechanisms for multiple activity modes of VTA dopamine neurons. Front Comput Neurosci 9:95
49. Canavier CC (1999) Sodium dynamics underlying burst firing and putative mechanisms for the regulation of the firing pattern in midbrain dopamine neurons: a computational approach. J Comput Neurosci 6:49–69
50. Knowlton CJ, Kutterer S, Roeper J et al (2018) Calcium dynamics control K-ATP channel mediated bursting in substantia nigra dopamine neurons: a combined experimental and modeling study. J Neurophysiol 119(1):84–95
51. Drion G, Bonjean M, Waroux O et al (2010) M-type channels selectively control bursting in rat dopaminergic neurons. Eur J Neurosci 31: 827–835
52. Yu N, Canavier CC (2015) A mathematical model of a midbrain dopamine neuron identifies two slow variables likely responsible for bursts evoked by SK channel antagonists and terminated by depolarization block. J Math Neurosci 5:5
53. Qian K, Yu N, Tucker KR et al (2014) Mathematical analysis of depolarization block mediated by slow inactivation of fast sodium channels in midbrain dopamine neurons. J Neurophysiol 112(11):2779–2790
54. Grace AA, Bunney BS (1986) Induction of depolarization block in midbrain dopamine neurons by repeated administration of haloperidol: analysis using in vivo intracellular recording. J Pharmacol Exp Ther 238:1092–1100
55. Shadlen MN, Newsome WT (1994) Noise, neural codes and cortical organization. Curr Opin Neurobiol 4:569–579
56. van Vreeswijk C, Sompolinsky H (1996) Chaos in neuronal networks with balanced excitatory and inhibitory activity. Science 274: 1724–1726
57. Brunel N (2000) Dynamics of sparsely connected networks of excitatory and inhibitory spiking neurons. J Comput Neurosci 8: 183–208
58. Bean BP (2007) The action potential in mammalian central neurons. Nat Rev Neurosci 8: 451–465
59. Hausser M, Stuart G, Racca C et al (1995) Axonal initiation and active dendritic propagation of action potentials in substantia nigra neurons. Neuron 15:637–647
60. Meza RC, López-Jury L, Canavier CC et al (2018) Role of the axon initial segment in the control of spontaneous frequency of nigral dopaminergic neurons in vivo. J Neurosci 38: 733–744
61. López-Jury L, Meza RC, Brown MTC et al (2018) Morphological and biophysical determinants of the intracellular and extracellular waveforms in nigral dopaminergic neurons: a computational study. J Neurosci 38: 8295–8310
62. Grace AA (1990) Evidence for the functional compartmentalization of spike generating regions of rat midbrain dopamine neurons recorded in vitro. Brain Res 524:31–41
63. Yu Y, Shu Y, McCormick DA (2008) Cortical action potential backpropagation explains spike threshold variability and rapid-onset kinetics. J Neurosci 28:7260–7272
64. McCormick DA, Shu Y, Yu Y (2007) Neurophysiology: Hodgkin and Huxley model--still standing? Nature 445:E1–E2; discussion E2–3
65. Knowlton CJ, Ziouziou TI, Hammer N et al (2021) Inactivation mode of sodium channels defines the different maximal firing rates of conventional versus atypical midbrain dopamine neurons. PLoS Comput Biol 17: e1009371

Chapter 6

Electrophysiological Characterization of Dopaminergic Neurons of the Rat Substantia Nigra Compacta

Marcelo Aguilar-Rivera, Rafael Ignacio Gatica, Victor Azócar, Darinka Buc, and José Antonio Fuentealba-Evans

Abstract

Dopamine neurons of the substantia nigra pars compacta (SNc) are involved in several physiological and pathological processes, ranging from motor control to decision-making. For example, a deficit in the activity of these neurons is associated with Parkinson's disease, while an increase in their activity has been associated with the compulsive drug seeking. Therefore, the study of the electrical activity of the SNc neurons is essential to continue studying the contribution of the nigrostriatal pathway to cognitive and behavioral processes. In this chapter, we describe a detailed protocol to carry out electrophysiological "single-unit recording" in dopamine neurons of the SNc in anesthetized rats. In addition, we provide electrophysiological criteria to classify different neuronal types and statistical tools to determine the effect of an acute injection of amphetamine on the activity of SNc neurons.

Key words Dopamine, Substantia nigra, Single cells, Tetrodes, Action potential, Firing rate, Firing pattern, Amphetamine

1 Introduction

Evidence suggests that dopaminergic neurotransmission is among the most important factors adjusting behavioral control [1]. Dysfunction in dopamine signaling has been associated with disorders sharing failure in behavioral control, such as drug addiction, attention-deficit/hyperactivity disorder, and obsessive-compulsive disorder [2, 3]. Two involved dopaminergic pathways have been described: the mesolimbic dopaminergic pathway that comprises dopamine (DA) neurons from the ventral tegmental area (VTA) to the nucleus accumbens (NAc) [4] and the nigrostriatal pathway involving DA neurons from the substantia nigra compacta (SNc) to the dorsal striatum [5, 6]. In this regard, electrophysiological recording in anesthetized rats show that the repeated exposure to drugs of abuse is accompanied by an increase in firing rate and burst

José Antonio Fuentealba-Evans and Pablo Henny (eds.), *Dopaminergic System Function and Dysfunction: Experimental Approaches*, Neuromethods, vol. 193, https://doi.org/10.1007/978-1-0716-2799-0_6,

activity in DA neurons of SNc and VTA [7, 8]. Thus, the objective of this chapter is to describe a protocol to carry out electrophysiological recordings of DA neurons in the SNc.

There has been a notorious evolution in the type of electrodes used to record cell units and field potentials in the alive brain: from single wire electrodes, made of different metal alloys that started being used early in the last century, to high channel density silicon probes widely used nowadays [9]. The electrophysiology recording technique is continually being improved not only to record more cell units from multiple sites but also to get a better fidelity of the recorded neural signal. Both low channels and high-density recordings have advantages and disadvantages regarding electrode hardness vs brain damage, or impedance vs noise, and price vs usability, so the trade-off should be evaluated in virtue of the biological questions under research. For instance, in some cases inexpensive wire-based electrodes are good enough to explore correlational hypothesis about brain activity on anesthetized preparations where live brain tissue is recorded only once. However, other metal alloys such as platinum iridium (PtIr) are more convenient to explore a causal hypothesis that involves the manipulation of neural activity, for instance, by electric stimulation [10] through the same electrodes used to record action potentials. The recording of brain activity at the single cell level and from multiple sites is challenging since it involves the use of small manipulators to control single or multiple electrode bundles with the help of a mechanical system named microdrive. Such microdrive as that presented in this chapter allows us to control the electrode depth into the brain tissue. Importantly, a microdrive can be configurated to target more than one brain area, using for simplicity, a single bundle array. In this case, the tip of each tetrode or some tetrodes subsets can record at different depths into the brain due to different wire length. In this way, the same bundle will be composed by two subsets of electrodes where the only difference between subsets is the tetrode length. In this chapter, the main aim is to discuss affordable electrophysiological approaches to study DA neurons of SNc in anesthetized animals.

2 Materials and Drive Components

Before starting to build a microdrive, it is necessary to define the tetrode configuration associated with the biological research questions to be researched and the target brain areas to be recorded. In this regard, it is necessary to define the rostral-caudal, medial-lateral, and dorso-ventral stereotaxic coordinates using an atlas such as the Paxinos and Watson atlas [11] or the Swanson's brain atlas [12]. Both atlases are specific for rat strains and a weight of 290 g, so to increase the targeting precision, the stereotaxic coordinates should follow such weight.

Usually, single tetrode arrays with a specific spatial configuration, for instance, 4 columns by 2 rows, with a total of 8 tetrodes (32 channels), are easy to build and will be optimum for targeting most of the brain areas, including those associated with the mesolimbic and nigrostriatal dopaminergic pathways. However, it is always possible to create more customized tetrode arrays, with different and specific spatial configurations, depending on the hypothesis under consideration and the skills of the researchers building such arrays.

We present a step-by-step guide to build tetrodes, tetrode arrays, and microdrives that allow for the control of tetrode positioning on the dorsal ventral axis. Also, we discussed some of the most standard analytics to quantify changes in the firing rate as well as how to qualitatively measure the way that dopaminergic neurons fire action potentials.

In a general sense, five different steps can be identified from wires selection to manipulator assembly as discussed below. In short, we refer to these steps as (1) selecting wires, (2) assembling manipulator, (3) soldering the tubing array, (4) twisting tetrodes, and (5) electrode interfacing using a printed circuit board (PCB).

A list of materials that includes the type of material and preferred sellers are listed in Table 1. Figures 1, 2, 3, 4, 5, and 6 present the most standard techniques and analysis in neuroscience to record neural activity and then study firing rate patterns in a quantitative way.

2.1 Wire Selection

There is a well-known relationship between the metal alloy of which a wire is made and its softness or stiffness. Stiff wires such as tungsten could be preferred to target deep brain areas as they can succeed in reaching all the way down (~8 mm) to the deepest tissue, for instance, when recording DA neurons from the VTA [13, 14]. Here we presented an alternative option based on cheaper wires, such as NiCr, that attain good enough stiffness based on twisting to successfully reaching deep brain area such as the SNc, when these types of wires are tied and twisted together.

Additionally, there are other electrode features such as the inverse relationship between area and impedance of an electrode that could also determine the recording quality. Electrode impedance is one of the most relevant features as described in the classic electrophysiological literature [15], as well as in a comprehensive discussion on this topic by Roger Lemon in 1984 on his book *Methods for Neuronal Recording in Conscious Animal* [16]. On the Chapter 5 entitled "Microelectrodes and Microdrives," Roger discussed the electrical and biological properties of microelectrodes based on the detailed previous work of David Robinson [15]. Both readings, highly recommended, describe different electronic and biological parameters such as impedance, capacitance, and the resistance associated with the brain–electrode interface, as well as

Table 1
Materials and reagents for tetrode electrodes and manipulator

Material	Company	Contact
Micromanipulator		www.nervetools.com
PCB		www.nervetools.com
Stainless steel tubing, 27G, 0.016″ OD, 0.002″ wall, 0.012″ ID (guiding cannulae array)	MicroGroup	https://microgroup.com/
Stainless steel tubing, 32G, 0.009″ OD, 0.002″ wall, 0.004″ ID (thinnest cannula to protect each tetrode)	MicroGroup	https://microgroup.com/
18-8 Stainless steel machine screw, plain finish, pan head, slotted drive, 1/8 length, fully threaded, 0-80 threads	Amazon, Grainger, McMaster-Carr	www.amazon.com www.grainger.com/ www.mcmaster.com/
Tetrode wire 0.0005″ (tetrode wire)	Sandvik Co.	https://www.materials.sandvik/en/
Small gold pins	Neuralynx	https://neuralynx.com/
36 position dual male nano-miniature connector A79022-001	Omnetics Connector Co.	https://www.omnetics.com/
PFA-coated stainless steel wire, 0.003″ bare, 0.0055″ coated (ground wire)	A-M Systems	https://www.a-msystems.com/
Gold plating solution	Neuralynx	https://neuralynx.com/

The main components are stainless steel and 3D printed plastic materials. There are several vendors around the globe from where to buy such materials and also some open sources from where to download, for instance, the 3D files for the manipulators. Depending on the type of neural potentials and brain areas, some materials are preferred for electrode twisting (i.e., stereotrodes and tetrodes), as discussed in the main section

the metal's electrode alloy, length, electric insulation, and shape of specific wires that determine the impedance and therefore the signal-to-noise ratio of electrophysiological recordings. In summary, the surface area, mostly related to the diameter of the exposed tip of an insulated wire, and its associated impedance are two key features to take into consideration on the fabrication of custom-made electrode array. Such features determine the noise and the fidelity of the neural recordings, and they can have a specific impact on the recording of single units and field potentials.

Using a thick wire, between 50 and 100 μm, is preferred to record field potentials since the thicker the wire the more cell

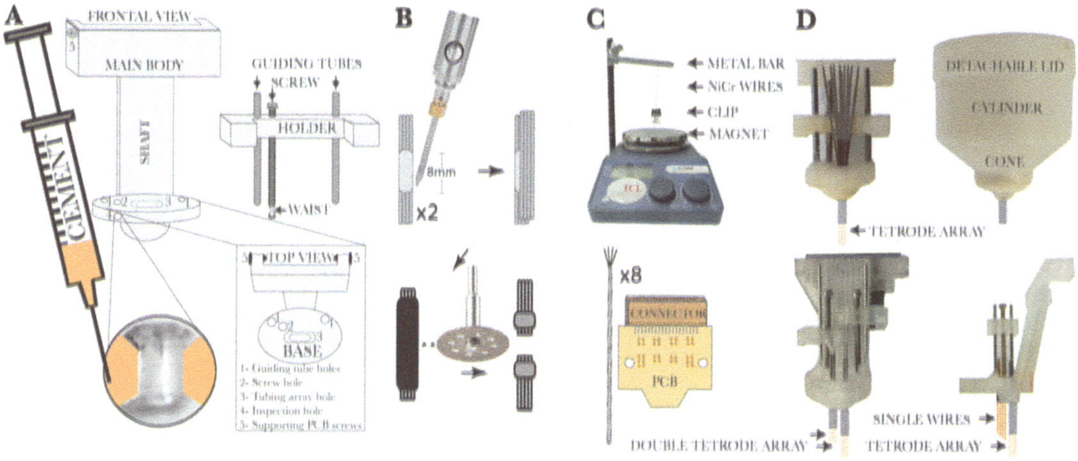

Fig. 1 Assembly of single and multiple electrode arrays. (**a**) 3D plastic parts that allow to hold vertically the printed board circuit and to control the electrode array positioning in the vertical axis. The drive is a simple mechanism based on a modified 00-80 stainless steel screw as discussed in the main text, two 21G cannulae glued to the base by any cyanoacrylate, and a horizontal plastic holder. To fill the space around the waist screw with dental cement, it is possible to use a 1-mL syringe and a thick needle (18 or 19G). (**b**) Assembly of stainless steel 27G cannulae array to spatially organize electrodes (i.e., 4 × 2 tetrodes) and guide them down into the brain. With the help of a soldering iron, eight cannulae are soldered (4 × 2 cannulae) side by side using tin (gray area). Then the array is cut in half to get two cannulae arrays. For this, the cannulae array is protected by a shrink tube (black area), thus preserving the recently soldered cannulae array. (**c**) Tetrode twisting process showing how to hang two wires from a horizontal bar placed over a magnetic dish. After bending the pair of wires around the bar, a loop is formed having four wires that will be twisted together to form a tetrode. Also, the custom PCB is depicted. (**d**) Different examples of the electrodes array configuration. Top panels depicted the eight tetrode arrays used in our recordings from the SNc (as in Figs. 3, 4, 5, and 6). The bottom panels show two other examples of tetrode configurations to target multiple brain areas. At the left two bundles of four tetrodes each, one to reach the amygdala and the other for the insular cortex. At the right, a multiple electrode array is optimized to record action potentials using thin tetrodes (12 μm) and field potentials using thicker (~100 μm of diameter) single wires

activity can be recorded. This allows for a good proxy of the averaged brain activity of hundreds to thousands of neurons [17] around the electrode tip. On the other hand, the use of thin wires optimizes the recording of action potentials from single cells since tiny wires can get closer than thicker wires to the neural bodies. An electrode with a small tip area will record changes on the membrane potential over a smaller distance than an electrode with a course tip, increasing the selectivity of such a thin electrode and the chance to well isolate action potentials sourced by single neurons. This proximity involves a small electric distance between the wire tip and the cell body or axon, increasing the spatial resolution that favors the recording of well-defined waveforms whose peak amplitude is "away" (usually >2 standard deviations) from the averaged background noise. Thus, the use of thin wires optimizes the spike sorting that is a common analytic based on the neural waveform as described later in this chapter (Fig. 3). However, a potential

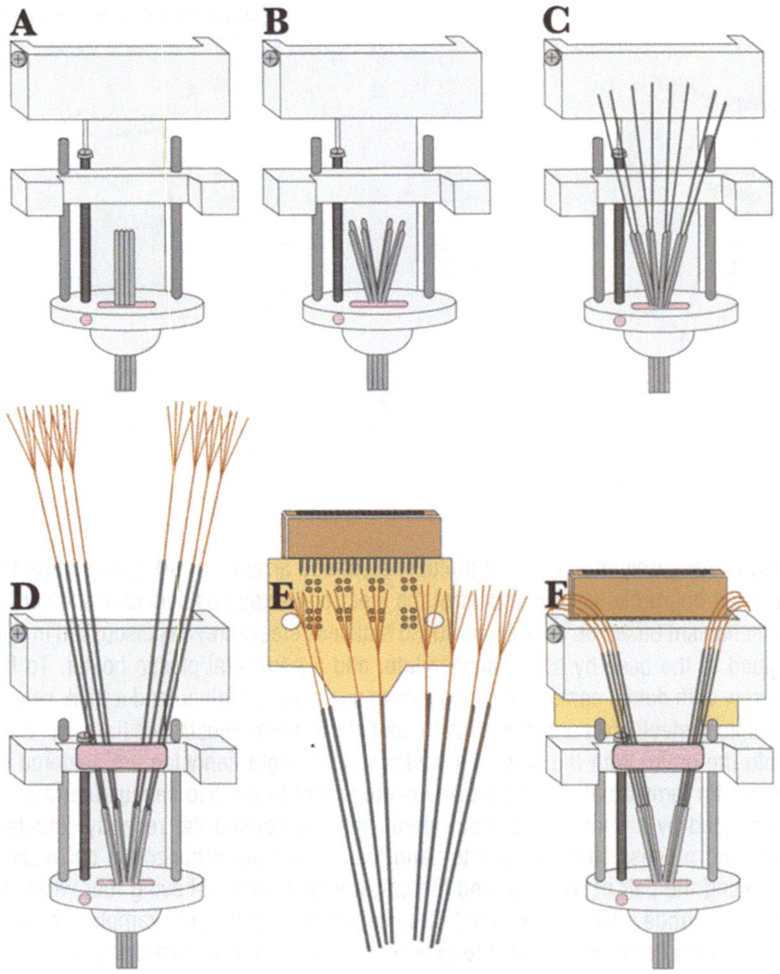

Fig. 2 Step-by-step assembly of a custom single electrode array. (**a**, **b**) The modified screw and cannulae are installed in the plastic base of the manipulator. Since the screw is trapped to the base by dental acrylic around the waist, it can spin on its own axes, and only the horizontal plastic part moves up or down. The guiding cannulae array also is installed in the base. The tip of the cannulae array should protrude ~6 mm from the bottom end of a manipulator to see under the manipulator and over the brain (i.e., during cutting the dura), for instance, when recording under anesthesia or during implantation of a manipulator for chronic recordings. The tetrode length has to stay in the case and should be able to protrude up to approximately 9 mm from the tip of the cannulae array to reach deep brain areas as the dopaminergic neural tissue. (**c**) The thinnest cannulae (32G) should be installed inside of the 27G cannulae array. (**d**) Next, each tetrode is installed one by one inside the thinnest 32G cannulae. The horizontal holder has to be in the upper position, so the tetrode can move all the way down, ~9 mm here, after tetrode is glued to the holder. For this, glue each tetrode to each 32G cannula using loctite, as well as cement the 32G cannulae using dental cement to the horizontal plastic holder (pink area). (**e**) The PCB is placed vertically behind the main plastic part, and each wire of a tetrode is interfaced to it using a gold pin. For simplicity, the main body and microdrive are not shown on the draw at the right. (**f**) The PCB is kept in place by lateral screws, but in addition, it can be secured to the connector holder by a small amount of dental cement

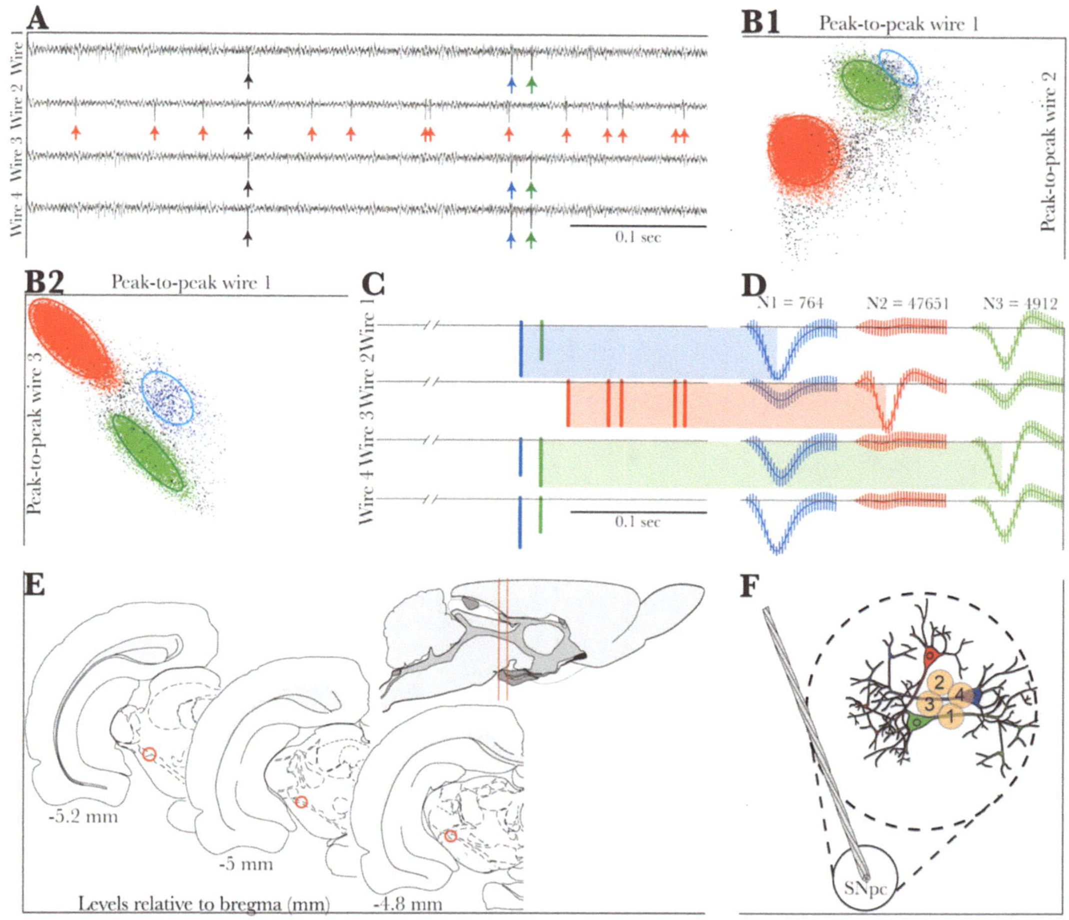

Fig. 3 Example of a canonical tetrode recording and spike sorting of single neurons. (**a**) Four traces depict a raw recording of high-frequency (300–5000 Hz) electrical neural activity from each wire of a tetrode. The multiple color arrowheads show three potential single neurons, while black arrows depict a single noise event that appears with equal amplitude at the four channels simultaneously. (**b1**) Scatter plots of the peak-to-peak amplitude of the action potential waveforms recorded in wire 1 versus those recorded in wire 2, while **b2** depicted spike activity recorded by wire 1 vs wire 3. Both scatter plots depicted the amplitude waveforms from the SNc record shown in **a**. Two clusters clearly emerge in the first plot **b1**, while the third cluster in blue is not evident away and separated from the cluster in green. However, the scatter plot **b2** shows complementary information where clusters in red, green, and blue are not overlapped, disambiguating that the action potentials in clusters in blue come from a different neuron than the red and green clusters. This is confirmed with the averaged waveform depicted in **d**. (**c**) Pictorial representation of the sorted spike trains extracted from the traces shown in **a**. (**d**) The extracted spikes from the clusters shown in **b**. Shaded area connected to **c** represents the same spike. The neuron 1 (blue) is captured clearly by wires 1, 3, and 4. Neuron 2 (red) shows as a well-defined waveform in wire 2, while neuron 3 (green) is evident in wire 3 and also in wires 1 and 4. Vertical lines around the averaged waveform represent ± 1 standard deviation. (**e**) Pictorial representation of rat brain coronal cuts on the rostral-caudal axis (modified from [12]). Red circles bound the area where tetrodes were lowered. (**f**) Scheme that represents the electric distance from the three neurons to each wire of the tetrode analyzed in this figure. The amplitude differences on the averaged waveform showed in **d** can be understood as the distance of the four tetrode channels to the neurons around. Every time a cell fires, action potentials are recorded simultaneously by the four channels that are part of the tetrode

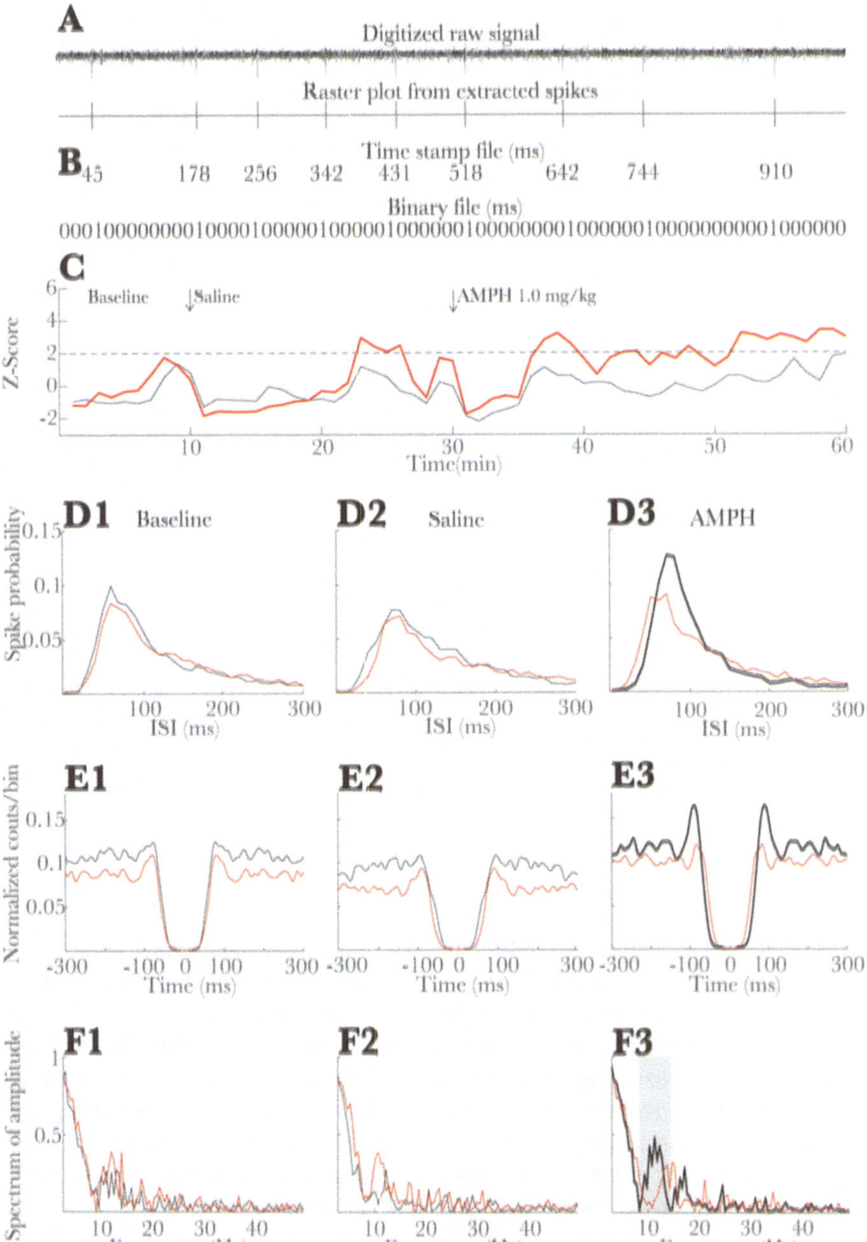

Fig. 4 Spikes extraction and firing rate characterization. (**a**) Top trace depicts a digitized raw signal of high-frequency (300–5000 Hz) neural activity as action potentials. It is possible to see the single spike events as a sharp vertical deflection. The bottom trace shows a raster plot of the extracted spike as a vertical line that matches temporally the single spikes events on top. (**b**) The top and bottom series are the timestamps and binary files, respectively, from the raster plot in **a**. (**c**) Changes on firing rate that are expressed as a *Z*-score from the baseline for a sequence of bins (60 s each bin) before and after injections. Red and black traces represent two different neurons. It is possible to observe a steady increment of the firing rate after AMPH 1.0 mg/kg ip that is significant (permutation test: $p < 0.05$) only for the neuron represented by the red trace. Interestingly, AMPH did not significantly change the firing rate on the second neuron (black trace). (**d**) Histograms of the interspike interval (ISI) during baseline, post-saline, and post-AMPH. The same number of spikes ($N = 3700$) was used to compute each histogram. It is possible to see that AMPH injection increased

drawback of thin wires (10–12 μm) is a high impedance (>2 MOhm) with a putative detriment of the quality signal since an increase of the noise. It is possible to reduce electrode impedance (even <1 MOhm) by metal deposition through gold plating, or by grinding the electrode tip [18], with the help of a diamond wheel spinned by a Dremel.

One of the low impedance reasonably priced wires on the market is the 50-μm tungsten wire (by California Fine Wire™). This wire is insulated with a single polyimide layer (SML) of coating around the tungsten metal bar. Given that polyimide has an excellent thermal stability, the heat treatment of such coating after twisting has a minimal effect on preserving the characteristic tetrode shape. Many researchers still prefer tungsten wires when the main goal is to record field potentials and spikes. Several wire alloys have been used to record both field and spikes activity from rats, although an ~65-μm diameter looks like the superior limit to successfully record action potentials and to get well-isolated waveforms [19].

Other researchers prefer to use alloys that are coated with materials that keep the shape of the twisted electrodes after activation by heat. One example is electrodes insulated with polyimide resin coating such as polyamic acid solution (PAC) as on the Sandvik™ wires. When this resin is exposed to heat, it helps to preserve the tetrode shape after twisting. Additionally, this type of coating reduces the biological response during chronic implants. Another option is to use a Heavy Formvar (HFV) with a secondary heat-activated bond coat from California Fine Wire™ and A-M systems™ that also helps to keep tetrodes twisted after treating the wire coating with heat. A nickel-chromium (NiCr) wire with such a heat-activated coating is a good combination since it retains the tetrode shape after twisting and activation by heat. Moreover, NiCr wires are cheaper than other alloys like tungsten or (platinum iridium) PlIr wires. Another benefit of using NiCr wires is the fact that is possible to find bars with diameter as thin as 10–12 μm, which is wanted to optimize single-cell recordings. We have gotten good quality recordings using NiCr alloy of 12 μm of diameter [20–22]. As discussed below, it is possible to get a good signal-to-noise ratio besides clearly identifying the waveforms from a single cell, where the functional concept of "unit" is further confirmed by functional measurements calculated from its firing such as the interspike interval (ISI).

Fig. 4 (continued) the chance that consecutive spikes are fired with a time delay under 100 ms for the neuron represented by the black trace. (**e**, **f**) Autocorrelograms and their respective power spectrums were calculated for the same neurons and periods of time as in **d**. This analysis confirms that amphetamine-induced change (gray shade area) on the firing mode only for the neuron represented by trace in black

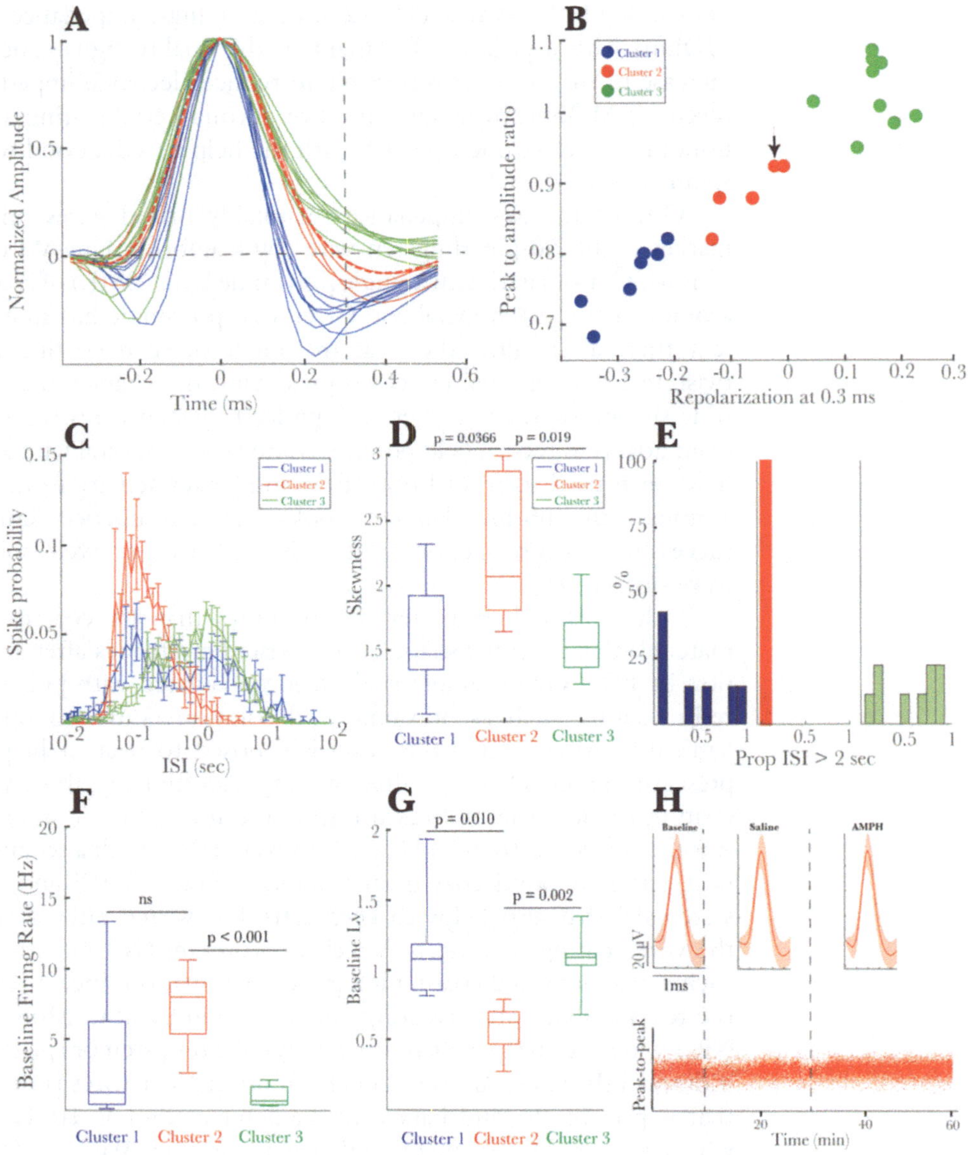

Fig. 5 Neural phenotypes classification based on waveform and spike firing. (**a**) Normalized waveforms of three different putative neural phenotypes. Since during repolarization waveform differs significantly among groups, the repolarization value at 300 us after the peak (dotted vertical line) was used for clustering as shown in the figures. (**b**) Scatter plot of repolarization values (as in **a**) versus the peak to amplitude ratio showing the separation of three neural phenotype clusters recorded from the SNc. The arrow in the second cluster in red depicts the neuron showed in **h**. (**c**, **d**) Averaged histograms of the interspike intervals (ISI) for the three phenotypes. It is possible to see that the cluster of putative dopaminergic neurons in red shows a distinctive distribution of ISI with a rise around 100 ms (10^{-1} s), having this a significant positive skewness as in **d**. The same number of spikes ($N = 200$) per neuron was used to compute each histogram before averaging them. This means there is a bigger probability these putative dopaminergic neurons fire action potentials with a short time among spikes ~100 ms (10^{-1} s). The chance that these putative dopaminergic neurons fire action potentials with a long time among spikes is low as depicted by the histogram in red of (**e**). (**f**, **g**) The cluster of putative dopaminergic neurons has a higher basal firing rate compared to cluster 3 and a smaller Lv than clusters 1 and 3 ($p < 0.05$, Mann-Whitney test). (**h**) Typical electrophysiological example showing a steady

When the aim is also to perturbate brain activity by electric microstimulation through the same electrodes used to record neural activity, platinum iridium (PlIr) alloy is a good option since it is less sensitive to the redox process associated with electric microstimulation than NiCr [23]. Additionally, this alloy has a lower impedance than NiCr and evokes minimal scarring in chronic implants [10], although PlIr alloy is an expensive type of wire.

2.2 Manipulator Assembly

The manipulator is a set of 3D printed plastic parts that contain and protect all its components, such as the tetrode array, the PCB, the microdrive, and the main body (Fig. 1). The microdrive is the main mechanical component of a manipulator that allows to move a tetrode array vertically, that is, dorsally and ventrally, into the brain in a very controlled way. This microdrive is composed of a modified screw, a couple of guide tubes, and a plastic horizontal holder where the tetrodes are glued (Fig. 1a). Next to the drive system, there is a shaft that serves as the physical support where the PCB is placed (Fig. 1a). The PCB is the physical interface between each wire of a tetrode (Fig. 1c) and the headstage amplifier. This printed circuit has 36 gold-plated vias where every one of them works as a channel (~330 μm in diameter) to insert one of the ends

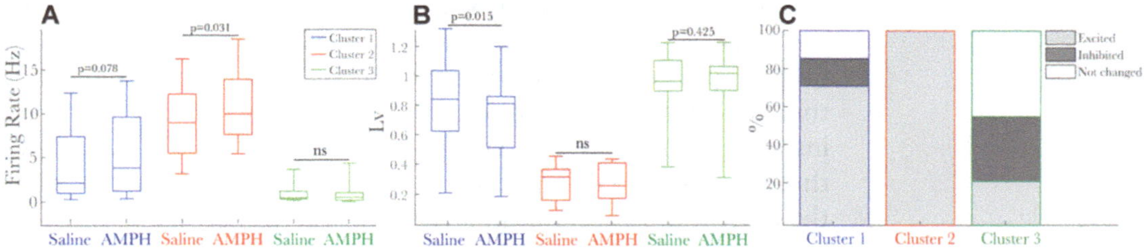

Fig. 6 Changes in the rate of action potentials on putative dopaminergic neurons induced by amphetamine injection. (**a**, **b**) Firing rate and Lv analysis. AMPH 1.0 mg/kg ip triggered a significant change in the number of spikes per second (**a**), but not on the firing pattern of putative dopaminergic neurons (**b**, depicted in red, $N = 6$). On the other hand, a decrease in irregular firing, without changes in firing rate, was found on non-dopaminergic neurons (depicted in blue, $N = 8$, Wilcoxon signed-rank test $p < 0.05$). Moreover, AMPH injection did not induce changes in firing rate or Lv of those non-dopaminergic neurons depicted in green. Data are represented as box and whiskers, showing the median (middle line), the 25th and 75th percentile in the box, and the lowest and highest values (whiskers). (**c**) AMPH population response analysis. After AMPH 1.0 mg/kg ip, neurons from clusters 1 (non-dopaminergic) and 2 (dopaminergic) were mainly excited, while both excitation and inhibition were mainly found in cluster 3 (non-dopaminergic). Data is shown as stacked bars

Fig. 5 (continued) waveform amplitude of action potential discharges of a neuron (pointed by the arrow in **b**) through the whole 1 h of recording. Based on the cloud density, it is possible to see an increment of firing rate after amphetamine injection. Boxplot depicts median and percentile 25th and 75th, while whisker depicted minimum and maximum of every wire that is part of the tetrode. The other end of the same tetrode is inserted in the brain after opening the dura. Also, three of these vias are to interface the ground electrodes, and one via works as the interface for reference. This PCB has soldered a military-grade connector (by Omnetics) for 32 channels plus 3 grounds and 1 reference channel. This board allows to electronically interface the whole tetrode array placed in the brain and the ground usually placed on the skull, to a headstage pre-amplifier. The microdrive allows to control the tetrode positioning during acute recordings under anesthesia or several days after the implantation surgery when the animals are fully recovered from surgery. The microdrive is composed of two stainless steel guiding tubes (21G) that are glued to the main plastic body floor (Fig. 1a). If such tubes are not available in the laboratory, it is also possible to use the stainless steel portion of 21G needles after cutting the bevel portion and rounding both ends. Such a 21G tube vertically guides the sliding of the movable horizontal plastic holder where the electrodes are glued. The use of a modified screw, that has a waist in the lower portion of the shaft, is justified by the need to trap the screw, avoiding it to move up or down but still letting it spin on its own axis moving the electrode holder instead. The waist on the screw is made with the help of a Dremel, 1 mm upper from the bottom edge of the screw. This is a stainless steel narrow cheese head slotted 0–80 screw, and the waist is ~400-μm deep so the diameter is ~700 μm in the deepest portion of the waist, while the unmodified remaining thread portion is 1.5 mm of diameter. Pouring acrylic cement, for instance, using a 1-mL syringe and a thick needle, i.e., 18G, through the lateral inspection hole helps to trap the screw when the cement dries around the waist (Fig. 1a, frontal view). Thus, the screw can spin in a clock and counterclockwise, moving up or down, respectively, the horizontal plastic holder along with the tetrode array glued on it, to different deepness into the brain. During the screw planting and before the dental acrylic cure, the screw must be rotated penetrating a tinny oil machine drop. If it was correctly done, the screw will rotate without getting out of piece. The movement on the vertical axis of this system is equivalent to 312 μm for each full turn screw and 78 μm when the screw is spinned only a quarter of a full turn. In summary, the microdrive is a simple but critical component to move tetrodes.

2.3 Tube Array Soldering

To build an array of cannulae that guide the movement of electrodes down into the brain, eight stainless steel guide cannulae 27G of 4 cm length are needed. It is also possible to use the shaft of 27G long needles to prepare such array. Solder together the four cannulae side by side by spreading tin on the top surface and covering 8 mm distal on each side from the middle of the array. Repeat the process with the remaining four cannulae (Fig. 1b). The new two half must be soldered together using the other face that is free of soldering (Fig. 1b). After soldering both halves together, you will have a single array of eight cannulae in total and organized as four by two cannulae. Before cutting the cannulae array in the half, it is necessary to cover the array using a heat shrink tube of 3 cm length. This will protect the array from damage when it is being cut in the middle to get the two halves of four by two cannulae (Fig. 1b). After cutting the array, it is necessary to remove the heat-shrinking tube and reopen the holes of each cannula in the array that resulted in clogging by cutting them by abrasion using a Dremel. With the help of a thin needle (i.e., 32G), deburr each opening to make it smooth and free of debris as well as clean inside each cannula using a stainless steel 34G bar.

2.4 Tetrode Twisting

Cut two pieces of your selected wire 20 cm long and align them together by one end of each thread. Bent the couple of wires on the middle keeping all tips together and making a loop around the horizontal main shaft of a tetrode spinning holder (Fig. 1c). Leave the free tetrode tips facing down of a custom tetrode twisting. Alternatively, if you do not have a spinning tetrode tool, it is possible to use any hot plate stirrer with a rotary magnet on its base (Fig. 1c). When this is the case, a metallic binder clip or metal hinge clip must be hanged from the free tips just on the opposite end to the loop (Fig. 1c). The distance between the clamp and the surface of the stir plate should be ~1.5 cm. This way, the rotary magnet under the surface of the steering plate will spin the metallic binder clip, and the four wires that compose a tetrode will spin too. Immediately after the binder clip and four wires start spinning, count approximately 90 full turns clockwise if working with NiCr wire. When the total number of turns is reached, stop the spinning tool or steering plate and allow the relaxation of the twisted wires, either by allowing it to spin in the opposite direction—this is approximately 30 turns when working with a tetrode twister—or by allowing the metallic binder to spin anti-clockwise until it stops by itself. Finally, use a heat gun to activate the wire coating that allows to fix and keep the characteristic twisted shape of a tetrode when it is removed from the horizontal bar where it was hanging over the spinner. Be careful not to apply excessive or moderate heat for too long or at a to short distance, which could melt the wire insulation, causing a short-circuit between the four channels of a tetrode. Move the heat gun up and down around the tetrode at medium temperature and a distance of 5 cm approx. The total number of spins will depend on the wire length, the type of alloy, and how much twist does the tetrode need. Also, some researchers would prefer to lightly modify the order of the steps, for instance, when a twisted tied tetrode is aimed, the heat can be applied before anticlockwise spinning the braided tetrode.

2.5 Electrode Interfacing Using a Custom-Printed Circuit Board

Using Altium Designer 2014, a custom double layer PCB was designed as the interface between electrodes and the headstage pre-amplifier. This PCB is built based on a standard FR-4 fiber glass of 600 μm of thickness, 1 Oz of copper, ENEPIG finish surface, and a green, yellow, or any other color available for the solder mask. On this PCB there are plated vias of 330 μm of diameter and an annular ring of 100 μm thick in both sides to improve the electric connection between the wire placed inside the via and the gold pin. Thirty-two vias are for the eight tetrodes, and the remaining vias are for the three ground connections and one for a reference channel. Typically, the ground wire consists of a perfluoroalkoxy alkane (PFA)-coated stainless steel, 0.003" bare, soldered to a 00–90 short stainless steel screw implanted in the skull touching the dura but without penetrating it. A military-grade connector (Omnetics Connector Corporation, model A79022-001) was soldered to this PCB using standard low-temperature solder paste (CHIPQUIK, SMDLTLFPT5) in a custom DIY reflow oven based on a commercial controller (i.e., Reflowster or Reflow Master). Alternatively, it is possible to use lead-free tin wire resin core and a soldering iron to hook the connector to the PCB by hand under the magnifier, although this approach is slower than the reflow soldering. If soldering by hand is still preferred, get help by using any tape, although polyimide is preferred given its thermal stability, or by using a cross-lock tweezer to hold the connector close to the PCB and proceed to soldering it. It is advised to cover the connector holes with tape to prevent to molten tin drops in the holes accidentally blocking them. After soldering, run a visual inspection to check if there are no shorts between the pins of the connector or shorts between PCB pads. With the help of a multimeter, check the proper continuity between each via and the corresponding connector pin. An overview of the connector attached to a PCB is shown in Fig. 1c.

3 Electrode Manipulator Assembly

3.1 Installing Cannulae Array Guide

First step is to mount the cannula array using cyanoacrylate and then dental acrylic in the center of the manipulator floor (Figs. 1a, d, 2a, b). After dental acrylic has thoroughly dried, separate the cannulae soldered free end (Fig. 2b), from the other cannulae to facility the interface with the 32G stainless-steel tubes (Fig. 2c). Before gluing the 32G cannulae to the movable system, turn the screw to rise the movable holder to the higher position. That will define the maximum distance that the tetrodes could be lowered (Fig. 2c). Glue the 32G to the holder using dental acrylic (Fig. 2d). Finally, slide each tetrode into the respective 32G tube of the eight cannulae array and glue each tetrode to the top portion of its respective 32G cannula using cyanoacrylate glue (Fig. 2d).

3.2 Interfacing Each Single Wire of a Tetrode with PCB

Once the electrode array is glued to the holder (Fig. 2d–f), it is time to proceed with the electrical contact between the wires and the PCB vias as mentioned above. For that, pass each wire of a tetrode through each corresponding via in the PCB (Fig. 2e). Write down to which set of vias (four vias per set) is each tetrode connected. It is preferable to pick them in a predetermined order according to a standard map used by your laboratory. This map must contain wires linking PCB vias information as well as identifying which electrode corresponds to which channel in the connector. This step ensures to associate the recorded signals to the targeted brain areas. After passing each wire per via, use small gold pins to make a strong electrical connection between the metal bar of each wire and the gold-plated wall and annular ring of a via. Finally, organize the wires around the connector (Fig. 2f) allowing free access to the holes on the top of the connector. Then fix the PCB to the microdrive by tightening the lateral screws and secure the PCB to the main body of the manipulator using dental cement (Fig. 2f).

3.3 Gold Plating

As mentioned before, the relationship between the surface area of the electrode tip, its impedance, and the signal-to-noise ratio is well known. Using a NiCr wire with a diameter between 12 and 17 μm, the impedance is over 3 MOhm when tested at 1 KHz before plating. For this reason and to reduce the wire impedance, each tetrode tip wire must be gold plated. A bigger signal-to-noise ratio will maximize the ability to record and discriminate action potentials sourced by single units. To achieve this, spin counterclockwise the screw of the microdrive system to put out from each guide tubing the tetrodes. Then, securely mount the microdrive on a clamp that will allow to submerge the wire tips into the gold plating solution. Fill an Eppendorf tube with a gold solution (i.e., Neuralynx™) and other with distilled water. Proceed to place the electrode tips in into the gold solution. We usually apply a DC current of 1.0 μA to each channel aiming to set the wire impedance around 250–350 kΩ using an isolated stimulator. If there is any single channel that has not been plated below 350 kΩ, repeat the process. Once all the tip wires have been gold plated, so the impedance is in an acceptable range, remove the electrode array from the gold plating solution and lower the tip of the tetrode array into distilled water to rinse off the excess gold particles. After rinsing the tetrode tips with water, it is convenient to use 70% alcohol and 100% alcohol as a final step. These two extra steps associated with the use of alcohol will help to keep the tetrode tips clean and remove the distilled water and moisture.

When gold plating and cleaning are done, pull up the electrode array to their top position leaving the tips immediately inside the guiding cannula. To do this, turn clockwise the drive screw. Thus, the microdrive and electrodes are ready for implantation.

3.4 Improving Signal-to-Noise Ratio

It is possible to use a tiny faraday cage inside a custom manipulator to electromagnetically insulate the electrodes and PCB from external electromagnetic interference or noise coming, for instance, from the alternating current (50 or 60 Hz depending on the country) associated with the power line. The main idea is to avoid that extra noise to interact with the metal portion of the implant such as electrodes and PCB that can work as antennas and translate such noise into currents that contaminate the action and field potentials recorded from the alive brain. Thus, the goal is to protect the electric neural signals just before they are amplified at the headstage level; leading to amplifications of biological signals with less electromagnetic interference. For this purpose, the manipulator can be covered on its internal wall with copper tape or copper mesh that is easily found in some hardware stores or online. After installing the Faraday cage shield, it has to be electrically connected to the ground at the PCB level. To achive this, a wire is glued to the internal wall of the manipulator using a copper tape or mesh, and connected to the other end of the ground PBC by a mechanical connection such as a gold pin or by soldering the wire directly into the PCB via.

4 Electrophysiological Setup

We built an amplification system combining custom software and commercial hardware. Thus, the amplification stages pertain to one commercial brand, while the analog to digital converter was acquired from a second-party hardware developer. The custom set of amplification and digitizing systems were interfaced with a Windows PC (Windows 10 Pro) where a custom-made software was wrote for data acquisition, spike extraction, and spike sorting.

Using the manipulator showed in the previous section, SNc recordings were performed using a 32-channel headstage (μPA 32, Multi Channel Systems) that allows a 10× pre-amplification of the signal. Then, the signals were amplified again (500×) on a second stage by a 32-channel amplifier (PGA 32, Multi Channel Systems) and band-pass filtered (1–5000 Hz). These data were sent to a USB National Instruments analog to digital converter (NI USB-6363, National Instruments), connected to a PC running a custom-made software written by Dr. Pedro Maldonado at the Universidad of Chile, Santiago, Chile. This software named "Get Data" was written on C with the help of LabWindows/CVI (National Instruments Inc.; version 10.0.0 (360), 2010) employing advanced signal analysis libraries (version 17.0, 2017). This custom software allows the offline visualization of each tetrode channel and then the storage of the digitized neural activity with a maximal temporal resolution of 30 KHz and a voltage precision of 16 bits. In this regard, when neural data was sampled at 30 kHz, this allows 32 data points for 1.07 ms time window to represent each action potential waveform. Extracellular waveforms that exceeded a voltage threshold of two standard deviations from the average noise were extracted. The action potential waveforms were extracted and represented by 32 voltage data points, having 10 data points to the voltage value data points pre-threshold and 22 data points for the post-threshold.

For our following data, two adult male Sprague Dawley rats were used. All procedures were in strict accordance with the guidelines published in the "NIH Guide for the Care and Use of Laboratory Animal" (8th Edition) and the principles presented in the "Guidelines for the Use of Animals in Neuroscience Research" by the Society for Neuroscience. When we recorded neural electric activity on rodents under anesthesia, we use urethane (1.5 g/kg i.p) since it does not modify normal firing patterns of DA neurons [24, 25]. The rat was then positioned in a stereotaxic apparatus (Stoelting Co). A craniotomy on the following coordinates was performed, relative to bregma: anteroposterior = −4.8 to −5.2 mm, mediolateral = −3.0 to −2.0 mm. The size of the craniotomy must be wide enough to position the tube array that is ~1.2 mm × 0.7 mm. After removing the dura mater, lower the tube array until it touches the brain, without perforating it. It is a good practice to use some type of sealing such as agar around the cannulae array to protect the brain underneath. To lower the tetrodes, use the screw from the microdrive. Tetrodes are first lowered at a 312 μm every 10 s rate (one full turn of the screw), until reaching the beginning of the targeted region. SNc is located in the following coordinates: dorsoventral = −7.6 to −8.0 mm. To search for single units, lower the tetrodes at a ~38 μm rate (1/8 turn of screw). If it is possible, try to find single unit activity in all the tetrodes (or at least in a half of them) and then wait 30–45 min until the single unit activity seems to be stable. More details about single unit identification and extraction are described on the following section.

5 Qualitative and Quantitative Analysis of the Discharge Pattern of SNc Neurons

Changes on the rate (spikes/second) and mode of discharging action potentials (i.e., bursty versus a tonic pattern) of neurons in response to amphetamine (AMPH) and other psychostimulants have been reported. In this regard, preclinical studies have shown consistent evidence of neuroplastic changes in the mPFC [26] and the dorsolateral striatum (DLS) [20, 21] induced by AMPH and cocaine [27]. However, the data regarding the effect of AMPH on SNc neurons is sparse, and the preliminary data we show in this chapter might contribute to fill the gap. In this regard, we discussed quantitative and qualitative changes on the firing rate of putative DA neurons recorded from the SNc on anesthetized rats after an acute AMPH injection.

Using "Get Data" software, it is possible to visualize the raw signal of the four channels of a tetrode in an offline mode as in the Fig. 3a. It is possible to see how the amplitude of the spikes pertaining to a neuron is different among channels at the same time point; however, the noise event shows up simultaneously with the same amplitude in all the four channels of a tetrode. The spike extraction was made on "Get data," while the clustering was made with the help of the software "Spike Sorter" also wrote by Dr. Pedro Maldonado (1999) using LabWindows/CVI. In both cases we used the Windows version of such software. Waveforms were sorted mainly using two-dimensional plots of the peak-to-peak amplitude of action potentials for pairs of two channels from the same tetrode (Fig. 3b). Also, it is possible to use principal component analysis of the spike waveforms to confirm the clustering defined on the scatter plots. Basically, we searched through different combinations of the channels pertaining to the same tetrode where the clusters did not overlap. Isolated clusters were used to obtain the average waveforms $\pm$ standard deviation based on all the spikes in a cluster pertaining to each single units recorded (Fig. 3d). Additionally, each timestamp associated with each spike when neurons discharge action potentials was extracted (Fig. 3c). Using the timestamps, we computed standard analytics per neurons described on the coming sections. As depicted on Fig. 3f and discussed on some of the early seminal works, (i) the electric distance from voltage membrane to electrode tip, (ii) the relative position of electrode tip in relationship to the neural body, and (iii) the surface area of electrode tip [15, 16], as well as (iv) the spatial relationship among channels that are part of an tetrode [28, 29], determine the spatial resolution and waveforms of the recorded action potentials. Also, it is important to note that when the extracellular recordings involve cells with a wide waveform as neuron 1 depicted on Fig. 3d, further analysis can be optimized by increasing the time window and ideally the number of samples for such time windows, to capture the whole action potential and details of waveform, respectively. For DA neurons, a time window of 1.4 ms has been used previously in tetrode recordings [30].

Following spike sorting, we proceeded with the analysis of firing parameters of SNc neurons. Here we analyzed the firing rate and firing pattern measures such as the interspike interval (ISI), autocorrelograms, and their respective power spectrums. Neurons with a percentage of ISI < 2 ms greater than 1% were considered as multi-units and excluded for further analysis to avoid false positives about firing patterns. Twenty-one neurons were finally used in the analytics discussed in this chapter.

A representative data analysis for single units is showed in Fig. 4, using two neurons from our data of SNc single-cell recordings (red and black trace in Fig. 4c–f). The main data needed to proceed to single-unit analysis is the timestamp or its respective binary file with a 1 ms time resolution. To obtain the spike timestamp associated with each single cell, the digitized signal (Fig. 4a, top) was analyzed by a clustering procedure shown in Fig. 3b–d. The time when each of these spikes was found was consolidated in a timestamp file (Fig. 4b, top; a raster plot was created from the timestamps depicted on the bottom). Finally, the timestamp file can be transformed to a binary file (Fig. 4b, bottom). The binary file is a continuous series of zeros and ones (in a millisecond scale), which allows to identify the occurrence of a spike in a logical (true or false) matter: a one represents that a spike is found in that millisecond, whereas a zero show the absence of a spike.

In our current experiments, we used the following recording protocol to study the effect of an acute AMPH injection: 10 min of basal activity, 20 min of post-saline injection, and 30 min of post-AMPH 1.0 mg/kg injection. Figure 4c shows the time course of firing rate during the 60 min of recording from two SNc neurons (bin size = 1 min). Firing rate was calculated dividing the total number of spikes in a period divided by the total time of that period. Then, data was plotted as a *Z*-score, to better visualized how the firing rate deviate from the basal in response to AMPH. This *Z*-score can be obtained as follows:

$$\frac{X \text{ firing rate} - \text{mean baseline firing rate}}{\text{standard deviation baseline firing rate}}$$

where X is the firing rate on a specific time, in this case for a bin size of 1 min. The *Z*-score vs time plot showed a sustained increment (*Z*-score > 2) on firing rate of the red trace neuron but not in the firing rate of the neuron depicted by the black trace. While *Z*-score can be used to study the statistical significance of the firing rate changes after a stimulus such as AMPH or lithium chloride, we recommend using a more robust statistical method like the non-parametric permutation test [21, 22]. First, baseline and post-AMPH firing rate of an individual neuron were divided in 1-min bins. Then, the mean firing rate of the baseline and post-AMPH bins was calculated, and the difference between them (original mean) was computed. Then, to estimate the empirical distribution of the test statistic under the null hypothesis, we randomly mixed the binned firing rate of baseline and post-AMPH and then computed the difference of the mean firing rate in the same manner (new mean firing rate). This permutation process was repeated 1000 times, and a distribution using all the new means firing rate differences was obtained. Finally, by comparing the test statistic computed using the original mean to the estimated empirical distribution, we calculated the p value and rejected the null hypothesis when the probability associated with test statistic was less than a significance level of 0.05. Using this method, the red trace neuron was classified as an excited neuron to AMPH (i.e., the neuron increased its firing rate after AMPH injection) and the neuron represented by the trace in black as neuron that did not change its firing rate in response to AMPH.

The firing pattern of a neuron can be visualized using ISI histograms (Fig. 4d). The ISI is defined as the difference in time between adjacent spikes fired by a single cell. It is computed with the timestamp data file (or transforming the binary file to timestamp). Following, this data is plotted in histograms, with the number of counts in each bin of time and normalized by the total number of ISIs (spike probability). The red and black trace neurons showed a similar ISI histogram during baseline and post-saline injection, indicating that their firing pattern did not change during these periods. During post-AMPH, a change in the ISI histogram distribution was found in the black trace neurons, showing a higher spike probability than the red trace neuron between 80 and 100 ms. This suggest a change in firing mode of the black trace neurons after AMPH injection that was not accompanied by a modification in firing rate. It is important to note that the total amount of spikes ($N = 3700$) was the same for each histogram.

A complementary analysis for firing pattern is the autocorrelogram. It is defined as the correlation of a spike train with itself but with a lag. Normalized autocorrelogram was calculated using the binary file as input for the *xcorr* MATLAB function using the following parameters: maxlag = 1024 (range of ± 1024 ms) and scaleopt = "coeff" (this specifies normalization). This function calculates the normalized correlation between the binary data and the binary data with a lag (lag = 0 is omitted, because it is always 1). Autocorrelogram curves were smoothed using a 25 ms hamming window (MATLAB; *filter* function). As in ISI analysis, our results showed similar autocorrelogram curves for the two neurons during baseline and post-saline (Fig. 4e). The lower probability values in the red trace neurons around 100 ms are because these neurons showed longer lag times between spikes. Importantly, during post-AMPH, the black trace neuron shows an increment in the normalized number of spikes between 80 and 100 ms as observed in Fig. 4d3. This confirms that a specific pattern of neuronal activity emerges in response to AMPH as depicted for the black trace neuron.

Last, the oscillatory nature of the autocorrelogram can be analyzed using a fast Fourier transform (MATLAB; *fft* function). Plotting the spectrum of amplitude vs frequency, it is possible to find the characteristic frequency at which the recorded SNc neurons respond to AMPH. In our example results, we found that the main difference between neurons was around 11 Hz of oscillatory activity after AMPH injection (Fig. 4f). The black trace neuron shows an increase between 9 and 14 Hz in response to AMPH (gray shade area), a result that is expected considering the data found in the ISI and autocorrelograms, where the spike firing was more evident within 80–100 ms. Overall, these analyses allow a fine dissection of the firing signatures of a neuron in response to drug administration.

6 Classification of Dopaminergic Neurons and Effect of an AMPH Injection

The activity of DA neurons has been widely studied (for review, *see* ref. 31). Many of the studies in DA neurons have been performed using glass pipettes [32–34] that are optimal to record single units but have a disadvantage: only one cell at a time is recorded per pipette. In this sense, the use of tetrodes has the advantage that with one single tetrode, several single units can be recorded (as showed in Fig. 3). However, we must consider that around the SNc, not only DA neurons are found but also GABAergic neurons and interneurons (for review, *see* ref. 35), evidencing the need to use several electrophysiological parameters to differentiate DA neurons. In electrophysiological recordings using pipettes, the following parameters were defined to classify a neuron as dopaminergic: (i) recording location; (ii) action potential duration >2.2 ms (unfiltered), >1.1 ms (filtered; 300–5000 Hz); (iii) firing rate <10 Hz; and (iv) bursty firing pattern, with a burst episode starting with an ISI < 80 ms and burst ending when ISI > 180 ms [36]. Here in this chapter, we explore the classification of DA neurons based on extracellular recordings using tetrodes.

For the analysis of phenotypes to identify putative DA neurons, we adapted some parameters that allowed us to classify striatal medium spiny neurons from anesthetized [20] and freely moving [21] recordings. The classification protocol we used considers the spike waveform and timestamp analyses (firing rate and ISI) of each neuron. First, each averaged waveform was normalized by their peak amplitude value and plotted for each of the 21 neurons (Fig. 5a) analyzed in this chapter. Differences on the waveforms after the peak were found among three potential clusters. This measure here is called repolarization at 0.3 ms (vertical black dotted line in Fig. 5a; [37]). To divide the waveforms in different subpopulations, we can combine the repolarization at 0.3 ms, for instance, with the peak-to-amplitude ratio (PAR, [38]) that is defined as follows: peak value/peak to valley value. Figure 5b shows a scatter plot of PAR vs repolarization at 0.3 ms for the 21 SNc neurons analyzed in this chapter. To obtain the clusters of neurons based on the repolarization vs PAR data, the MATLAB *kmeans* function was used. Three clusters were obtained. As defined before, the action potential duration of DA neurons is higher than 1.1 ms in filtered signals like in tetrodes [36]. In this sense, neurons from the second (red) and third (green) clusters were candidates to be defined as putative DA neurons, as they show large action potential durations. Given the sharp and more symmetric action potential waveform, neurons from the first cluster depicted in blue (Fig. 5a) could be considered as a population of putative GABAergic neurons [35].

To confirm which of the two clusters showing large action potential correspond to the DA neurons, firing parameters were also analyzed. In this regard, the ISI histograms were obtained from each neuron and then plotted as the averaged ISI for each one of the three clusters (Fig. 5c, time in logarithmic scale to show large ISI values) vs spike probability. The first (blue) and third (green) clusters showed a similar distribution of their ISI values, with a large proportion of their being higher than 1 s (10^0 s). Interestingly, the ISI distribution of the second cluster depicted in red (Fig. 5c) showed a peak around 0.1 s (10^{-1} s). To analyze if the ISI histograms were significantly different between clusters, we obtained the skewness value of the ISI distributions (MATLAB; *skewness* function). Based on extracellular tetrode recordings, it has been previously defined that DA neurons show an ISI skewness value around 2 [39, 40]. Indeed, the second neural cluster depicted in red shows a skewness median value around 2. This means the cluster of putative DA neurons shows a biased distribution to short ISI, whereas skewness values close to 0 show a uniform distribution. Moreover, the proportion of ISIs higher than 2 s (number of ISIs >2 s/total number of ISIs) was the lowest for the cluster of putative DA neurons and more uniform for the first and third clusters (Fig. 5e). Thus, these results strongly suggest that the second cluster depicted in red corresponds to a set of DA neurons recorded from the SNc (Fig. 3e). The ISI histogram and ISI skewness reported here are similar to those previously described for DA neurons [34, 39, 40]. While it is not clear to which neuronal phenotype the neurons from the third cluster could correspond, even considering the long waveform and irregular firing pattern, it is safer to leave these neurons as an unidentified phenotype.

When looking to more common parameters as the basal firing rate (Fig. 5f), we also observed that the cluster of putative DA neurons showed a median activity of 8 Hz as reported for DA neurons [31].

Burst activity has been commonly analyzed on DA neurons using the following criteria [34]: a burst begins when two spikes are found within 80 ms and ends when an ISI larger than 160 ms is observed. Also, the coefficient of variation (defined as the standard deviation of the ISI divided by the mean of the ISIs) has been used for the same purpose [24]. However, this method has the caveat that is dependent on the firing rate. To address this and to avoid using an arbitrary threshold as the Grace and Bunney method [34], we measured the local variation index (Lv). This is an index about firing mode and was calculated as follows [41]:

$$\mathrm{Lv} = \frac{3}{n-1} \sum_{i=1}^{n-1} \left(\frac{I_i - I_{i+1}}{I_i + I_{i+1}} \right)^2$$

where I_i and I_{i+1} correspond to the i-th and i + 1st ISIs and n are the number of ISIs. Then, Lv index allows to analyze the instantaneous variability of ISIs, with the advantage that this index is less sensitive to changes on firing rate compared to the coefficient of variation. Thus, Lv range from values under to around one (1) means a regular to irregular pattern, respectively, while values over 1 point to a firing mode type burst. For our current data, we found that the cluster of putative DA neurons showed a lower Lv than the other two clusters ($p < 0.05$, Mann-Whitney test, Fig. 5g). This result is consistent with that reported in the literature for DA neurons with a regular firing rate [24].

In summary, based on extracellular tetrode recordings, single cells can be defined as putative DA neurons if they meet the following criteria based on the data discussed so far:

- PAR between 0.8 and 0.95 (*Y*-axis, Fig. 5b)
- Repolarization at 0.3 ms between −0.13 and 0 ms (*X*-axis, Fig. 5b)
- Median skewness around 2 (*Y*-axis, Fig. 5d)
- High % of low chance of ISIs >2 s (*Y*-axis, Fig. 5e)
- Basal firing rate ~8 Hz (*Y*-axis, Fig. 5f)
- Basal Lv ~0.5 (*Y*-axis, Fig. 5g)

After attempting the classification of SNc cells as DA neurons based on the criteria described above, then we studied the effect of an acute AMPH 1.0 mg/kg ip injection on the sample of SNc neurons (Fig. 6). We found that an acute AMPH injection induced an increment on the firing rate (Fig. 6a, cluster 2, $p < 0.05$, Wilcoxon signed-rank test) of all the five putative SNc DA neurons (Fig. 6c, cluster 2, $p < 0.05$, permutation test). The putative GABAergic neurons in the first cluster were also mainly excited in response to AMPH, while the unidentified third cluster of neurons showed a mixed response to AMPH. Putative DA neurons did not show a change on their firing pattern in response to AMPH ($p > 0.05$, Wilcoxon signed-rank test, Fig. 6b), while putative GABAergic neurons adjust their firing in response to AMPH. Overall, these results indicate that putative SNc DA neurons are mainly excited by AMPH but did change their firing pattern. Also, we only record a small number of neurons with five putative SNc DA neurons that constitute a very preliminary result.

Remarks

In this chapter, we discussed the extraction of waveform and firing parameters to characterize putative neural phenotypes from the recording of single unit on anesthetized rats. The flexibility of this method makes it possible to use exactly the same analyses on free-behaving animals, altogether with pharmacological and optogenetic manipulations to increase robustness, and more importantly to explore causal hypothesis about the contribution of different neural phenotypes to physiological and behavioral disorders. Thus, it should be possible to better understand how the neural signature evolves in correlation with the casual drug intake to habits. It is well known that the behavioral sensitization by AMPH is a good approach to better understand the evolution of drug addiction. In this regard, we started using the Lv index on medium spiny neurons activity recorded from the dorsolateral striatum of anesthetized [20] and freely moving rats [21]. In these studies we found a higher Lv, meaning that a burst firing pattern predicted the susceptibility to express AMPH locomotor sensitization. Then, after a repeated AMPH administration, a challenge of this psychostimulant on sensitized rats was associated with a decrease in the Lv index. Dorsolateral striatum neurons showed a regular tonic likewise firing pattern in response to AMPH [21]. This experimental approach can be extended to the recording of multiple sites, for instance, from where DA neurons are placed to the brain areas targeted by their projections.

The scope of our approach is electrode agnostic. Given that the main parameters used to identify different neural phenotypes are the spike waveform and their timing, our approach is valid for intra- and extracellular recordings, while the minimum condition of capturing the action potential waveforms and timestamps from single units is achieved. Importantly, we also here described an easy and versatile electrode array configuration that allows such goal based on basic laboratory tools and materials. In this regard, the custom microdrive described in this chapter is a cheap and very accessible solution for electrophysiological recordings on anesthetized or freely moving animals. The main advantage of such a physical platform is its flexibility to control electrode positioning of multiple electrode bundles that are part of the same manipulator and can target several brain areas simultaneously.

Bibliography

1. Cools R (2008) Role of dopamine in the motivational and cognitive control of behavior. Neuroscientist 14(4):381–395
2. Robbins TW, Gillan CM, Smith DG, de Wit S, Ersche KD (2012) Neurocognitive endophenotypes of impulsivity and compulsivity: towards dimensional psychiatry. Trends Cogn Sci 16(1):81–91. https://doi.org/10.1016/j.tics.2011.11.009
3. Escobar AP, Casanova JP, Andrés ME, Fuentealba JA (2020) Crosstalk between kappa opioid and dopamine systems in compulsive behaviors. Front Pharmacol 11:57. https://doi.org/10.3389/fphar.2020.00057. eCollection 2020. Review
4. Fallon JH, Moore RY (1978) Catecholamine innervation of the basal forebrain. IV. Topography of the dopamine projection to the basal forebrain and neostriatum. J Comp Neurol 180:545–580
5. Fallon JH, Riley JN, Moore RY (1978) Substantia nigra dopamine neurons: separate populations project to neostriatum and allocortex. Neurosci Lett 7:157–162

6. Haber SN, Fudge JL, McFarland NR (2000) Striatonigrostriatal pathways in primates form an ascending spiral from the shell to the dorsolateral striatum. J Neurosci 20(6):2369–2382. https://doi.org/10.1523/JNEUROSCI.20-06-02369.2000
7. Pérez-Valenzuela EJ, Andrés Coke ME, Grace AA, Fuentealba Evans JA (2020) Adolescent exposure to WIN 55212-2 render the nigrostriatal dopaminergic pathway activated during adulthood. Int J Neuropsychopharmacol 23(9):626–637. https://doi.org/10.1093/ijnp/pyaa053
8. Marinelli M, Cooper DC, Baker LK, White FJ (2003) Impulse activity of midbrain dopamine neurons modulates drug-seeking behavior. Psychopharmacology 168(1–2):84–98. https://doi.org/10.1007/s00213-003-1491-1
9. Buzsáki G, Costas AA, Christof K (2012) The origin of extracellular fields and currents—EEG, ECoG, LFP and spikes. Nat Rev Neurosci 13(6):407–420. https://doi.org/10.1038/nrn3241
10. della Valle E, Koo B, Patel PR, Whitsitt Q, Purcell EK, Chestek CA, Weiland JD (2021) Electrodeposited platinum iridium enables microstimulation with carbon fiber electrodes. Front Nanotechnol 3:782883. https://doi.org/10.3389/fnano.2021.782883
11. Paxinos G, Charles W (2007) The rat brain in stereotaxic coordinates, 6th edn. Elsevier
12. Swanson LW (2004) Brain maps: structure of the rat brain, 3rd edn. Elsevier
13. Jo YS, Heymann G, Zweifel LS (2018) Dopamine neurons reflect the uncertainty in fear generalization. Neuron 100(4):916–925.e3. https://doi.org/10.1016/j.neuron.2018.09.028
14. Heymann G, Jo YS, Reichard KL, McFarland N, Chavkin C, Palmiter RD, Soden ME, Zweifel LS (2020) Synergy of distinct dopamine projection populations in behavioral reinforcement. Neuron 105(5):909–920.e5. https://doi.org/10.1016/j.neuron.2019.11.024
15. Robinson DA (1968) The electrical properties of metal microelectrodes. Proc IEEE 56(6):1065–1071. https://doi.org/10.1109/PROC.1968.6458
16. Lemon R (1984) Methods for neuronal recording in conscious animals. Wiley
17. Buzsáki G (2004) Large-scale recording of neuronal ensembles. Nat Neurosci 7(5):446–451. https://doi.org/10.1038/nn1233
18. Kaltenbach JA, Gerstein GL (1986) A rapid method for production of sharp tips on preinsulated microwires. J Neurosci Methods 16(4):283–288. https://doi.org/10.1016/0165-0270(86)90053-1
19. Olds J, Disterhoft JF, Segal M, Kornblith CL, Hirsh R (1972) Learning centers of rat brain mapped by measuring latencies of conditioned unit responses. J Neurophysiol 35(2):202–219. https://doi.org/10.1152/jn.1972.35.2.202
20. Gatica RI, Aguilar-Rivera MÍ, Azocar VH, Fuentealba JA (2020) Individual differences in amphetamine locomotor sensitization are accompanied with changes in dopamine release and firing pattern in the dorsolateral striatum of rats. Neuroscience 427:116–126. https://doi.org/10.1016/j.neuroscience.2019.11.048
21. Gatica RI, Aguilar-Rivera M, Henny P, Fuentealba JA (2022) Susceptibility to express amphetamine locomotor sensitization correlates with dorsolateral striatum bursting activity and GABAergic synapses in the globus pallidus. Brain Res Bull 179:83–96. https://doi.org/10.1016/j.brainresbull.2021.12.005
22. Aguilar-Rivera M, Kim S, Coleman TP, Maldonado PE, Torrealba F (2020) Interoceptive insular cortex participates in sensory processing of gastrointestinal malaise and associated behaviors. Sci Rep 10(1):21642. https://doi.org/10.1038/s41598-020-78200-w
23. Manzur HE, Alvarez J, Babul C, Maldonado PE (2013) Synchronization across sensory cortical areas by electrical microstimulation is sufficient for behavioral discrimination. Cereb Cortex 23(12):2976–2986. https://doi.org/10.1093/cercor/bhs288
24. Brown MT, Henny P, Bolam JP, Magill PJ (2009) Activity of neurochemically heterogeneous dopaminergic neurons in the substantia nigra during spontaneous and driven changes in brain state. J Neurosci 29(9):2915–2925. https://doi.org/10.1523/JNEUROSCI.4423-08.2009. PMID: 19261887; PMCID: PMC4262786
25. Walczak M, Błasiak T (2017) Midbrain dopaminergic neuron activity across alternating brain states of urethane anaesthetized rat. Eur J Neurosci 45(8):1068–1077. https://doi.org/10.1111/ejn.13533
26. Aguilar-Rivera MI, Casanova JP, Gatica RI, Quirk GJ, Fuentealba JA (2015) Amphetamine sensitization is accompanied by an increase in prelimbic cortex activity. Neuroscience 288:1–9. https://doi.org/10.1016/j.neuroscience.2014.12.027
27. Steketee JD, Kalivas PW (2011) Drug wanting: behavioral sensitization and relapse to drug-seeking behavior. Pharmacol Rev 63(2):

348–365. https://doi.org/10.1124/pr.109.001933

28. Gray CM, Maldonado PE, Wilson M, McNaughton B (1995) Tetrodes markedly improve the reliability and yield of multiple single-unit isolation from multi-unit recordings in cat striate cortex. J Neurosci Methods 63(1–2):43–54. https://doi.org/10.1016/0165-0270(95)00085-2
29. Harris KD, Henze DA, Csicsvari J, Hirase H, Buzsáki G (2000) Accuracy of tetrode spike separation as determined by simultaneous intracellular and extracellular measurements. J Neurophysiol 84(1):401–414. https://doi.org/10.1152/jn.2000.84.1.401
30. Wang DV, Tsien JZ (2011) Conjunctive processing of locomotor signals by the ventral tegmental area neuronal population. PLoS One 6(1):e16528. https://doi.org/10.1371/journal.pone.0016528. PMID: 21304590; PMCID: PMC3029369
31. Marinelli M, McCutcheon JE (2014) Heterogeneity of dopamine neuron activity across traits and states. Neuroscience 282:176–197. https://doi.org/10.1016/j.neuroscience.2014.07.034
32. Grace AA, Bunney BS (1983) Intracellular and extracellular electrophysiology of nigral dopaminergic neurons--1. Identification and characterization. Neuroscience 10(2):301–315. https://doi.org/10.1016/0306-4522(83)90135-5. PMID: 6633863
33. Grace AA, Bunney BS (1984) The control of firing pattern in nigral dopamine neurons: single spike firing. J Neurosci 4(11):2866–2876. https://doi.org/10.1523/JNEUROSCI.04-11-02866.1984
34. Grace AA, Bunney BS (1984) The control of firing pattern in nigral dopamine neurons: burst firing. J Neurosci 4(11):2877–2890. https://doi.org/10.1523/JNEUROSCI.04-11-02877.1984
35. Dudman JT, Gerfen CR (2015) Chapter 17 – The basal ganglia. In: Paxinos, George. Academic Press, San Diego, pp 391–440
36. Ungless MA, Grace AA (2012) Are you or aren't you? Challenges associated with physiologically identifying dopamine neurons. Trends Neurosci 35(7):422–430. https://doi.org/10.1016/j.tins.2012.02.003
37. Vinck M, Bos JJ, Van Mourik-Donga LA, Oplaat KT, Klein GA, Jackson JC, Gentet LJ, Pennartz CMA (2016) Cell-type and state-dependent synchronization among rodent somatosensory, visual, perirhinal cortex, and hippocampus CA1. Front Syst Neurosci 9:1–25. https://doi.org/10.3389/fnsys.2015.00187
38. Yarom O, Cohen D (2011) Putative cholinergic interneurons in the ventral and dorsal regions of the striatum have distinct roles in a two choice alternative association task. Front Syst Neurosci 5:5–36. https://doi.org/10.3389/fnsys.2011.00036
39. Li W, Doyon WM, Dani JA (2011) Acute in vivo nicotine administration enhances synchrony among dopamine neurons. Biochem Pharmacol 82(8):977–983. https://doi.org/10.1016/j.bcp.2011.06.006. Epub 2011 Jun 12. PMID: 21684263; PMCID: PMC3162092
40. Doyon WM, Ostroumov A, Ontiveros T, Gonzales RA, Dani JA (2021) Ethanol produces multiple electrophysiological effects on ventral tegmental area neurons in freely moving rats. Addict Biol 26(2):e12899. https://doi.org/10.1111/adb.12899
41. Shinomoto S, Kim H, Shimokawa T, Matsuno N, Funahashi S, Shima K, Fujita I et al (2009) Relating neuronal firing patterns to functional differentiation of cerebral cortex. PLoS Comput Biol 5(7):e1000433. https://doi.org/10.1371/journal.pcbi.1000433

Part III

Models for Addiction and Neuropsychiatric Diseases

Chapter 7

Pulmonary Inhalation to Assess Effects of Coca Paste on Behavior and Dopamine Neurotransmission

Cecilia Scorza, José Pedro Prieto, Sara Fabius, and Martín Galvalisi

Abstract

In human, the pulmonary inhalation route is one of the most effective and rapid forms for drug delivery with therapeutic, medical, and recreational purposes. In addition to being a noninvasive method, the extensive irrigated lung surface turns extremely beneficial when drug delivery for pulmonary use is chosen to study the systemic effect of different drugs. Due to different issues, however, this route of administration is not very often used to evaluate the effects of drugs in animals; however, here we provided methodological information applying this route of administration combined with traditional approaches.

Coca paste (CP) is an illicit drug of abuse commonly used in several countries of South America, and likewise crack, it is classified as a smoked cocaine form. In humans, CP consumption induces a fast and strong dependence in addition to other physiological alterations. Yet, these other factors, which collaborate in defining the clinical profile of CP consumers, continue under study. Furthermore, until recently there was no preclinical data so far showing the effects induced by the pulmonary inhalation of CP using traditional behavioral and neurochemical tools, such as the open field test and in vivo brain microdialysis technique, respectively. In this work, the open field test was used for the acute motor and exploratory activity assessment. In vivo brain microdialysis is a widely known neurochemical approach used for monitoring neurotransmitter extracellular levels in discrete brain areas of freely moving rats. Considering that increases in dopamine neurotransmission induced by cocaine and other drugs of abuse in the nucleus accumbens are related to their rewarding properties, we designed a novel methodological procedure to measure the extracellular dopamine levels in the nucleus accumbens of freely moving rats during and after the passive inhalation of different volatilized CP samples. The pharmacological relevance of the major components of CP samples, cocaine and caffeine (main active adulterant), is highlighted on the behavioral and neurochemical outcomes. Our findings pointed out the relevance of the route of administration and the chemical composition of CP preparations in smoked cocaine abuse liability and place them as the main contributing factors.

Key words Cocaine, Caffeine, Addiction, Route of administration, Nucleus accumbens

1 Introduction

1.1 Cocaine Base Paste or Coca Paste: Definition

Cocaine is a psychostimulant alkaloid substance extracted from coca leaves (*Erythroxylum coca bush*) through a chemical process that usually ends with the obtainment of the cocaine hydrochloride.

José Antonio Fuentealba-Evans and Pablo Henny (eds.), *Dopaminergic System Function and Dysfunction: Experimental Approaches*, Neuromethods, vol. 193, https://doi.org/10.1007/978-1-0716-2799-0_7,

There are different forms of cocaine consumption. Since cocaine hydrochloride is hydrophilic, this compound represents its intranasal or intravenous form of consumption. However, other forms of cocaine are used through the pulmonary inhalation route. Crack and coca paste (CP), known as cocaine base paste as well, are the most known smoked cocaine forms [1]. Crack is a cocaine derivative presented in a free base form, and it is processed from cocaine hydrochloride with sodium bicarbonate, yielding solid crack cocaine [2]. In turn, CP is one of the earliest intermediate products obtained during any of the preliminary phases of cocaine alkaloid extraction [3, 4]. CP is an illicit drug of abuse frequently consumed in several South American countries [5–7]. In humans, its consumption produces an intense feeling of euphoria combined with psychomotor alterations. Frequent users of CP quickly develop a high level of dependence with high rates of relapse [8]. Behavioral alterations such as cognitive deficits, increased impulsivity, aggressiveness, and sleep disturbances are also seen. These behavioral traits shape a clinical profile shared by CP users which is differentiable from cocaine consumers [8].

There are similarities and differences between smoked cocaine (crack or CP) and cocaine hydrochloride. Although cocaine shows the same pharmacological action at the organism (i.e., blockade of dopamine transporter) regardless of whether it is crack, CP, or the hydrochloride form, the different route of administration (e.g., intranasal, intravenous, or smoked) is an essential factor for the abuse potential, pharmacokinetics profile, and biological and psychological effects of cocaine. Then, the route of administration and the chronicity of cocaine use establish a pattern of consumption and determine the potency of its effect. Moreover, we determined in animal studies that, in addition to the route of administration, the purity of cocaine and adulterants could also be collaborating to the physiological and behavioral effects after its exposure [9, 10]. It is also important to consider that when cocaine base is burnt for smoking, it generates anhydroecgonine methyl ester (AEME). This exclusive product of cocaine pyrolysis is used in forensic analysis as the analytical biological marker for crack cocaine consumption [11–14]. In animal studies, it has been described that AEME can impair cognitive functions [14], while in in vitro studies, it has been reported that AEME seems to be neurotoxic and can potentiate cocaine neurotoxicity [15, 16]. Although no data have been provided to confirm that hypothesis, both studies suggest that the effects of its pyrolysis product may aggravate the condition of cocaine addiction.

1.2 Coca Paste Chemical Composition: Cutting and Adulteration

Illegal substances seized by police and analyzed by forensic laboratories usually reveal cutting agents like diluents and adulterants in their chemical composition. Diluents are pharmacologically inactive and readily available substances, while adulterants are pharmacologically active substances, usually more expensive or less available than diluents [17]. This kind of information is commonly *reported* in forensic literature. However, being the adulterants substances that could imitate or boost the main drug (e.g., cocaine), it is imperative to take into consideration the pharmacological interactions between all these substances, which in turn can alter the behavioral output of the drug consumer. In this context, we have studied the chemical composition of seized CP samples in order to quantify the cocaine content and identify the presence of adulterants [9, 10, 18] and their stability (i.e., chemical integrity) after the volatilization process [18]. Providing this information was peculiar and novel, since no data of a seized CP samples had not been published before. Thus, we found that cocaine, phenacetin (antipyretic drug), and caffeine were the main compounds identified in the analyzed CP samples, whereas other compounds like aminopyrine, levamisole, lidocaine, and benzocaine were less frequently detected [18]. Interestingly, a similar profile of adulteration is also frequently reported for different drugs of abuse (e.g., heroin, amphetamine, cocaine hydrochloride) [19–21], where pulmonary inhalation is not the usual route of administration. Regarding this, it seems that a similar pattern of adulteration is applied even if the way of drug consumption is different, and such adulterants should be volatilizable in the case of smoking form of cocaine. We also found that cutting agents typically used for seized adulterated CP samples have been changing over time and the amount of the adulterants has increased [22], suggesting that the adulteration in illegal and seized drugs is not a random phenomenon [23] and seems to depend on the availability of the cutting substances.

1.3 Active Adulterant and Its Role in Cocaine Effects

Forensic and research data has revealed that caffeine is one of the most frequently found substances in seized samples of psychostimulants, including cocaine hydrochloride and even in smoked forms of cocaine, like crack and CP [9, 19, 24–27] among other class of drugs [28]. Although caffeine is a legal substance consumed worldwide to improve attention and overall performance, extensive caffeine use may lead to the development of mild psychological and physical dependence [29]. In humans, oral consumption is the main way to use caffeine; however, some informal data provided by websites (http://boingboing.net/2009/01/19/how-tomake-smokable.html) indicate that caffeine can also be consumed as a recreational drug using the pulmonary inhalation route. Although it has already been described that caffeine can be volatilized [30], we demonstrated in animal studies that volatilized caffeine is able to induce behavioral and neurochemical changes that are associated to the dopaminergic neurotransmission [31]. In addition to its per se effect, it has been suggested that caffeine adulteration is intended not only to increase the weight and volume but also to mimic or potentiate the stimulant and the reinforcing effects of cocaine

[17, 25]. In animal studies, we have reported that caffeine boosts several pharmacological effects induced by acute or repeated intraperitoneal or intravenous injections of cocaine. For example, we have shown that caffeine increases the acute stimulation motor effects of cocaine [9, 10] and enhances the reinforcing and motivational value of cocaine in the self-administration paradigm in rats [32]. Moreover, caffeine potentiates cocaine reward-associated learning in the conditioning place preference test [33] in mice and maintains the acute wake-promoting effect of cocaine in rats [34]. The repeated treatment with the combination of cocaine and caffeine also elicits the locomotor sensitization phenomenon, an effect that was not observed in cocaine-treated rats under the same protocol schedule [27]. The behavioral sensitization response induced by a repeated administration of seized caffeine-adulterated CP samples, containing specific ratios of cocaine and caffeine [18], was accompanied by changes in the expression of plasticity markers in the medial prefrontal cortex (mPFC) and nucleus accumbens (NAc) of mice, both brain regions highly linked with drugs addiction [35].

1.4 Pulmonary Inhalation Route and Drug Addiction

The respiratory tract of mammals has been used to deliver biologically active chemicals into the organism for therapeutic (e.g., asthma treatment), medical (e.g., anesthesia delivery), and recreational (e.g., cannabis or cocaine consumption) purposes [36]. The efficiency of the lungs resides in the large surface area and profuse blood supply, facilitating the rapid absorption of drugs into the systemic circulation. Because of this property, smokable forms of cocaine like crack has largely been preferred comparatively with other routes of recreational cocaine use, including nasal insufflation or intravenous injection of cocaine hydrochloride [1, 37]. Although this route of administration could be considered advantageous, taking into account its expected effect, heavy or regular inhalation of cocaine could have harmful effects on the lung among others clinical issues [38, 39]. It is widely accepted that the rate of drug delivery to the brain is important in determining the risk of developing addiction, i.e., formulations and routes of administration that result in the rapid entry of a drug into the brain increase the propensity to addiction [40]. In humans, smoke cocaine, or even when injected intravenously, is associated with a greater vulnerability to developing addiction than its intranasal use [1, 37]. Similarly, preclinical data indicate that differences in the rate of drug delivery can have important behavioral and neurochemical effects on animals [40, 41]. Although the potential for addiction is significantly increased when drugs are administered rapidly, why this occurs is unknown. It is worth mentioning that almost all of these studies were largely carried out using the intravenous route of administration; in contrast, only a few studies have been carried out after animal exposure to the drugs volatilization, especially to smoked cocaine [42, 43]. The legal difficulties of obtaining the illicit drugs to work in preclinical studies, the complexity of drug burning, and the selection of the rodent paradigms under which the animals would be expose to volatilized drugs constitute some issues that may hinder the obtainment of reliable data. We have been working for several years in overcoming those challenges focusing our investigations on the behavioral, molecular, and neurochemical effects induced by seized smoked cocaine samples in rodents. This chapter will summarize the behavioral and neurochemical effects induced by the passive exposition to the fume of seized CP samples and the role of its active components (cocaine and caffeine). Particularly, in this study we also reported the combination of the in vivo brain microdialysis technique with the passive inhalation of the fume of volatilized CP samples as a novel procedure to study the effects of this smokable form of cocaine on dopamine release in freely moving rats.

2 Materials

2.1 Animals

All the experiments included in this chapter were performed using Wistar male rats (IIBCE animal facilities, Montevideo) weighing 250–310 g. All animals were housed in groups of five in plastic cages ($50 \times 37.5 \times 21$ cm^3) and kept under controlled conditions (temperature 22 ± 2 °C, 12 h day-night cycle, lights on at 7:00 am) with food and water available ad libitum. All experimental procedures were performed following the "Guidelines for the Care and Use of Mammals in Neuroscience and Behavioral Research" (National Research Council (US) Committee on Guidelines for the Use of Animals in Neuroscience and Behavioral Research) [44] and approved by the IIBCE Bioethics Committee's requirements and carried out under the current national ethical regulations.

2.2 Drugs

Smoked cocaine or seized CP samples were supplied by the Technical Forensic Institute (Uruguay), with the authorization of the National Drug Board and the Ministry of Public Health (Uruguay), from seizures made by the local police after raid procedures. These are samples that could possibly be consumed by drug users. Cocaine hydrochloride, used as a standard solution in chemical analysis, was generously donated by Laboratory Verardo y Cia (Argentina). Caffeine was obtained from Sigma-Aldrich (Germany). Those CP samples included in this chapter (Table 1) were chosen after a chemical analysis previously performed [18]. The identification and quantification of cocaine and caffeine were performed using HPLC (Waters 1525) coupled with a diode array detector (Waters 2998) and an autosampler (Waters 717 plus; Waters Corporation, USA), equipped with a C18 column (100×4.6 mm, 5 μ, 100 Å, Luna, Phenomenex, USA).

Table 1
CP-seized samples

CP samples	Cocaine base (%)	Caffeine (%)
CP1	67.4 ± 1.2	14.0 ± 0.2
CP2	68.2 ± 2.0	0.0 ± 0.0
CP3	37.1 ± 0.0	41.9 ± 0.0
CP4	50.2 ± 1.0	0.0 ± 0.0

CP-seized samples chosen in this chapter as examples to show the cocaine and caffeine role in the behavioral and neurochemical effects induced by each volatilized CP sample

3 Methods

3.1 Procedure to Volatilize Drugs: Setup of Pulmonary Inhalation in Rats

The method applied to volatilize drugs reproduces at the lab the procedure used by consumers to smoke CP (or crack). Based on previous reports [43, 45], the sequence of procedures for the volatilization and pulmonary inhalation setup is illustrated in Fig. 1. The volatilization device (Fig. 1a) consisted of a stainless steel tube (length 10 cm, i.d. 0.6 mm) filled with steel wool used in order to keep the drug at the top of the tube. The tube was closely connected to a plastic syringe (30 mL) through which a vacuum system was set up. At the end of the tube, a lighter was turn on, and drugs were heated until the vapor generation started. At this point, the syringe piston was moved away, and the vapor immediately reached the syringe. Animals are exposed to drugs vapor into the inhalation chamber for 10 min (Fig. 1b) where a passive inhalation of drug takes place. After this volatilization procedure, behavioral (Fig. 1c) and neurochemical assays (Fig. 1d, e) were performed.

3.2 Procedure to Study the Stimulant Effect of Drugs: Open Field Assay

After the exposition to the fume of volatilized drugs, animals were removed from the inhalation chamber and placed in an Open Field box (OF, a square box of 60×60 cm^2 with red 40-cm-high acrylic sides), in a quiet experimental room with controlled temperature (22 ± 2 °C). A camera placed above the OF and connected to a computer equipped with the video-tracking EthoVision XT 7.0 software (Noldus, Netherlands) was used to automatically record motor parameters (Fig. 1c). We specifically measured the horizontal locomotor activity defined as the total distance moved in meters (m) and velocity of movement. Locomotor and exploratory (number of rearings) activities were registered for 30 min. During all the experiments, the inhalation chamber and the OF were cleaned with alcohol 30% before placing the following rat.

To evaluate the behavioral effect induced by CP samples on animal motor behavior, we acutely treated rats with CP samples with different chemical composition (*see* Table 1) and following the

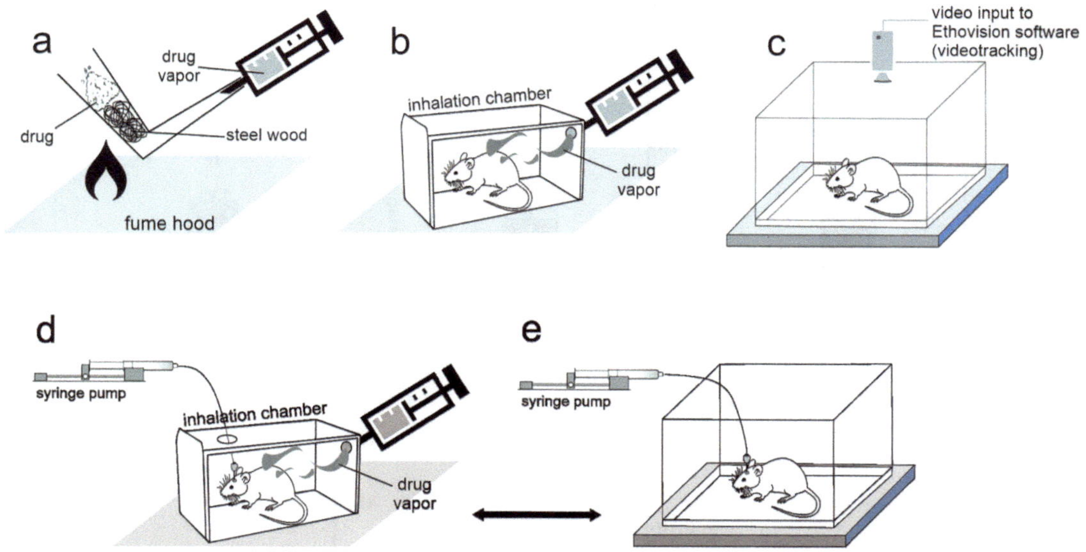

Fig. 1 Illustrate the sequence of steps to volatilize CP seized samples or cocaine pure (**a**), the passive exposition to vapor drug in the inhalation chamber (**b**), and the behavioral evaluation on the open field (**c**) and automatically recording motor parameters by the video tracking software. Images show cannulated animal (NAcc) placed into the inhalation chamber (**d**) where the dialysis sample collections continue during the CP volatilization. After that, animals were removed from the inhalation chamber and returned to the dialysis cage and were not disconnected during the dialysis experiment (**e**). Control animals were subjected to the same procedure although only the steel wool was heated

methods described above. The acute stimulant effect was evaluated by changes in the following parameters: distance moved, velocity of movement, and number of rearings (exploratory activity). Results showed that CP1- and CP3-treated animals exhibited a statistically stimulant effect significantly higher compared to control group, while the effect induced by CP2 was lower and not significantly different compared to the control group. Distance moved profile was similarly accompanied by changes in the velocity of movement and rearing behavior (Fig. 2a–i). We can appreciate how the presence of caffeine in samples with high or low content of cocaine (CP1 and 3, respectively) substantially contributes to the acute stimulant effect induced by smoked cocaine under the pulmonary inhalation route. This result is in accordance with our previous data of different seized CP samples that were intraperitoneal administered [9, 10], confirming that both cocaine and caffeine also exert their pharmacological effects after pulmonary inhalation. Since it has been reported that the severity of dependence to drugs of abuse is influenced by the route of administration [37], we can assume that smoking seized CP samples adulterated with caffeine could induce a greater stimulant effect and increase its reinforcing effects and consequently its abuse liability. Evidence that caffeine was able to increase the reinforcing and motivational value of cocaine in the self-administration paradigm in rats supports this hypothesis

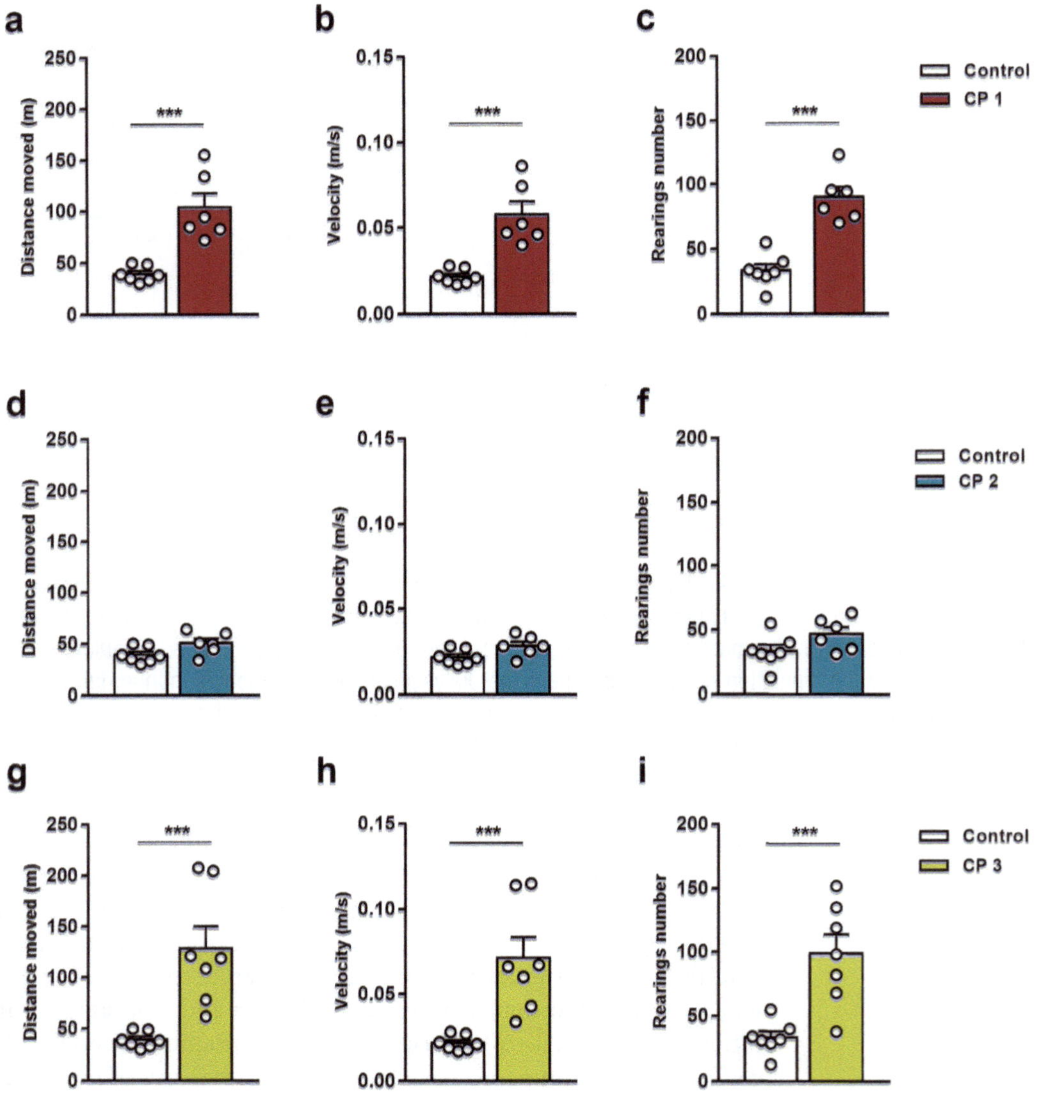

Fig. 2 Acute stimulant effect induced by volatilized CP seized samples. Distance moved (m), velocity of movement (m/s), and number of rearings in CP-treated animals (50 mg) or control group (0). Each parameter was evaluated immediately after its volatilization (10 min exposed). Mean ± SEM. Student *t*-test. * = versus control group; *** = $p < 0.001$. $N = 6$ per group

[32]. In this case, the intravenous route was used, which implies another fast way for the entry of the drug into the brain.

It is well established that after the repeated exposure to drugs of abuse, particularly psychostimulants, there is a progressive and persistent increase in this drug-induced hyperactivity. This phenomenon proved that repeated drug exposure causes the brain motivational circuitry to become persistently hyperresponsive or sensitized to drugs and drug-associated stimuli [46]. As far as we know, there were no data available in rodents showing the

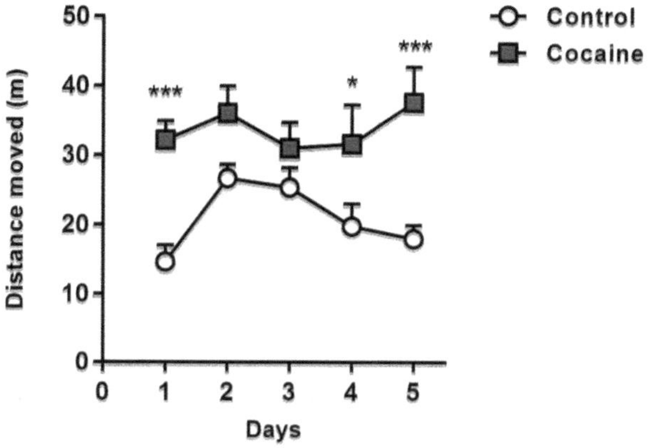

Fig. 3 Repeated stimulant effect induced by volatilized cocaine pure (25 mg) or control group. Distance moved (*m*) was evaluated immediately after its volatilization (10 min). Mean ± SEM. Two-way ANOVA followed by Newman-Keuls. * = versus control group. * = $p < 0.05$, *** = $p < 0.001$. $N = 8$ per group

persistent hyperactivity induced by the exposition to smoked cocaine. By using of the volatilization device and all procedures illustrated above (Fig. 1a, b), we carried out the evaluation of the stimulant effect induced by the repeated exposition to pure smoked cocaine (without adulterants). Animals were treated for 5 days with volatilized cocaine and placed into OF each day. Distance moved was recorded automatically for 30 min. As expected, results showed that a low dose (25 mg) of volatilized cocaine induced a persistent increase of animal motor activity along 5 days (Fig. 3). Again, these data indicate that cocaine can exert its pharmacological effects after pulmonary inhalation and after a repeated schedule of administration. It remains to be determined if the presence of caffeine can boost this behavioral effect.

3.3 Procedure to Study the Neurochemical Effect of Drugs: In Vivo Intracerebral Microdialysis Technique

It is widely known that the preferential release of dopamine in the nucleus accumbens shell has been associated to the reinforcing effects of psychostimulants, *like cocaine* [47, 48]. *Since the effect on dopamine release induced by CP-seized samples in neurochemical in vivo models was lacking*, we evaluated the neurochemical changes induced by seized CP samples on the dopamine neurotransmission using the *brain* microdialysis technique in freely moving animals. *Thus, we implemented a novel procedure which combined the in vivo microdialysis technique with the CP volatilization method.* We acutely treated rats with CP samples with different chemical composition (Table 1), following the methods described above. Anesthetized rats with a combination of ketamine (90 mg/kg, i. p.) and xylazine (5 mg/kg, i.p., Konsol Kning S.A Laboratory) were placed in a stereotaxic frame in a flat-skull position. A dialysis probe (MD-2200, 2 mm membrane length, BAS, USA) was implanted into the NAcc shell following these coordinates (in mm): AP, 1.9; L, 1.0; DV, 7.6 [49]. After surgery, animals were individually housed and monitored while recovered. Microdialysis experiments were conducted at 18–20 h after surgery in a freely moving condition [31]. On the experimental day, each animal was located in the dialysis cage and connected with tubings to the perfusion micropump. Artificial cerebrospinal fluid (aCSF: 125-mM NaCl; 1.26-mM $CaCl_2$; 2.5-mM KCl; 1.18-mM $MgCl_2$) was perfused at a constant flow rate of 1.5 μL/min. Dialysate samples were collected every 20 min during all the experiment (30 μL each). After a stabilization period (100 min), four fractions were collected to obtain basal values before the CP volatilized samples exposure. At this moment and in order to expose animals to the CP fume, each animal was placed for a moment (10 min) into the inhalation chamber where the dialysis samples collection continued during the drug volatilization (Fig. 1d, e). After that, animals were removed from the inhalation chamber and returned to the dialysis cage (Fig. 1e). To avoid air entrance to dialysis system, animals were not disconnected during the dialysis experiment. Control animals were subjected to the same procedure although only the steel wool was heated. Dopamine and the metabolite 3,4-dihydroxyphenylacetic acid (DOPAC) concentration in dialysate samples were immediately determined by high-performance liquid chromatography (HPLC) equipped with a degassing unit (DGU-20A5R; Shimadzu, USA), a binary HPLC pump (1525; Waters, USA), an electrochemical detector with a temperature controller (2465, Waters, USA), and a C18 column (3-μm particle size; Phenomenex, USA). The mobile phase consisted of citric acid (0.15 M), sodium octyl sulfate (0.6 mM), 3% acetonitrile, and 1.5% tetrahydrofuran at pH 3.0 and was infused at a flow rate of 0.3 mL/min. Oxidation potential was set of +650 mV versus Ag/AgCl. The analytical column and the electrochemical flow cell were maintained at 30 °C. At the end of the experiments, brains were processed for histological procedures in order to confirm the localization of each dialysis probe.

Results showed that the control group dopamine and DOPAC levels were stable along the entire experimental session. On the other hand, the exposure to CP1 and CP4 elicited a significant increase in the extracellular dopamine levels (Fig. 4a, b). The effect of both samples was evident immediately after its exposure and reached statistical significance after 20 min; however, the effect of CP1 was greater than that observed for CP4. The extracellular level of DOPAC was significantly changed by CP1 during 80 min (Fig. 4c), whereas the effect of CP4 was less pronounced and short-lasted (Fig. 4d). This is the first reported demonstration that a smoked form of cocaine, with or without caffeine (as adulterant), increases dopamine extracellular levels in the

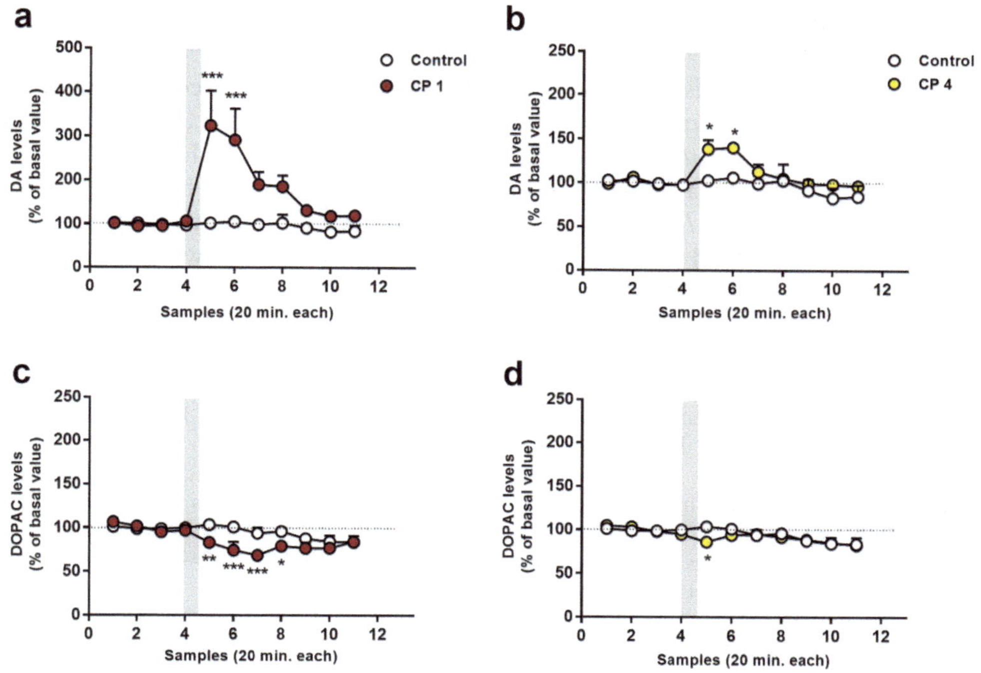

Fig. 4 Effect of CP1 and CP4 on the dopamine (DA) and DOPAC extracellular levels in the NAc shell. Animals were exposed to 50 mg of volatilized caffeine (CP, closed circles) or control animals (open circles) during 10 min. Gray bar indicates the volatilization period (10 min). Mean ± SEM. Two-way ANOVA followed by Newman-Keuls. *versus control * $= p < 0.05$, ** $= p < 0.01$, *** $= p < 0.001$. $N = 5$ per group

nucleus accumbens shell. This finding suggests that the presence of caffeine in smoked forms of cocaine may potentiate its positive reinforcing effect and finally its abuse liability.

4 Conclusions

Our study revealed that the pulmonary inhalation route is a valid option to study behavioral and neurochemical effects induced by the volatilization of smoked cocaine samples with or without active adulterants. Passive exposure to fume of drugs increased acute and chronic motor activity and activated dopamine transmission in the NAcc shell. Accordingly, the model of exposure to smoked drugs effectively reproduced systemic changes in an animal model. From a methodological point of view, here we provided a novel procedure to measure dopamine changes in behave animals under the CP effects using the pulmonary inhalation route. In addition, this study contributes to the knowledge of caffeine as an active adulterant in interaction with cocaine. In conclusion, this study offers new insights into pulmonary drug delivery in the psychostimulant addiction field.

Acknowledgments

This study was partially supported by Smoked Cocaine in South Cone Countries Grant CICAD-OEA/USINL, *Premio Concursable Junta Nacional de Drogas* (Uruguay), and PEDECIBA (Uruguay). Sara Fabius is a postgraduate fellow supported by *Agencia Nacional de Investigación e Innovación* (ANII), Uruguay; Martín Galvalisi and José Pedro Prieto had postgraduate fellowships from ANII.

References

1. Hatsukami D, Fischman M (1996) Crack cocaine and cocaine hydrochloride. Are the differences myth or reality? J Am Med Assoc 276:1580–1588
2. Siegel RK (1982) Cocaine smoking. J Psychoactive Drugs 14(4):271–359
3. Castaño GA (2000) Cocaínas fumables en Latinoamérica. Adicciones 12:541–550
4. Paly D, Jatlow P, Van Dyke C, Jeri FR, Byck R (1982) Plasma cocaine concentrations during cocaine paste smoking. Life Sci 30:731–738
5. Castro de la Mata R (1989) Aspectos farmacológicos de la pasta básica de cocaína. In: León FR, Castro de la Mata R (eds) Pasta básica de cocaína: un estudio multidisciplinario. CEDRO. http://www.cedro.org.pe
6. United Nations Office on Drugs and Crime, UNODC 1400 Vienna, Austria. World Drug Report 2015. Available at: https://www.unodc.org/documents/wdr2015/World_Drug_Report_2015.pdf
7. Pascale A, Hynes M, Cumsille F, Bares C (2014) Consumo de pasta base de cocaína en América del Sur: revisión de los aspectos epidemiológicos y médico-toxicológicos. Comisión Interamericana para el Control del Abuso de Drogas. Organización de los Estados Americanos
8. Pascale A, Negrin A, Laborde A (2010) Pasta base de cocaína: experiencia del Centro de Información y Asesoramiento Toxicológico [Cocaine base paste: experience from the Montevideo Poison Control Center]. Adicciones 22(3):227–231
9. López Hill X, Prieto J, Meikle M, Urbanavicius J, Prunell J, Abin Carriquiry A, Umpiérrez E, Scorza C (2011) Coca-paste seized samples characterization: chemical analysis, stimulating effect in rats and relevance of caffeine as a major adulterant. Behav Brain Res 221:134–141
10. Prieto JP, Meikle M, López Hill X, Urbanavicius J, Abin Carriquiry A, Prunell G, Scorza MC (2012) Relevancia del adulterante activo cafeína en la acción estimulante de la pasta base de cocaína. Rev Psiquiatr Urug 76: 35–48
11. Jacob P 3rd, Lewis ER, Elias-Baker BA, Jones RT (1990) A pyrolysis product, anhydroecgonine methyl ester (methylecgonidine), is in the urine of cocaine smokers. J Anal Toxicol 14: 353–357
12. Kintz P, Cirimele V, Sengler C, Mangin P (1995) Testing human hair and urine for anhydroecgonine methyl ester, a pyrolysis product of cocaine. J Anal Toxicol 19:479–482
13. Toennes SW, Fandino AS, Kauert G (1999) Gas chromatographic-mass spectrometric detection of anhydroecgonine methyl ester (methylecgonidine) in human serum as evidence of recent smoking of crack. J Chromatogr B Biomed Sci Appl 735:127–132
14. Gomes EF, Lipaus IFS, Martins CW, Araújo AM, Mendonça JB, Pelição FS, Lebarch EC, de Melo Rodrigues LC, Nakamura-Palacios EM (2018) Anhydroecgonine methyl ester (AEME), a product of cocaine pyrolysis, impairs spatial working memory and induces striatal oxidative stress in rats. Neurotox Res 34(4):834–847
15. Garcia RC et al (2012) Neurotoxicity of anhydroecgonine methyl ester, a crack cocaine pyrolysis product. Toxicol Sci 128:223–234
16. Udo MSB, da Silva MAA, de Souza Prates S, Dal'Jovem LF, de Oliveira Duro S, Faião-Flores F, Garcia RCT, Maria-Engler SS, Marcourakis T (2021) Anhydroecgonine methyl ester, a cocaine pyrolysis product, contributes to cocaine-induced rat primary hippocampal neuronal death in a synergistic and time-dependent manner. Arch Toxicol 95(5): 1779–1791. https://doi.org/10.1007/s00204-021-03017-z
17. Broséus J, Gentile N, Esseiva P (2016) The cutting of cocaine and heroin: a critical review. Forensic Sci Int 262:73–83
18. Abin Carriquiry JA, Martínez Busi M, Galvalisi M, Minteguiaga M, Prieto JP, Scorza

C (2018) Identification and quantification of cocaine and active adulterants in coca-paste seized samples: useful scientific support to health care. Neurotox Res 34:295–304

19. Evrard I, Legleye S, Cadet-Taırou A (2010) Composition, purity and perceived quality of street cocaine in France. Int J Drug Policy 1(5): 399–406
20. Solimini R, Rotolo MC, Pellegrini M, Minutillo A, Pacifici R, Busardò FP, Zaami S (2017) Adulteration practices of psychoactive illicit drugs: an updated review. Curr Pharm Biotechnol 18:524–530
21. Kudlacek O, Hofmaier T, Luf A, Mayer FP, Stockner T, Nagy C, Holy M, Freissmuth M, Schmid R, Sitte HH (2017) Cocaine adulteration. J Chem Neuroanat 83–84:75–81
22. Galvalisi M (2016) Efectos neuroquímicos y comportamentales inducidos por pasta base de cocaína inhalada en ratas. Master thesis. Available in https://www.colibri.udelar.edu.uy/jspui/handle/20.500.12008/10168
23. Behrman A (2008) Luck of the draw: common adulterants found in illicit drug. J Emerg Nurs 34(1):80–82
24. Pawlik E, Mahler H (2011) Smoke analysis of adulterated illicit drug preparations. Toxichem Krimtech 78:200–210
25. Cole C, Jones L, Mcveigh J, Kicman A, Syed Q, Bellis M (2011) Adulterants in illicit drugs: a review of empirical evidence. Drug Test Anal 3: 89–96
26. Fukushima AR, Carvalho VM, Carvalho DG, Diaz E, Bustillos JO, Spinosa HD et al (2014) Purity and adulterant analysis of crack seizures in Brazil. Forensic Sci Int 243C:95–98
27. Prieto JP, Galvalisi M, López Hill X, Meikle MN, Abin Carriquiry JA, Scorza C (2015) Caffeine enhances and accelerates the expression of sensitization induced by coca paste indicating its relevance as a main adulterant. Am J Addict 24(5):475–481
28. Vanattou-Saifoudine N, McNamara R, Harkin A (2012) Caffeine provokes adverse interactions with 3,4-methylenedioxymethamphetamine (MDMA, 'ecstasy') and related psychostimulants: mechanisms and mediators. Br J Pharmacol 167:946–959
29. Morelli M, Simola N (2011) Methylxanthines and drug dependence: a focus on interactions with substances of abuse. Handb Exp Pharmacol 200:483–507
30. Gostic T, Klemenc S, Stefane B (2009) A study of the thermal decomposition of adulterated cocaine samples under optimized aerobic pyrolytic conditions. Forensic Sci Int 187:19–28
31. Galvalisi M, Prieto JP, Martínez M, Abin-Carriquiry JA, Scorza C (2017) Caffeine induces a stimulant effect and increases dopamine release in the nucleus accumbens shell through the pulmonary inhalation route of administration in rats. Neurotox Res 31(1): 90–98
32. Prieto JP, Scorza C, Serra GP, Perra V, Piras G, Galvalisi M, Abin Carriquiry JA, Valentini V (2016) Cocaine motivational value is enhanced when co-administered with caffeine: relevance of adulterants in reinforcement. Psychopharmacology 233(15–16):2879–2889
33. Muñiz JA, Gomez G, González B, Rivero-Echeto MC, Cadet JL, García-Rill E, Urbano FJ, Bisagno V (2016) Combined effects of simultaneous exposure to caffeine and cocaine in the mouse striatum. Neurotox Res 29(4): 525–538
34. Schwarzkopf N, Lagos P, Falconi A, Scorza C, Torterolo P (2018) Caffeine as an adulterant of coca paste seized samples: preclinical study on the rat sleep-wake cycle. Behav Pharmacol 29(6):519–529
35. Prieto JP, López Hill X, Urbanavicius J, Sanchez V, Nadal X, Scorza C (2020) Cannabidiol prevents the expression of the locomotor sensitization and the metabolic changes in the nucleus accumbens and prefrontal cortex elicited by the combined administration of cocaine and caffeine in rats. Neurotox Res 38(2):478–486
36. Sanders M (2007) Inhalation therapy: an historical review. Prim Care Respir J 16(2):71–81
37. Gossop M, Griffiths P, Powis B, Strang J (1992) Severity of dependence and route of administration of heroin, cocaine and amphetamines. Br J Addict 87:1527–1536
38. Tashkin DP (2001) Airway effects of marijuana, cocaine, and other inhaled illicit agents. Curr Opin Pulm Med 7(2):43–61
39. Pascale A, Negrín A, Ormaechea R (2011) Preliminary study of the effect on the lungs due to the consumption of cocaine base paste. Arch Bronconeumol 47(2):108
40. Samaha AN, Robinson TE (2005) Why does the rapid delivery of drugs to the brain promote addiction? Trends Pharmacol Sci 26: 82–87
41. Minogianis EA, Shams WM, Mabrouk OS, Wong JT, Brake WG, Kennedy RT, du Souich P, Samaha AN (2019) Varying the rate of intravenous cocaine infusion influences the temporal dynamics of both drug and dopamine concentrations in the striatum. Eur J Neurosci 50(3):2054–2064

42. Hueza IM, Ponce F, Garcia RC, Marcourakis T, Yonamine M, Mantovani Cde C, Kirsten TB (2016) A new exposure model to evaluate smoked illicit drugs in rodents: a study of crack cocaine. J Pharmacol Toxicol Methods 77:17–23
43. Galvalisi G, Prieto JP, Martinez M, Abin-Carriquiry JA, Scorza C (2015) Smoked cocaine: chemical analysis of seized samples and the role of caffeine in its central actions. In: IBRO 9th world congress, Rio de Janeiro. http://ibro.info/events/meetings/
44. National Research Council (US) Committee on Guidelines for the Use of Animals in Neuroscience and Behavioral Research (2003) Guidelines for the care and use of mammals in neuroscience and behavioral research. National Academies Press, Washington, DC
45. Herculiani PP, Pires-Neto RC, Bueno HM, Zorzetto JC, Silva LC, Santos AB, Garcia RC, Yonamine M, Detregiachi CR, Saldiva PH, Mauad T (2009) Effects of chronic exposure to crack cocaine on the respiratory tract of mice. Toxicol Pathol 37(3):324–332
46. Vanderschuren L, Pierce RC (2010) Sensitization processes in drug addiction. Curr Top Behav Neurosci 3:179–195
47. Di Chiara G, Imperato A (1988) Drugs abused by humans preferentially increase synaptic dopamine concentrations in the mesolimbic system of freely moving rats. Proc Natl Acad Sci U S A 85:5274–5278
48. Volkow ND, Fowler JS, Wang GJ, Baler R, Telang F (2009) Imaging dopamine's role in drug abuse and addiction. Neuropharmacology 56:3–8
49. Paxinos G, Watson C (2005) The rat brain in stereotaxic coordinates, 5th edn. Academic Press, Sydney

Chapter 8

Ethanol Intake on Preclinical Models: Methods of Assessment and Effects on Dopamine Signaling

Ricardo M. Pautassi, Aranza Wille-Bille, Macarena S. Fernández, and María Carolina Fabio

Abstract

Alcohol (ethanol) use is associated with significant health problems. This prompted efforts to recreate ethanol intake in murine animal models. The present chapter discusses the advantages and disadvantages of several methods to assess ethanol self-administration in these animal models. We then describe in detail a variant of a drinking-in-the-dark procedure, which exposes rats to ethanol drinking in a single bottle of ethanol (concentration range 5–10%), 2 h a day, three times a week, thus mimicking some aspects of the intermittent access procedure. The involvement of the dopaminergic system in ethanol self-administration, and some consequences of ethanol self-administration in this transmitter system, is then discussed.

Key words Ethanol, Self-administration, Rat, Mouse, Dopamine

1 Introduction

1.1 Why Is It Important to Measure Ethanol Intake? How Preclinical Models Recapitulate Its Features?

Alcohol (ethanol) use is a normative behavior, ubiquitously found in most countries worldwide [1–4]. Yet, for many its use is associated with significant health problems. This risk is higher in those who incur in heavy episodic drinking, a pattern involving the ingestion of four to five standard drinks in a relatively short timeframe (usually $\leq$2 h), or for those who began drinking or had their first drunkenness episode early in life (e.g., $\leq$14 years) [5]. Heavy episodic drinking is associated with greater likelihood of exhibiting several acute negative consequences of ethanol use (i.e., ethanol-related injuries or car crashes) as well as greater possibility of being diagnosed with an alcohol use disorder (AUD) [6]. On the other hand, individuals who begin drinking early in life usually exhibit more ethanol-related negative consequences, more use of other drugs (e.g., marihuana), and greater risk of being diagnosed with an AUD, when compared to peers that began drinking later in life [7, 8].

José Antonio Fuentealba-Evans and Pablo Henny (eds.), *Dopaminergic System Function and Dysfunction: Experimental Approaches*, Neuromethods, vol. 193, https://doi.org/10.1007/978-1-0716-2799-0_8,

Several studies focus on factors that increase or decrease (vulnerability and protective factors, respectively) the ingestion of ethanol. These factors, in turn, can be classified as distal or proximal, according to how close in time are to the actual consumption of ethanol. Distal factors are usually unspecific; that is, they affect a wide variety of behaviors or diseases. For instance, having suffered early life maltreatment (e.g., child abuse) puts the individual at risk for several problems, including ethanol abuse [9, 10]. Likewise, protective distal factors (e.g., paternal monitoring) deter from the likelihood of engaging into a wide array of negative situations (e.g., criminal behavior or risky sexual behavior [11, 12]), including substance use [13]. Proximal factors are positioned at the very end of the casual chain leading to ethanol use and are close in time to the actual drinking behavior. For instance, the motives to drink ethanol [14] are cognitions (i.e., "by drinking at this party I will enhance my abilities to interact with other people") that ignite the actual drinking behavior.

Some of the distal factors, such as personality factors (e.g., impulsivity, extroversion, sensation-seeking), exert their action via proximal factors [15]. Among those, drinking to reduce negative affect (i.e., so-called drinking to cope) or drinking to promote fitting in a social group ranks among the most studied proximal factors that regulate ethanol use [16]. Sometimes these proximal factors are thought to fully explain how distal factors exhibit their effect; this is called mediation in the clinical literature, whereas it is also possible that proximal factors moderate (e.g., exacerbate or inhibit) the effect of distal factors.

No matter how good these theoretical models are, they require actual measurement of ethanol ingestion. Measuring ethanol use in humans is, however, subjected to several methodological issues. As many behaviors, it is subjected to recall and telescopic biases, an issue that becomes more important when we ask about drinking in specific timeframes. The telescoping effect occurs when temporally remote events are perceived closer in time than they actually are, whereas more recent events are thought to have occurred at a more distant time [17].

The diversity of ethanol-containing beverages and modes of mixing them is also another hurdle. To reduce variability, the researchers have created standard guidelines to measure ethanol ingestion in so-called standard drinks or units. A standard drink usually contains about 14 g of absolute ethanol (albeit the definition can change in different countries or regions) and is the quantity that can be usually found in a can of beer of 330–350 cm^3, or in a serving of wine or liquor. This information, along with representative images of beverages, can be provided to study participants to facilitate their reporting on actual, daily life, drinking patterns. Researchers have also created "ethanol consumption diaries" [18], allowing participants to register their ethanol ingestion in a daily basis, thus reducing recall and telescopic biases.

The use of animal models of ethanol intake [19] provides greater accuracy in the measurements and the possibility of imposing a variety of experimental manipulations, including putative pharmacological or environmental therapeutics, or the use of unique biological specimens, such as rats or mice genetically modified to express (or not) proteins of interest [20]. Of course, this experimental complexity comes at the cost of a much reduced external validity. Despite this drawback, the field of animal models of ethanol ingestion or effects has provided significant information toward understanding of the neurobiological consequences of acute and chronic ethanol use [21].

1.2 Aim of the Chapter

In this chapter we will review the different procedures available to model ethanol intake in rodents (mainly in rats) and then will describe, with methodological detail, a variant of the drinking-in-the-dark (DID) procedure used in our lab [22], useful to assess age-related differences in ethanol intake and neurobiological determinants of this behavior. We will then focus on the effects of ingested alcohol on transmitters systems, notably dopamine (DA).

1.3 Different Procedures to Model Ethanol Intake in Rodents

Several studies administer ethanol to rats or mice via intraperitoneal or intragastric procedures, or via vapor. These passive exposure models are well tailored to assess the long-term consequences of ethanol exposure. It has been shown, for instance, that intermittent exposure to binge-like ethanol dosing during the adolescence of the rat (e.g., 2.5 g/kg every other day between postnatal day (PD) 25 and 42) induces alterations in glutamatergic and dopaminergic transmission and in the neural pruning of the latter system [23] and alters the functionality of the hypothalamic-pituitary-adrenal axis, responsible for the stress response [24]. These studies are useful to demonstrate persistent changes associated with ethanol exposure but employ routes of administration that exert behavioral and neural effects quite different from those induced by the self-administration. For example, a seminal study indicated that self-administration, but not injections, of ethanol increased glucose accumulation in brain areas associated with positive reinforcement [25].

Perhaps the simplest way to measure ethanol ingestion in rodents is to make the drug available in a bottle, in a 24-h basis. When the drug is the only fluid available, this setup is referred to as "forced access" and can be used to promote dependence or a high level of drug exposure. For instance, in a series of studies [26–28], pregnant dams were given ethanol as the sole fluid 22 h a day (10% v/v, sweetened with 64 mg/L sucralose), throughout pregnancy and during the first week of nursing. Two hours per day the dams were given water, to prevent dehydration. The dams drank, on average, 6.6 ± 0.4 g/kg of ethanol a day, and this pattern was associated with behavioral and neural alterations in the offspring.

The rats derived from the ethanol-exposed dams exhibited heightened ethanol consumption themselves, which was coupled with greater persistence of aversively motivated memories and, relevant for the purposes of the present chapter, heightened levels of 3,4-dihydroxyphenylacetic acid (DOPAC, a metabolite of dopamine) in the nucleus accumbens, a brain area associated with the processing of reward [26, 29].

The offspring of the studies described in the preceding paragraph [26–28] were not tested in a forced choice procedure; instead, they had the possibility to choose between an ethanol and a water bottle. This so-called two-bottle choice procedure is widely popular as it recapitulates the voluntary nature of ethanol self-administration in humans. Its use in rodents can be traced to an early paper [30] that tested how procedural factors affected ethanol intake in rats. The report indicated that progressive exposure to increasing concentration of ethanol (i.e., "acclimation") and the possibility to actively choice ethanol over water had facilitating effects upon ethanol acceptance. Yet, it was the availability of ethanol in an intermittent basis (as opposed to a continuous 24-h exposure) that had the biggest impact on ethanol intake and preference vs. water. The procedure was then recaptured by an influential study [31], which showed that rats exposed to 20% ethanol in intermittent 24-h sessions initially mostly rejected the ethanol solution. Specifically, in the first sessions, these rats drank circa 2 g/kg/day of ethanol and exhibited 18% preference for ethanol vs. water. This implies that of the total amount of fluid consumed (i.e., ethanol + water, measured in ml), less than 20% had been obtained from the ethanol bottle. As the intake sessions progressed, however, they exhibited high levels of ethanol intake and preference (approximately 5.25 g/kg/day and 50% of predilection vs. water). Intriguingly, treatment with naltrexone or acamprosate (drugs of choice for the clinical treatment of AUD) significantly decreased ethanol consumption in rats exposed to the intermittent access to 20% ethanol but left mostly unaffected the ethanol drinking in counterparts given continuous access to 10 or 20% ethanol.

It is not clear why the intermittent schedule yields better outcomes, in terms of increased ethanol consumption, when compared to the continuous exposure. It is possible that the intermittency (i.e., the day off) allows development of withdrawal or hangover effects [32]. This, in turn, favors drinking to cope with the negative affect of hangover [32]. Also, intermittent and continuous exposure to a given stimulus is usually associated with sensitization and habituation, respectively, of the behavioral responses elicited by that stimulus. It is notable that intermittent ethanol is preferred, over continuous exposure, in studies that aim at promoting ethanol-induced behavioral sensitization. The latter is the progressive enhancement in ethanol-induced motor activation, observed after repeated ethanol administration, which is usually considered a proxy for the appetitive effects of the drug [33].

Before the intermittent ethanol access procedure became a fixture of the preclinical work, another widely used method was to acclimate the rats to drink alcohol by offering a mixture of alcohol and sucrose, with the latter solution being progressively removed from the mixture as sessions progressed. This "sucrose-fading" or sucrose-substitution initiation procedure [34] received, however, several critics, raising doubts on what was the driving force of the drinking. Specifically, it is possible that the rats acquire sucrose-ethanol association that promote greater ethanol acceptance for reasons unrelated to the pharmacological effects of the drug but instead related to hedonic shifts toward the taste of ethanol. A study, however, critically appraised this possibility [35]. After the initial training in which the sucrose mixture was faded, the rats received ethanol intubations closely followed by the administration of the prototypical emetic agent lithium chloride. The pairing, of the postingestive effects of ethanol and the illness induced by lithium chloride injection, led to a significant reduction in seeking behavior for ethanol [35]. This result is consistent with the notion that the pharmacological effects of ethanol are responsible for maintaining the ethanol drinking behavior in the sucrose-fading paradigm.

The sucrose-fading procedure has been also modified to include operant procedures. By operant we imply that the rats are not just given the alcohol to drink but, instead, have to perform a previously defined, arbitrary, behavior such as pressing a lever or nose poke a hole, to have access to the liquid. The operant version of the sucrose-fading procedure involves completing a behavioral requirement (for instance, pressing a lever four times) to access a tube, from which the animal can drink for 20 min [34]. The advantage of this procedure should be obvious: it combines in a single task appetitive and consummatory phases of the ethanol self-administration behavior.

Operant ethanol self-administration is also widely employed and, unlike regular access to ethanol from bottle, can be used to assess discrimination of the interoceptive context induced by ethanol, "willingness to work" for the ethanol concentrations, and cue-induced relapse. We will briefly review the last two applications. Access to ethanol in an operant self-administration procedure is dependent upon completion of a previously defined, arbitrary, behavior, say pressing a lever. Moreover, those requirements can be upregulated, for instance, after a day of training the rat, or mouse can be asked to press twice or thrice to have access to the ethanol solution. By checking the "breaking point" at which the behavior is no longer emitted, the researchers can assess the effects of treatments upon the motivation to work for ethanol. A study that used this so-called progressive ratio [36] found that a selective cannabinoid receptor antagonist lowered the breaking point to self-administer ethanol. The researchers also found that the rats decreased responding as ethanol access was withheld during extinction sessions. Yet the rats exhibit renewed interest in self-administering ethanol after being exposed to an odorant cue that had signaled ethanol access.

The latter study [36] illustrates the methodological subtleties allowed by the use of operant ethanol self-administration. Also interesting is that the latter, with some modifications, can be applied in infant or adolescent rats or mice, to assess ontogenetic patterns of ethanol intake and acceptance. In those cases, an intraoral cannula is implanted in the animal's cheek and used to deliver the solution (for a review, see [37]), upon completion of a certain behavior, such as nose poke or depression of a floor plate. In one of those studies [38], the researchers put 1-day-old rats in a restrictor vest. The pups could activate, via head or forelimb movements, a touch-sensitive sensor, which in turn caused the intraoral delivery of milk or ethanol. Head or forelimb responses for ethanol or milk were above 100 in a 15-min session, significantly higher than the 50–60 responses emitted in yoked control rats, which received the reinforcer no matter the emission of the responses.

The intraoral cannula can also be connected to a pump that delivers the ethanol solution in a pulsating manner, independent of the execution of a specific behavior. This procedure is usually employed to assess orofacial responsiveness to ethanol [39]. Rats emit a stereotypical set of facial responses when stimulated with appetitive or aversive taste stimuli. Appetitive stimuli induce, among other, tongue protrusions and paw licking, whereas the aversive stimuli induce head shaking, gaping, and forelimb flailing. Ethanol, a complex taste with sweet and bitter components that at high concentrations also stimulates the trigeminal nerve, is associated with a mix of appetitive and aversive orofacial responses [40]. In infant rats or mice, the continuous infusion of ethanol into the oral cavity (via an intraoral cannula) serves to mimic the single-bottle ethanol intake tests conducted in adults. Although seemingly forced, the rats have control over the amount of ethanol ingested, as they can emit passive drips, head shaking, or other rejective behaviors, if needed [41].

Ethanol exposure in infancy is highly prevalent and exerts lingering consequences. Ethanol tasting was shown to occur in circa 60% of Argentinean children aged 8 to 12 years [42], findings congruent with those reported in the USA [43]. Moreover, those children or adolescent that report an early age of first of contact with ethanol (usually defined as having a full drink at age 14–15 or before) are more likely to engage in hazardous drinking behaviors, or to experience ethanol-related disorders later on [44]. Overall, these findings indicate that assessing ethanol acceptance in infant rats or mice is of outmost importance to understand the etiology and trajectories of ethanol use.

One problem of the intraoral infusion method is that, when applied to animals older than 21 days old, the placement of the cannula requires invasive surgery with administration of anesthesia (described in [45]). This prompted the development of methods to measure ethanol self-administration across the ontogeny, with minimal variations in methodology. One of those had its heyday during the early 2000s and was known as consume off the floor (COF). The test is conducted in a terrarium featuring a floor made of stone. Ethanol-soaked wipes line the stone, which is heated via an external source. The rats (to our knowledge the procedure was never applied to mice) are then placed into the terrarium and can lick the ethanol directly from the floor. The procedure has been shown to induce relatively high levels of ethanol ingestion, particularly in young females, who exhibited blood ethanol concentrations of circa 300 mg/dl, when stimulated with 30% at PD 22. Levels of ingestion decrease in a roughly monotonic basis as age of testing increases, a finding consistent with the notion that predisposition for ethanol intake peaks at adolescence [46].

The COF is particularly suited to analyze the effects of early exposure to ethanol upon later acceptance of the drug. Sanders and Spear [47] gave infant rats access to ethanol or just water, via COF or via intubations, every other day from postnatal day 2 to 10. When tested at postnatal day 12, those pups that had been exposed to the drug exhibited heightened ethanol drinking than those treated with vehicle. Despite its advantages, the COF has some limitations. For instance, it is possible that a significant fraction of the blood ethanol concentration achieved by the rats is due to inhalation of the vapor or transdermal absorption [46].

We will now move on to describe in detail a variant of a DID procedure we regularly use in our lab. The DID [48] is a variant of a restricted access ethanol intake procedure, in which rats or mice have access to ethanol for a limited time every day, anywhere between 15 min and 2 h, usually beginning 2 to 3 h into the dark phase of the circadian cycle [49]. An advantage of these restricted access procedures is that they make more likely that the levels of ethanol ingested are pharmacologically relevant (usually attaining or surpassing 100 mg/dl/2 h) and thus may affect dopamine or other neurotransmitter signaling.

The DID variant we employ exposes Wistar rats to ethanol drinking in a single bottle of ethanol (concentration range 5–10%), 2 h a day, three times a week, thus mimicking some aspects of the intermittent access procedure. As it will be described, drinking takes place between 700 and 900 PM, following preliminary data indicating that this strain shows, in a vivarium with a 12-h light period with lights on at 700 AM, a peak of alcohol drinking shortly after vivarium lights go off and, then, another a few hours before lights are turned on.

2 Assessing Ethanol Intake in Wistar Rats via an Adapted Version From the DID Paradigm: Materials

The procedure we will describe has been tested in rats, adolescent or adults, yet it should be possible to run mice as well on it. No special housing or handling of the rats is necessary prior to the procedure (*see* **Note 1**). During the tests each home cage is divided in two halves via a divider made of white high-impact acrylic. Thus, the rats should be able to smell and hear, but not touch, each other. During the test the regular lid of each cage is replaced by a special lid featuring two opposite feeder-like sections, which serve to put the bottles and food pellets. A representative image of this setup can be observed in Fig. 1.

Ethanol is provided in a solution made of tap water and 8 or 10% 96-proof ethanol, previously weighed in a scale. The scale, depending on the precision, can be also used to weigh the bottles that will be employed during the tests. The ethanol or water solutions are poured into standard 100-ml glass bottles capped with open-ended (OE) metal spouts (*see* **Note 3**).

Fig. 1 (**a**) Representative photograph depicting the arrangements made to a regular rat home chamber, during the variant of the drinking-in-the-dark procedure described in the present chapter. The procedure exposes rats (mice could be tested as well) to ethanol drinking in a single bottle of ethanol (concentration range 5–10%), 2 h a day, three times a week, thus mimicking some aspects of the intermittent access procedure. During the tests, each home cage is divided in two halves via a divider made of white high-impact acrylic, and the regular lid of each cage is replaced by a special lid featuring two opposite feeder-like sections, which serve to put the bottles and food pellets. (**b**) Same as (**a**), yet from a lateral view

3 Assessing Ethanol Intake in Wistar Rats via an Adapted Version of the DID Paradigm: Procedure

The procedure employs water restriction to motivate the rats to drink ethanol. The deprivation is relatively mild (i.e., rats are given 50% of the water they would have drank under ad libitum conditions) and helps avoid other alternatives to motivate drinking, such as a sucrose-substitution procedure or exposing the rats for several weeks to ethanol. Thus, on day 1 of the procedure, the rats, up to this moment housed in same-sex pairs, are weighed at about 200 PM and moved to separates halves of the home cage, which is now divided in two halves and equipped with a special lid. Each rat is provided a water bottle previously weighed to the nearest 0.01 g, and food is provided ad libitum. The next day (Experimental day 2), the water bottle is removed at about 2 PM and weighed, so as to provide a measure of ad libitum water intake, per each individual rat. The animals are immediately weighed and provided a new water bottle, containing 50% of the water they had consumed the day before (corrected by weight of the animal, *see* **Note 4**).

On Experimental day 3, at about 630 PM, the rats are weighed. By 645 PM—i.e., about 15 min before the lights go off—the water bottle (that should be empty by now, as the rats had been offered only 50% of their ad libitum water requirement) is replaced by an ethanol-containing bottle which had been previously weighed (pre-intake score). Two hours later the ethanol bottle is removed and weighed again (post-intake score) and replaced by water bottles, and the divider is removed to resume pair-housing condition. The difference in weight of the ethanol bottle (i.e., pre-intake score minus post-intake score) provides a raw measure of ethanol intake than can be used to calculate absolute ethanol intake (g/kg) scores. The rats then have ad libitum access to food and water until 2 PM of the next day (Experimental day 4), in which they are again provided 50% of the water volume they would usually consume, given their weight. The next day (Experimental day 5) is a recurrence of the procedures described for Experimental day 3. The procedure can be now be repeated as many times as needed, albeit in our laboratory we tend to expose the rats to ethanol every other day on weekdays (i.e., Monday, Wednesday, Friday).

Some methodological subtleties are worth mentioning. Leakage can often take place yet can be accounted for by placing an ethanol bottle in an empty cage and measuring its content before and after the 2-h period. The difference between the pre- and the post-testing period can then be used to adjust intake values in the experimental cages. The addition of photoelectrical cells or lickometers to measure each individual drinking bout can be also a welcomed addition [50]. Acquiring video files of the behavior can be a cost-effective alternative.

Our published findings [22] indicate that adolescent rats exposed to this procedure consume approximately 2.5–4.0 g/kg of ethanol when tested across 3 weeks, significantly more than adult counterparts who rarely exceed 1.5 g/kg. Moreover, the ethanol drinking of the adolescents can be blocked by the administration of the opioid antagonist naloxone prior to the test, which provides translational value to the model [22].

4 Ethanol Self-Administration Modulates Dopaminergic System

Among the neurotransmitter systems, dopamine is an important substrate for the actions of ethanol in the brain. The activation of dopaminergic circuits has been implicated in several physiological functions, including feeding, locomotion and coordinated movements, hormone secretion, behavioral activation, and impulse control [51, 52], and early studies made it clear dopamine's mediational role in the reinforcing properties and rewarding effects of abuse drugs [53–57]. A wide range of data suggests that alterations of this system are associated with compulsive alcohol seeking and consumption [58–60] and may underlie the predisposition to and the development of ethanol abuse [61, 62].

Electrophysiological and microdialysis studies have revealed that ethanol exerts its action on DA systems by modulating the firing activity of DA neurons in VTA, both in vivo [63–65] and in vitro [66–68]. Some regional differences have been recently observed in the electrophysiological activity of VTA after systemic alcohol administration in freely moving rats. In one study ethanol induced a more effective stimulation in DA neurons of the lateral VTA than in those situated in medial VTA, as found in both firing rate and in burst activity [69]. The ethanol-induced increase in VTA activity consequently increases the concentration of dopamine and its metabolite (homovanillic acid and 3,4-dihydroxyphenylacetic acid) in several target areas, such as the nucleus accumbens (AcbN), striatum, amygdala, hippocampus, and medial prefrontal cortex (mPFC) [70–75]. This effect has been reported after intraperitoneal or intravenous (systemic) injections of moderate doses of ethanol [76–79] or directly into VTA [80, 81].

Voluntary consumption of ethanol in rodents enhances dopamine release in AcbN [82, 83]. This facilitatory effect of ethanol does not seem to depend on a specific route of self-administration, as it is found after different protocols (e.g., operant self-administration involving lever press, nose poke, or wheel turn, among other actions) and—perhaps more important—is also observed after self-administration of the drug directly into VTA [84]. Another work further dissected the area associated with this effect, as the DA release was found after self-administration into a neuronal subpopulation of the posterior, but not the anterior, VTA [85].

It is important to assess if these effects have translational value. In humans, the effects of self-administered ethanol have been the subject of countless experiments (e.g., to assess its detrimental effects upon car-driving abilities). Yet it is notable that just a few of these directly analyzed the effects of ethanol on the dopamine system. One of those studies [86] reported greater DA release in AcbN after a single ethanol administration, using positron emission tomography. However, a hypodopaminergic state has been reported after chronic ethanol exposure followed by a withdrawal phase in rodents [87, 88], or in patients with AUD [89, 90]. At least in animal models, it seems that the greater willingness to drink ethanol, usually observed after a withdrawal period [91], is driven by the need to compensate a DA deficit associated with long-term exposure to the drug [89, 91]. Notably, there seems to be certain ontogenetic periods in which the chronic exposure to the drug is associated with greater, rather than diminished, DA activity. Specifically, studies with adolescent rats [23, 92] indicate heightened basal DA levels in the AcbN after i.p. ethanol administration [3 g/kg, 2 days on, 2 days off in [23]; 0.75 g/kg daily from postnatal day 30 to 50 in [92]]. Notably, ethanol self-administration at roughly the same ages (i.e., PDs 30 to 60) was associated with heightened extracellular levels of DA, as measured in AcbN via microdialysis, in response to a challenge dose of i.p. ethanol (2.5 g/kg). The duration of the DA peak at adulthood was also longer in those rats that had been exposed to ethanol as adolescents, than in controls peers [93]. The latter study tested ethanol's effects in selectively bred, alcohol-preferring (P) rats, which have been widely employed to dissect ethanol-DA interactions [94].

Several variations in DA basal levels have been observed between strains selectively bred for high or low ethanol preference. The latter involve rats that are mated as a function of exhibiting high or low intake and preference for free choice drinking of ethanol. A study [95] showed that the P rats, that usually drink around five times more ethanol than non-preferring (NP) rats in a free-choice protocol, exhibit lower basal DA concentration in mPFC, compared with outbred Wistar rats, and also exhibit a more potent releasing of DA in AcbN after the self-administration of ethanol [82]. Other study [96], in turn, observed in another alcohol-preferring strain (Sardinian, Sp) higher basal dopamine content at AcbN, when compared to either non-preferring Sardinian counterparts (Snp) or non-selected Wistar rats.

Interestingly, several studies reported that the increase in dopamine concentration in AcbN during ethanol self-administration sessions can occur before a pharmacologically relevant brain ethanol concentration is achieved. Specifically, these studies report two phases of DA release in AcbN, in subjects with a history of operant self-administration. The first takes place after mere exposure to the self-administration environment, before the actual access to the ethanol solution, while the other occurs during the ethanol access, yet sometimes even before peak brain ethanol concentrations are achieved [97–100].

These results, which suggest that a significant fraction of ethanol-induced DA release at AcbN is driven by classical conditioning involving the contextual cues surrounding the training, are consistent with studies conducted in P rats, in which an increase in DA concentration at AcbN occurs during anticipation of self-administration [101]. It is interesting that similar context-driven DA release was observed in the mPFC of animals exposed for a week to a sweetened ethanol solution, yet not in response to a sweetened non-ethanol solution [102]. This supports the notion that ethanol's pharmacological effects are key to support these putative conditioned effects, also shown in [103]. Overall, these results suggest that the first peak in dopamine extracellular concentration in different mesolimbic brain areas (specially in AcbN and mPFC) is triggered by sensory cues that were present during training and predict the rewarding effects of ethanol [82, 103–105].

5 DA Receptors, Agonists, and Antagonists

Once dopamine is released in the neuron terminals, it interacts with G-protein membrane receptors that mediate the response in target neurons. The activation of dopamine receptors (DR) involves alterations in second messengers, through the cascade of cAMP-dependent protein kinase (PKA), leading to changes in the permeability of ion channels and membrane potential. The most abundant DA receptors are D1-like and D2-like receptors, and their activation seems to have opposite effects. D2 are expressed presynaptically, and its activation decreases DA synthesis and release [106], therefore inhibiting DA neurotransmission. This effect is supported by studies where local blockade of D2 receptors modulates VTA neuron firing rates [107]. On the other hand, D1-like receptors, expressed in target neurons (postsynaptically), have lower affinity to DA and are activated by high concentrations of DA [51, 52]. It is important to note that D2, postsynaptic, positive cells are also present in the striatum.

It has been suggested that alterations in the expression, function, or activity of D1 or D2 DA receptors predispose subjects to AUD. If so, these receptors might represent an efficient target in the treatment of AUD [108]. Epidemiological studies showed that patients diagnosed with AUD exhibit lower levels of DRD2 in the brain [109]. Generally, it has been shown that microinjections of DA agonists directly into AcbN enhance ethanol intake, whereas the use of antagonists decreases the consumption of the drug [110, 111]. However, some studies showed that some D1 and

D2 receptor agonists (e.g., SKF 38393, SKF 82958, quinpirole) reduce ethanol intake [108] or preference [112], in a dose-response fashion.

Altogether, these results support the notion that plastic changes in the mesocorticolimbic dopaminergic system play an important role in the reinforcing effects of ethanol. It is important to remark, however, that that DA modulates ethanol's reinforcing effects, seeking and intake, by interacting with several other transmitters, notably the endogenous opioid system, the GABAergic system, and the serotoninergic system.

Serotonin (5-HT) is a monoamine neuromodulator involved in the early formation of brain circuits and in the regulation of several physiological functions, including mood and anxiety-related behaviors, and it is also involved in AUD [113, 114]. Decreased 5-HT in dorsal raphe nucleus seems to induce impulsive or compulsive behavior in mice, which in turn are associated with drug-seeking behavior [115]. Moreover, 5-HT neurons encode reward signals and interact with glutamatergic and dopamine neurons in the mesolimbic dopaminergic pathway [116], modulating rewarding effects of drugs and natural reinforcers.

6 Concluding Comments

The different methods to assess ethanol self-administration in rodents have, compared to methods that administer the drug via injections or intubations, a set of disadvantages. Many have substantial technical complexity, and all of them share the impossibility to exactly titrate level of exposure across individuals. Yet, none of them outweigh the ecological richness of modeling the route of administration used by humans as well as its voluntary nature. Thus, self-administration methods should be considered as the ultimate benchmark upon which treatments or factors that affect liability for AUD should be tested. The DID variant we described in the present chapter is a simple and cost-effective alternative to measure ethanol intake or assess the effectiveness of treatments that can reduce such behavior.

7 Notes

1. The rats are, as described, briefly isolated during the tests, which take place in the home cages. Thus, we recommend that weaning is performed at postnatal day 21 and followed by same-sex pair housing. A few days before commencement of the intake tests, we recommend that each cage is divided in half, via a plexiglass or high-impact acrylic divider, so that the rats are briefly exposed to the same confinement conditions they

will experience during the tests. Also, handling is indeed required if the procedure involves i.p. or i.g. administration of substances prior to the test itself.

2. The divider can be made of plexiglass; in such case, it should be made opaque via a sticky tape and could be provided with holes covered with wire mesh. The latter arrangement reduces the social isolation of the rats. Special care should be taken to secure the cage lid.
3. We tend to use open-ended (OE) metal spouts, connected to a rubber base, which is then pressed into the bottle. Yet there are plenty of alternatives. Sipper tubes can be used instead of bottles, and these have the advantage of, most of the time, having a stamp with consumption levels marked. Another decision is to use either OE or ball bearing-containing ball-point (BP) sipper tubes or metal spouts. BP are much likely to get blocked, particularly if sugar-based solutions are used, yet it has been shown that its use can dramatically increase the level of ethanol ingestion, particularly in adolescent rats [117].
4. If on Experimental day 1 the rat weighed 750 and 775 grams, respectively, and on Experimental day 1 drank 25 ml of water, the rat should be given a bottle containing 12.92 ml of water (i.e., [775*25/750]/2). Historical values (i.e., weight and drinking scores from previous batches of animals) can be also used to calculate the amount of water to be offered.

Acknowledgments and Funding Sources

Funding was provided by grants PICT 2018-00597 to RMP and SECYT Consolidar 2018 to RMP and MCF. We would like to thank Mr. Leonardo Marengo for the input provided concerning the chapter.

References

1. Sornpaisarn B, Shield K, Manthey J, Limmade Y, Low WY, Van Thang V et al (2020) Alcohol consumption and attributable harm in middle-income South-East Asian countries: epidemiology and policy options. Int J Drug Policy 83:102856
2. Cremonte M, Biscarra MA, Conde K, Cherpitel CJ (2018) Epidemiology of alcohol consumption and related problems in Latin American countries: contributions of psychology. Int J Psychol 53(4):245–252
3. Rehm J, Room R, van den Brink W, Jacobi F (2005) Alcohol use disorders in EU countries and Norway: an overview of the epidemiology. Eur Neuropsychopharmacol J Eur Coll Neuropsychopharmacol 15(4): 377–388
4. Popova S, Rehm J, Patra J, Zatonski W (2007) Comparing alcohol consumption in central and eastern Europe to other European countries. Alcohol Alcohol 42(5): 465–473
5. Morean ME, Kong G, Camenga DR, Cavallo DA, Connell C, Krishnan-Sarin S (2014) First drink to first drunk: age of onset and delay to intoxication are associated with adolescent alcohol use and binge drinking. Alcohol Clin Exp Res 38(10):2615–2621

6. Stolle M, Sack PM, Thomasius R (2009) Binge drinking in childhood and adolescence: epidemiology, consequences, and interventions. Dtsch Arztebl Int 106(19):323–328
7. Vera B, Pilatti A, Pautassi R (2019) ELSA cohort 2014: association of age of first drink and progression from first drink to drunkenness on alcohol outcomes in Argentinean college freshmen. Am J Drug Alcohol Abuse 46: 1
8. Pilatti A, Read JP, Pautassi RM (2017) ELSA 2016 Cohort: alcohol, tobacco, and Marijuana use and their association with age of drug use onset, risk perception, and social norms in Argentinean College Freshmen. Front Psychol 8:1452
9. Schuckher F, Sellin T, Fahlke C, Engstrom I (2018) The impact of childhood maltreatment on age of onset of alcohol use disorder in women. Eur Addict Res 24(6):278–285
10. McPhie ML, Weiss JA, Wekerle C (2014) Psychological distress as a mediator of the relationship between childhood maltreatment and sleep quality in adolescence: results from the Maltreatment and Adolescent Pathways (MAP) longitudinal study. Child Abuse Negl 38(12):2044–2052
11. Dittus PJ, Michael SL, Becasen JS, Gloppen KM, McCarthy K, Guilamo-Ramos V (2015) Parental monitoring and its associations with adolescent sexual risk behavior: a meta-analysis. Pediatrics 136(6):e1587–e1599
12. Wertz J, Nottingham K, Agnew-Blais J, Matthews T, Pariante CM, Moffitt TE et al (2016) Parental monitoring and knowledge: testing bidirectional associations with youths' antisocial behavior. Dev Psychopathol 28(3): 623–638
13. Tornay L, Michaud PA, Gmel G, Wilson ML, Berchtold A, Surís JC (2013) Parental monitoring: a way to decrease substance use among Swiss adolescents? Eur J Pediatr 172(9): 1229–1234
14. Cox WM, Klinger E (1988) A motivational model of alcohol use. J Abnorm Psychol 97(2):168–180
15. Kenney SR, Anderson BJ, Stein MD (2018) Drinking to cope mediates the relationship between depression and alcohol risk: different pathways for college and non-college young adults. Addict Behav 80:116–123
16. Wardell JD, Kempe T, Rapinda KK, Single A, Bilevicius E, Frohlich JR et al (2020) Drinking to cope during COVID-19 pandemic: the role of external and internal factors in coping motive pathways to alcohol use, solitary drinking, and alcohol problems. Alcohol Clin Exp Res 44:2073–2083
17. Janssen SMJ, Chessa AG, Murre JMJ (2006) Memory for time: how people date events. Mem Cogn 34(1):138–147
18. Lindner P, Johansson M, Gajecki M, Berman AH (2020) Using alcohol consumption diary data from an internet intervention for outcome and predictive modeling: a validation and machine learning study. BMC Med Res Methodol 20(1):111
19. Koob GF (2005) Development of animal models of excessive drinking, protracted abstinence, and cue-induced relapse for medications development. Neuropsychopharmacology 30:S40–S
20. Gallegos S, San Martin L, Araya A, Lovinger DM, Homanics GE, Aguayo LG (2021) Reduced sedation and increased ethanol consumption in knock-in mice expressing an ethanol insensitive alpha 2 subunit of the glycine receptor. Neuropsychopharmacology 46(3): 528–536
21. Bell RL, Hauser SR, Liang T, Sari Y, Maldonado-Devincci A, Rodd ZA (2017) Rat animal models for screening medications to treat alcohol use disorders. Neuropharmacology 122:201–243
22. Salguero A, Suarez A, Luque M, Ruiz-Leyva-L, Cendán CM, Morón I et al (2020) Binge-like, naloxone-sensitive, voluntary ethanol intake at adolescence is greater than at adulthood, but does not exacerbate subsequent two-bottle choice drinking. Front Behav Neurosci 14:50
23. Pascual M, Boix J, Felipo V, Guerri C (2009) Repeated alcohol administration during adolescence causes changes in the mesolimbic dopaminergic and glutamatergic systems and promotes alcohol intake in the adult rat. J Neurochem 108(4):920–931
24. Logrip ML, Rivier C, Lau C, Im S, Vaughan J, Lee S (2013) Adolescent alcohol exposure alters the rat adult hypothalamic-pituitary-adrenal axis responsiveness in a sex-specific manner. Neuroscience 235:174–186
25. Porrino LJ, Williams-Hemby L, Whitlow C, Bowen C, Samson HH (1998) Metabolic mapping of the effects of oral alcohol self-administration in rats. Alcohol Clin Exp Res 22(1):176–182
26. Plaza W, Gaschino F, Gutierrez C, Santibanez N, Estay-Olmos C, Sotomayor-Zarate R et al (2020) Pre- and postnatal alcohol exposure delays, in female but not in male rats, the extinction of an auditory fear conditioned memory and increases alcohol

consumption. Dev Psychobiol 62(4): 519–531

27. Fernandez MS, Carrizo J, Plaza W, Haeger P, Pautassi RM (2019) Prenatal ethanol exposure potentiates isolation-induced ethanol consumption in young adult rats. Alcohol 75:39–46
28. Di Doi P, Fernandez MS, Haeger Soto P, Pautassi RM (2019) Effects of isolated or enriched housing at adolescence upon ethanol intake and anxiety responses, in rats exposed to prenatal ethanol. J Fetal Alcohol Spectrum Risk Prevent 1(1):e1–e12
29. Bardo MT (1998) Neuropharmacological mechanisms of drug reward: beyond dopamine in the nucleus accumbens. Crit Rev Neurobiol 12(1–2):37–67
30. Wise RA (1973) Voluntary ethanol intake in rats following exposure to ethanol on various schedules. Psychopharmacologia 29(3): 203–210
31. Simms JA, Steensland P, Medina B, Abernathy KE, Chandler LJ, Wise R et al (2008) Intermittent access to 20% ethanol induces high ethanol consumption in Long-Evans and Wistar rats. Alcohol Clin Exp Res 32(10):1816–1823
32. Varlinskaya EI, Spear LP (2004) Acute ethanol withdrawal (hangover) and social behavior in adolescent and adult male and female Sprague-Dawley rats. Alcohol Clin Exp Res 28(1):40–50
33. Camarini R, Pautassi RM (2016) Behavioral sensitization to ethanol: neural basis and factors that influence its acquisition and expression. Brain Res Bull 125:53–78
34. Samson HH, Sharpe AL, Denning C (1999) Initiation of ethanol self-administration in the rat using sucrose substitution in a sipper-tube procedure. Psychopharmacology 147(3): 274–279
35. Samson HH, Cunningham CL, Czachowski CL, Chappell A, Legg B, Shannon E (2004) Devaluation of ethanol reinforcement. Alcohol 32(3):203–212
36. Economidou D, Mattioli L, Cifani C, Perfumi M, Massi M, Cuomo V et al (2006) Effect of the cannabinoid CB1 receptor antagonist SR-141716A on ethanol self-administration and ethanol-seeking behaviour in rats. Psychopharmacology 183(4): 394–403
37. Pautassi RM, Miranda-Morales RS, Nizhnikov M (2015) Operant self-administration of ethanol in infant rats. Physiol Behav 148: 87–99
38. Bordner KA, Molina JC, Spear NE (2008) Analysis of ethanol reinforcement in 1-day-old rats: assessment through a brief and novel operant procedure. Alcohol Clin Exp Res 32(4):580–592
39. Pautassi RM, Arias C, Molina JC, Spear N (2008) Domperidone interferes with conditioned disgust reactions but not taste avoidance evoked by a LiCl-paired taste in infant rats. Dev Psychobiol 50(4):343–352
40. Kiefer SW, Badia-Elder NE (1997) Taste reactivity in high-alcohol-sensitive and low-alcohol-sensitive rats. Alcohol 14(3):225–229
41. Pautassi RM, Godoy JC, Spear NE, Molina JC (2002) Early responsiveness to stimuli paired with different stages within the state of alcohol intoxication. Alcohol Clin Exp Res 26(5): 644–654
42. Pilatti A, Godoy JC, Brussino S, Pautassi RM (2013) Underage drinking: prevalence and risk factors associated with drinking experiences among Argentinean children. Alcohol 47(4):323–331
43. Donovan JE, Molina BS (2014) Antecedent predictors of children's initiation of sipping/tasting alcohol. Alcohol Clin Exp Res 38(9): 2488–2495
44. Pautassi RM, Bravo AJ, Acquas E, Pilatti A (2020) Editorial: is early onset of alcohol use associated with later alcohol use? Front Behav Neurosci 14:133
45. Pautassi RM, Myers M, Spear LP, Molina JC, Spear NE (2008) Adolescent but not adult rats exhibit ethanol-mediated appetitive second-order conditioning. Alcohol Clin Exp Res 32(11):2016–2027
46. Truxell EM, Molina JC, Spear NE (2007) Ethanol intake in the juvenile, adolescent, and adult rat: effects of age and prior exposure to ethanol. Alcohol Clin Exp Res 31(5): 755–765
47. Sanders S, Spear NE (2007) Ethanol acceptance is high during early infancy and becomes still higher after previous ethanol ingestion. Alcohol Clin Exp Res 31(7):1148–1158
48. Boehm SL 2nd, Moore EM, Walsh CD, Gross CD, Cavelli AM, Gigante E et al (2008) Using drinking in the dark to model prenatal binge-like exposure to ethanol in C57BL/6J mice. Dev Psychobiol 50(6):566–578
49. Thiele TE, Crabbe JC, Boehm SL (2014) "Drinking in the Dark" (DID): a simple mouse model of binge-like alcohol intake. Curr Protoc Neurosci 68:9 49 1–12
50. Panksepp JB, Rodriguez ED, Ryabinin AE (2017) Sweetened ethanol drinking during social isolation: enhanced intake, resistance to genetic heterogeneity and the emergence of a distinctive drinking pattern in adolescent mice. Genes Brain Behav 16(3):369–383

51. Beaulieu JM, Gainetdinov RR (2011) The physiology, signaling, and pharmacology of dopamine receptors. Pharmacol Rev 63(1): 182–217
52. Tritsch NX, Sabatini BL (2012) Dopaminergic modulation of synaptic transmission in cortex and striatum. Neuron 76(1):33–50
53. Wise RA (2004) Dopamine, learning and motivation. Nat Rev Neurosci 5(6):483–494
54. Wise RA (1998) Drug-activation of brain reward pathways. Drug Alcohol Depend 51(1–2):13–22
55. Koob GF (1992) Drugs of abuse: anatomy, pharmacology and function of reward pathways. Trends Pharmacol Sci 13(5):177–184
56. Weiss F (2000) Neuroadaptive changes in neurotransmitter systems mediating ethanol-induced behaviors. In: Noronha A, Eckardt M, Warren K (eds) Review of NIAAA's Neuroscience and Behavioral Research Portfolio. NIAAA Research Monograph No. 34. NIH Pub. No. 00–4579. National Institute on Alcohol Abuse and Alcoholism, Bethesda, MD, pp 261–313
57. Gonzales RA, Job MO, Doyon WM (2004) The role of mesolimbic dopamine in the development and maintenance of ethanol reinforcement. Pharmacol Ther 103(2): 121–146
58. Risinger FO, Freeman PA, Rubinstein M, Low MJ, Grandy DK (2000) Lack of operant ethanol self-administration in dopamine D2 receptor knockout mice. Psychopharmacology 152(3):343–350
59. Kehagia AA, Murray GK, Robbins TW (2010) Learning and cognitive flexibility: frontostriatal function and monoaminergic modulation. Curr Opin Neurobiol 20(2): 199–204
60. Sinha R (2013) The clinical neurobiology of drug craving. Curr Opin Neurobiol 23(4): 649–654
61. Li TK (2000) Pharmacogenetics of responses to alcohol and genes that influence alcohol drinking. J Stud Alcohol 61(1):5–12
62. Repo E, Kuikka JT, Bergstrom KA, Karhu J, Hiltunen J, Tiihonen J (1999) Dopamine transporter and D2-receptor density in late-onset alcoholism. Psychopharmacology 147(3):314–318
63. Gessa GL, Muntoni F, Collu M, Vargiu L, Mereu G (1985) Low doses of ethanol activate dopaminergic neurons in the ventral tegmental area. Brain Res 348(1):201–203
64. Yim HJ, Gonzales RA (2000) Ethanol-induced increases in dopamine extracellular concentration in rat nucleus accumbens are accounted for by increased release and not uptake inhibition. Alcohol 22(2):107–115
65. Foddai M, Dosia G, Spiga S, Diana M (2004) Acetaldehyde increases dopaminergic neuronal activity in the VTA. Neuropsychopharmacology 29(3):530–536
66. Brodie MS, Shefner SA, Dunwiddie TV (1990) Ethanol increases the firing rate of dopamine neurons of the rat ventral tegmental area in vitro. Brain Res 508(1):65–69
67. Bunney EB, Appel SB, Brodie MS (2001) Electrophysiological effects of cocaethylene, cocaine, and ethanol on dopaminergic neurons of the ventral tegmental area. J Pharmacol Exp Ther 297(2):696–703
68. Brodie MS, Pesold C, Appel SB (1999) Ethanol directly excites dopaminergic ventral tegmental area reward neurons. Alcohol Clin Exp Res 23(11):1848–1852
69. Doyon WM, Ostroumov A, Ontiveros T, Gonzales RA, Dani JA (2021) Ethanol produces multiple electrophysiological effects on ventral tegmental area neurons in freely moving rats. Addict Biol 26(2):e12899
70. Pierce RC, Kumaresan V (2006) The mesolimbic dopamine system: the final common pathway for the reinforcing effect of drugs of abuse? Neurosci Biobehav Rev 30(2): 215–238
71. Khatib SA, Murphy JM, McBride WJ (1988) Biochemical evidence for activation of specific monoamine pathways by ethanol. Alcohol 5(4):295–299
72. Murphy JM, McBride WJ, Gatto GJ, Lumeng L, Li TK (1988) Effects of acute ethanol administration on monoamine and metabolite content in forebrain regions of ethanol-tolerant and -nontolerant alcohol-preferring (P) rats. Pharmacol Biochem Behav 29(1):169–174
73. Schier CJ, Dilly GA, Gonzales RA (2013) Intravenous ethanol increases extracellular dopamine in the medial prefrontal cortex of the Long-Evans rat. Alcohol Clin Exp Res 37(5):740–747
74. Imperato A, Di Chiara G (1986) Preferential stimulation of dopamine release in the nucleus accumbens of freely moving rats by ethanol. J Pharmacol Exp Ther 239(1):219–228
75. Carr DB, O'Donnell P, Card JP, Sesack SR (1999) Dopamine terminals in the rat prefrontal cortex synapse on pyramidal cells that project to the nucleus accumbens. J Neurosci 19(24):11049–11060
76. Heidbreder C, De Witte P (1993) Ethanol differentially affects extracellular monoamines

and GABA in the nucleus accumbens. Pharmacol Biochem Behav 46(2):477–481
77. Yoshimoto K, McBride WJ, Lumeng L, Li TK (1992) Alcohol stimulates the release of dopamine and serotonin in the nucleus accumbens. Alcohol 9(1):17–22
78. Yoshimoto K, Ueda S, Kato B, Takeuchi Y, Kawai Y, Noritake K et al (2000) Alcohol enhances characteristic releases of dopamine and serotonin in the central nucleus of the amygdala. Neurochem Int 37(4):369–376
79. Yan QS (1999) Extracellular dopamine and serotonin after ethanol monitored with 5-minute microdialysis. Alcohol 19(1):1–7
80. Ding ZM, Rodd ZA, Engleman EA, McBride WJ (2009) Sensitization of ventral tegmental area dopamine neurons to the stimulating effects of ethanol. Alcohol Clin Exp Res 33(9):1571–1581
81. Ding ZM, Oster SM, Hall SR, Engleman EA, Hauser SR, McBride WJ et al (2011) The stimulating effects of ethanol on ventral tegmental area dopamine neurons projecting to the ventral pallidum and medial prefrontal cortex in female Wistar rats: regional difference and involvement of serotonin-3 receptors. Psychopharmacology 216(2):245–255
82. Weiss F, Lorang MT, Bloom FE, Koob GF (1993) Oral alcohol self-administration stimulates dopamine release in the rat nucleus accumbens: genetic and motivational determinants. J Pharmacol Exp Ther 267(1): 250–258
83. Gonzales RA, Weiss F (1998) Suppression of ethanol-reinforced behavior by naltrexone is associated with attenuation of the ethanol-induced increase in dialysate dopamine levels in the nucleus accumbens. J Neurosci 18(24): 10663–10671
84. Rodd-Henricks ZA, McKinzie DL, Crile RS, Murphy JM, McBride WJ (2000) Regional heterogeneity for the intracranial self-administration of ethanol within the ventral tegmental area of female Wistar rats. Psychopharmacology 149(3):217–224
85. Rodd ZA, Melendez RI, Bell RL, Kuc KA, Zhang Y, Murphy JM et al (2004) Intracranial self-administration of ethanol within the ventral tegmental area of male Wistar rats: evidence for involvement of dopamine neurons. J Neurosci 24(5):1050–1057
86. Boileau I, Assaad JM, Pihl RO, Benkelfat C, Leyton M, Diksic M et al (2003) Alcohol promotes dopamine release in the human nucleus accumbens. Synapse 49(4):226–231
87. Diana M, Pistis M, Muntoni A, Gessa G (1996) Mesolimbic dopaminergic reduction outlasts ethanol withdrawal syndrome: evidence of protracted abstinence. Neuroscience 71(2):411–415
88. Diana M, Pistis M, Carboni S, Gessa GL, Rossetti ZL (1993) Profound decrement of mesolimbic dopaminergic neuronal activity during ethanol withdrawal syndrome in rats: electrophysiological and biochemical evidence. Proc Natl Acad Sci U S A 90(17): 7966–7969
89. Melis M, Spiga S, Diana M (2005) The dopamine hypothesis of drug addiction: hypodopaminergic state. Int Rev Neurobiol 63: 101–154
90. Volkow ND, Wang GJ, Telang F, Fowler JS, Logan J, Jayne M et al (2007) Profound decreases in dopamine release in striatum in detoxified alcoholics: possible orbitofrontal involvement. J Neurosci 27(46): 12700–12706
91. Weiss F, Parsons LH, Schulteis G, Hyytiä P, Lorang MT, Bloom FE et al (1996) Ethanol self-administration restores withdrawal-associated deficiencies in accumbal dopamine and 5-hydroxytryptamine release in dependent rats. J Neurosci 16(10):3474–3485
92. Badanich KA, Maldonado AM, Kirstein CL (2007) Chronic ethanol exposure during adolescence increases basal dopamine in the nucleus accumbens septi during adulthood. Alcohol Clin Exp Res 31(5):895–900
93. Sahr AE, Thielen RJ, Lumeng L, Li TK, McBride WJ (2004) Long-lasting alterations of the mesolimbic dopamine system after peri-adolescent ethanol drinking by alcohol-preferring rats. Alcohol Clin Exp Res 28(5): 702–711
94. Bell RL, Rodd ZA, Smith RJ, Toalston JE, Franklin KM, McBride WJ (2011) Modeling binge-like ethanol drinking by peri-adolescent and adult P rats. Pharmacol Biochem Behav 100(1):90–97
95. Engleman EA, Ingraham CM, McBride WJ, Lumeng L, Murphy JM (2006) Extracellular dopamine levels are lower in the medial prefrontal cortex of alcohol-preferring rats compared to Wistar rats. Alcohol 38(1):5–12
96. Leggio B, Masi F, Grappi S, Nanni G, Gambarana C, Colombo G et al (2003) Sardinian alcohol-preferring and non-preferring rats show different reactivity to aversive stimuli and a similar response to a natural reward. Brain Res 973(2):275–284
97. Doyon WM, Anders SK, Ramachandra VS, Czachowski CL, Gonzales RA (2005) Effect of operant self-administration of 10% ethanol plus 10% sucrose on dopamine and ethanol concentrations in the nucleus accumbens. J Neurochem 93(6):1469–1481

98. Doyon WM, York JL, Diaz LM, Samson HH, Czachowski CL, Gonzales RA (2003) Dopamine activity in the nucleus accumbens during consummatory phases of oral ethanol self-administration. Alcohol Clin Exp Res 27(10):1573–1582
99. Katner SN, Kerr TM, Weiss F (1996) Ethanol anticipation enhances dopamine efflux in the nucleus accumbens of alcohol-preferring (P) but not Wistar rats. Behav Pharmacol 7(7):669–674
100. Katner SN, Weiss F (1999) Ethanol-associated olfactory stimuli reinstate ethanol-seeking behavior after extinction and modify extracellular dopamine levels in the nucleus accumbens. Alcohol Clin Exp Res 23(11): 1751–1760
101. Melendez RI, Rodd-Henricks ZA, Engleman EA, Li TK, McBride WJ, Murphy JM (2002) Microdialysis of dopamine in the nucleus accumbens of alcohol-preferring (P) rats during anticipation and operant self-administration of ethanol. Alcohol Clin Exp Res 26(3):318–325
102. Doherty JM, Schier CJ, Vena AA, Dilly GA, Gonzales RA (2016) Medial prefrontal Cortical Dopamine responses during operant Self-Administration of Sweetened Ethanol. Alcohol Clin Exp Res 40(8):1662–1670
103. Carrillo J, Gonzales RA (2011) A single exposure to voluntary ethanol self-administration produces adaptations in ethanol consumption and accumbal dopamine signaling. Alcohol 45(6):559–566
104. Day JJ, Roitman MF, Wightman RM, Carelli RM (2007) Associative learning mediates dynamic shifts in dopamine signaling in the nucleus accumbens. Nat Neurosci 10(8): 1020–1028
105. Schultz W (1998) Predictive reward signal of dopamine neurons. J Neurophysiol 80(1): 1–27
106. Onali P, Olianas MC, Bunse B (1988) Evidence that adenosine A2 and dopamine autoreceptors antagonistically regulate tyrosine hydroxylase activity in rat striatal synaptosomes. Brain Res 456(2):302–309
107. White FJ, Wang RY (1984) Pharmacological characterization of dopamine autoreceptors in the rat ventral tegmental area: microiontophoretic studies. J Pharmacol Exp Ther 231(2):275–280
108. Cohen C, Perrault G, Sanger DJ (1999) Effects of D1 dopamine receptor agonists on oral ethanol self-administration in rats: comparison with their efficacy to produce grooming and hyperactivity. Psychopharmacology 142(1):102–110
109. Volkow ND, Wang GJ, Fowler JS, Logan J, Hitzemann R, Ding YS et al (1996) Decreases in dopamine receptors but not in dopamine transporters in alcoholics. Alcohol Clin Exp Res 20(9):1594–1598
110. Samson HH, Hodge CW, Tolliver GA, Haraguchi M (1993) Effect of dopamine agonists and antagonists on ethanol-reinforced behavior: the involvement of the nucleus accumbens. Brain Res Bull 30(1–2):133–141
111. Hodge CW, Samson HH, Haraguchi M (1992) Microinjections of dopamine agonists in the nucleus accumbens increase ethanol-reinforced responding. Pharmacol Biochem Behav 43(1):249–254
112. George SR, Fan T, Ng GY, Jung SY, O'Dowd BF, Naranjo CA (1995) Low endogenous dopamine function in brain predisposes to high alcohol preference and consumption: reversal by increasing synaptic dopamine. J Pharmacol Exp Ther 273(1):373–379
113. Bonnin A, Levitt P (2011) Fetal, maternal, and placental sources of serotonin and new implications for developmental programming of the brain. Neuroscience 197:1–7
114. Lin SH, Lee LT, Yang YK (2014) Serotonin and mental disorders: a concise review on molecular neuroimaging evidence. Clin Psychopharmacol Neurosci 12(3):196–202
115. Figee M, Pattij T, Willuhn I, Luigjes J, van den Brink W, Goudriaan A et al (2016) Compulsivity in obsessive-compulsive disorder and addictions. Eur Neuropsychopharmacol J Eur Coll Neuropsychopharmacol 26(5):856–868
116. Walsh JJ, Christoffel DJ, Heifets BD, Ben-Dor GA, Selimbeyoglu A, Hung LW et al (2018) 5-HT release in nucleus accumbens rescues social deficits in mouse autism model. Nature 560(7720):589–594
117. Doremus TL, Brunell SC, Rajendran P, Spear LP (2005) Factors influencing elevated ethanol consumption in adolescent relative to adult rats. Alcohol Clin Exp Res 29(10): 1796–1808

Chapter 9

Preclinical Models of THC and Nicotine Exposure During Adolescent Brain Development: Modeling Neuropsychiatric Phenotypes Related to Dopaminergic Transmission

Enzo Perez-Valenzuela and Steven R. Laviolette

Abstract

Adolescence is a gradual transition period between childhood and adulthood, characterized by greater sensitivity to rewarding stimuli. Consistently, demographic studies have shown that teenagers show a high prevalence of rewarding drugs use, mainly nicotine-containing products and cannabis. Clinical researches have associated the adolescence consume of nicotine and cannabis with a higher vulnerability to develop neuropsychiatric diseases in adulthood such as depression, schizophrenia, and drug addiction. Despite these evidence, it is difficult to conclusively prove causal relationships with longitudinal and retrospective clinical research in humans. Accordingly, preclinical animal models are indispensable tools to determine the causal relationship between early neurodevelopmental drug exposure and psychiatric disease risk. Preclinical rodent models have been widely used to research the neurobiological mechanisms underlying the vulnerability of the adolescent brain because the similarities in behavioral patterns and brain maturation processes with human adolescents. Particularly, both human teenager and adolescent rodents show similar patterns of DA maturation, and dysregulations in these neuronal circuits may induce phenotypes associated with psychiatric diseases.

This chapter will describe the THC and nicotine adolescent exposure methods in rats that induce a long-term dysregulation of the dopaminergic system. Furthermore, this chapter will detail the experimental protocol used in our laboratory to test the electrical activity of dopaminergic neurons. In addition, these methodologies include preclinical testing protocols for several neuropsychiatric behavioral phenotypes related to social cognition and motivation, memory processing, and anhedonia-like behaviors.

Key words Adolescence, Dopamine, Nicotine, THC, Depression, Working memory, Social interaction

1 Introduction

1.1 Drugs of Abuse and Adolescence

Adolescence is a gradual transition period between childhood and adulthood, characterized by social behaviors strongly influenced by peer pressure, greater sensitivity to rewarding stimuli, and more resistance to aversive stimuli. Although these behavioral changes are necessary for new skill learning and the transition to

José Antonio Fuentealba-Evans and Pablo Henny (eds.), *Dopaminergic System Function and Dysfunction: Experimental Approaches*, Neuromethods, vol. 193, https://doi.org/10.1007/978-1-0716-2799-0_9,

independent adult life, they are associated with more risk-taking behaviors [1]. Consistently, global demographic studies have shown that teenagers comprise a segment of society with a high prevalence of drug experimentation and increased vulnerability to drug dependence disorders. Two of adolescents' most widely used drugs include nicotine-containing products and cannabis [2].

According to the World Health Organization global report on the prevalence of tobacco use in 2018, at least 43.8 million adolescents aged 13–15 represent approximately 12% of global tobacco consumption [2]. In addition, it has been reported that over 90% of smokers first become dependent on tobacco during adolescence [3]. Electronic cigarette (e-cigarette) use is increasingly pervasive among teenagers due to the perception that such products are less harmful than conventional cigarettes [4]. For instance, in the United States, it has been observed that approximately 4.3% of middle school students and 11.3% of high-school students have used e-cigarettes [5, 6].

Adolescents' global use of cannabis is higher than the general population, with an estimated annual prevalence of 4.7% vs. 3.9%, respectively. However, these differential trends have shown regional patterns. For example, the most significant difference is reported in Oceania, which the cannabis prevalence in teenagers is around 17.8% compared to 10% of annual prevalence in the general population. Similar trends are observed in the Americas, with a past-year use of approximately 12% in adolescents vs. 8.8% in the general population, and in Europe with cannabis use in teenagers vs. the general population reported at 11.7% vs. 6%, respectively [7]. Considering these emerging trends, the possible neuropsychiatric consequences of the neurodevelopmental exposure to cannabinoids and nicotine are becoming an urgent question for clinical and preclinical research.

1.2 Evidence for the Impact of Neurodevelopmental Nicotine and Cannabinoid Exposure in Psychiatric Disorders

Adolescence is a transitional period of brain development in which several major neuropsychiatric diseases emerge due to anomalies in brain maturation processes [8, 9]. While multiple environmental and genetic factors converge in the ontogenetic of these pathologies, these developmental anomalies may also be triggered by exposure to chronic stressors and/or extrinsic insults to the developing brain associated with certain drugs of abuse [10, 11].

Neurodevelopmental exposure to nicotine and cannabis has been associated with a higher vulnerability to develop neuropsychiatric diseases in adulthood. For example, longitudinal studies in adolescents have shown a reciprocal association between early cigarette smoking and the emergence of depressive symptoms [12–14]. Furthermore, it has been proposed that adolescent smoking is a predictor for the later life development of depression [15]. In addition, there is a well-established comorbidity between schizophrenia and heavy nicotine dependence [16, 17]. In particular, chronic smoking during adolescence has been associated with a higher risk of developing schizophrenia during adulthood. One self-report study has found that adolescent smoking increases the risk of psychotic events even after adjustment for other factors, such as other parental substance abuse, alcohol use, or illicit substance use [18]. In addition, it has been reported that the number of cigarettes smoked per day during adolescence increases the risk for schizophrenia hospitalization during adulthood [19].

In the case of cannabis, a significant association between suicidal behavior and regular use of cannabis has been observed. Interestingly, these associations are more robust in a younger population. Adolescent chronic cannabis users (14–15 years old) present more risk of suicidal ideation and suicide attempts compared to adult regular cannabis users (21 years old) [20]. In addition, the early use of cannabis has been associated with a higher risk of schizophrenia in adulthood. A watershed longitudinal retrospective study reported that in a cohort of 45,570 Swedish military conscripts, there was a higher relative risk of developing schizophrenia among heavier consumers of cannabis [21]. Interestingly, a prospective study showed that cannabis users by age 15 had more schizophrenia symptoms at age 26 than controls and cannabis users at age 18, even after controlling for the presence of pre-cannabis prodromal symptoms [22]. Furthermore, it has been reported that adolescent cannabis use is associated with both positive and negative symptoms of schizophrenia in young adulthood. Hence, the first use of cannabis below 16 years showed a higher effect than the first use after 15 years [23]. These studies strongly support the hypothesis that adolescence is a sensitive period wherein exposure to either tobacco or cannabis products may increase the relative risk of developing severe neuropsychiatric disorders in later life.

Despite extensive correlational evidence associating the consumption of tobacco and cannabis products with increased neuropsychiatric risk, it is impossible to conclusively prove causal relationships with longitudinal, retrospective clinical research in humans. Inherent experimental limitations include subject memory bias in self-reporting, differing social backgrounds, and most importantly, the high variability in doses, frequency, routes of administration, or ages of drug exposure onset in these sorts of clinical investigations. Accordingly, preclinical animal models are indispensable tools to determine the causal relationship between early neurodevelopmental drug exposure and psychiatric disease risk [24]. Indeed, such preclinical models allow for precise experimental control of the timing of drug exposure, dosing, and environmental consistency to identify causal factors involved in these comorbid phenomena.

1.3 Rodent Models of Neurodevelopment: Similarities Between Human Vs. Rodent Brain Development Trajectories

Preclinical rodent models have been widely used to research the neurobiological mechanisms underlying the vulnerability of the adolescent brain. Extensive evidence has demonstrated that adolescent rodents present similar behavioral patterns and brain maturation processes to those in human adolescents. At the behavioral level, adolescents show a higher sensitivity to rewarding stimuli than adult subjects in both humans and rodents [25]. For instance, in teenagers, peer interaction is more influential in decision-making during adolescence than during other developmental periods [26]. Similarly, adolescent rodents show more social preference and spend more time playing with peers than adult rats [27, 28]. In addition, it has been observed that the preference for sweet-taste food is higher during early adolescence than during emerging adulthood in both rodents and humans [29, 30]. In addition, both human and rodent adolescents display a similar preference to accept a smaller reward delivered sooner than a more significant reward delayed than adulthood, demonstrating increased propensities to impulsive behaviors [31, 32]. These studies support the notion that both humans and rodents exhibit similar patterns of affective processing and social development.

In neurodevelopmental terms, adolescence has been described as an adaptative period of high plasticity which synaptic connections are pruned and restructured to refine synaptic connectivity across multiple brain circuits. This myriad of anatomical, molecular, and morphological changes are present in rodents and humans. In humans, there are ongoing changes in gray matter density [33] and changes in the integrity of the white matter pathways [34]. In the case of reward processing circuitry, many adaptive changes occur in neurochemical signaling pathways. For example, there is a peak in the expression of D1 receptor mRNA in the prefrontal cortex in adolescents and young adults [35]. Imaging studies have observed a similar pattern in the maturation of structure and function of the mesocorticolimbic pathway [36–42]. In rats, it has been reported that there is an "inverted U shape" pattern in dopamine release, dopamine receptor expression and firing rates of dopamine neurons during ontogenetic development [43–47]. The similarity between adolescent patterns of dopamine (DA) maturation across humans and rodents further underscores the utility of adolescent rodents to gain insight into translational effects in the human brain.

Although the start of these changes is diffuse and depends on each subject, it has been suggested that adolescence begins before the onset of puberty and finishes after sexual maturation. Puberty in male humans is marked with the semenarche (first ejaculation), and female puberty begins with thelarche (first breast development) and menarche (first menstrual cycle). While in male rats, puberty is determined by the presence of balano-preputial separation (around PND 38) and the completion of spermatogenesis (around PND 60), in female rats it begins with the vaginal opening (around PND

35) and finishes with the first ovulation cycle (around PND 40). Although puberty and adolescence are not interchangeable terms, both are closely related, and the timing of puberty can modulate the changes observed during adolescence [10, 11, 48]. With this in mind, the treatment protocol used in our laboratory starts from PND 35 to ensure that every rat is in the mid-adolescence period at the beginning of the drug exposure period.

The rodent model of neurodevelopmental drug exposure has been used extensively to study the effects of cannabinoids and nicotine on the developing mammalian brain. In these studies, the primary psychoactive components of tobacco (nicotine) or cannabis (delta-9-tetrahydrocannabinol; THC) can be systemically administered to rats, allowing the experimenter to precisely control both dosage and temporal exposure profiles to these drugs during confined periods of adolescent brain development. These experimental models have revealed that repeated treatment with cannabinoid receptor type 1 (CB1) agonists (such as THC or WIN55212-2) during adolescence increases dopaminergic neurotransmission and modifies dopaminergic receptor expression in adulthood [49–56]. In addition, several studies have used adolescent nicotine exposure in rats to model various neuronal, molecular, and behavioral phenotypes induced by human adolescent nicotine exposure. Interestingly, these studies have similarly reported long-term dysregulation of mesocorticolimbic DA transmission patterns [57, 58].

This chapter will describe the THC and nicotine adolescent exposure methods in rats that induce a long-term dysregulation of the mesocorticolimbic dopaminergic system. Furthermore, this chapter will detail the experimental protocol used in our laboratory to test the electrical activity of dopaminergic neurons from ventral tegmental area (VTA). In addition, these methodologies include preclinical testing protocols for several neuropsychiatric behavioral phenotypes related to social cognition and motivation, memory processing, and anhedonia-like behaviors.

2 Materials

2.1 Adolescent THC Treatment

(i) THC.

THC must be stored at 4 °C and wrapped in aluminum foil because this drug is sensitive to light and heat. THC is a regulated drug that must be registered in a controlled substance registry in most jurisdictions. In addition, the acquisition of THC will most likely require Governmental licensing.

(ii) Cremaphor® (also named polyoxyl 35 castor oil, cremulphor, emulsogen, or Kolliphor).

Note: Cremaphor® is a nonionic solubilizer and emulsifier made by reacting castor oil with ethylene oxide. This detergent is a pale yellow and viscous oily liquid. It can be stored at room temperature.

(iii) Ethanol (THC is suspended in 100% ethanol solution).

(iv) Saline 0.9%.

(v) 1 ml Syringe, 1 inch, 25-gauge needles.

(vi) Nitrogen gas stream.

(vii) Pasteur pipet borosilicate glass cotton plugged.

Note: Pipettes deliver the nitrogen gas into the THC Cremaphor® solution and remove (burn-off) the ethanol. In this case, the cotton is to avoid contamination in the solution. The recommended size of the pipet is 5 ¾″ (14.6 cm).

(viii) Centrifuge tubes (50 ml).

(ix) Aluminum foil or dark glass bottles (to avoid light exposure to THC during storage).

2.2 Adolescent Nicotine Treatment

(i) Nicotine bitartrate.

(ii) Phosphate-buffered saline 0.9%.

2.3 Behavioral Tests

2.3.1 Sucrose Preference

(i) Sucrose.

(ii) Bottle.

2.3.2 Social Interaction and Memory Test

(i) Three-chambered box.

The rectangular three-chambered box comprises transparent Plexiglas walls and a black Plexiglas floor. The base is 120 cm L × 60 cm W. Each chamber is 40 cm L × 60 cm W × 35 cm H. The dividing walls have a plastic guillotine door in the middle to allow movement across sections. Each plastic guillotine door is 10 cm L × 35 cm H (*see* Fig. 3a).

(ii) Two rectangular wire cages.

The wire cages are 21 cm L × 10 cm W × 10 cm H, allowing nose contact between the bars. Each cage has a plastic plate to facilitate cleaning between tests (*see* Fig. 3a). The wire cage has a locked door to avoid escape by the "stranger" rats during testing.

(iii) Video Camera or webcam.

(iv) Video Software.

Debut video capture free version is the software used to record the videos using a webcam and a computer. The software is available using this link: https://www.nchsoftware.com/capture/index.html (free version for noncommercial use).

The software used for video analysis is BehaView® version 0.0.16 free software. The software is available using this link: http://www.pmbogusz.net/?a=behaview

2.3.3 Novel Object Recognition

(i) Black open field box.

The open field black box is made with black Plexiglas. The box is 80 cm L × 80 cm W × 80 cm H (*see* Fig. 4a).

(ii) Pair of objects.

The objects must be easily distinguishable and have good contrast with the black floor of the open field box. Our protocol uses a yellow plastic bottle/orange glass bottle or white glass bottle/brown plastic bottle (*see* Fig. 4b).

(iii) Video Camara or webcam.

(iv) Video Software.

Debut video capture free version is the software used to record the videos using a webcam and a computer. The software is available using this link: https://www.nchsoftware.com/capture/index.html (free version for noncommercial use).

The software used for video analysis is BehaView® version 0.0.16 free software. The software is available using this link: http://www.pmbogusz.net/?a=behaview

2.4 Single Unit Recording

(i) Urethane dose 1.4 g/kg.

Urethane anesthesia was chosen due to the extended half-life [59]. In addition, urethane does not modify dopaminergic transmission [60, 61].

(ii) Glass Capillaries 2 mm outside diameter, 4 inch long.

(iii) Vertical electrode puller.

(iv) Pontamine blue sky 6B.

(v) Digitalizer Digidata1440A.

(vi) Software Clampex® 10.5.

(vii) Amplifier MultiClamp® 700B.

(viii) Software Multiclamp® 700B Commander.

(ix) Current clamp and voltage clamp Headstage CV-7B.

(x) Iontophoresis Pump BAB-501.

(xi) Micro electrode tester.

(xii) Audio monitor.

(xiii) Heat pad.

(xiv) Surgery instruments.

(xv) Drill.

(xvi) Thermometer.

(xvii) Lubricating jelly.

(xviii) Marker.

(xix) Needle 34-gauge/67 mm long.

(xx) Saline 0.9%.

3 Methods

3.1 Adolescent THC Treatment

(i) THC solution preparation.

THC is chemically classified as a cannabinoid due to its terpene phenolic structure of 21 carbons atoms. This chemical structure gives THC a high affinity to lipids at the expense of their solubility in water. Frequently to facilitate the dissolution in saline for systemic administration, THC is mixed with several vehicles such as solvents dimethyl sulfoxide (DMSO), ethanol, detergents (e.g., Tween 80), or vegetable oils. Our laboratory uses a dissolution protocol using the vegetal oil Cremaphor®, ethanol, and nitrogen gas.

Protocols for dissolving THC

1. First, put Cremaphor in a centrifuge tube of 50 ml. Using a permanent marker, mark the tube with the volume used by the Cremaphor.
2. Add the required amount of THC ethanol solution to the centrifuge tube. Vortex the tube until it homogenizes the solution.
3. Now, the ethanol is removed from this solution using a continuous stream of nitrogen gas. Connect a nitrogen gas tank with a plastic tube connector and a glass Pasteur pipet to perform this procedure. Then, open the gas tank and set the pressure under 10 bpi. This pressure should be enough to safely remove the ethanol without creating large bubbles in the solution. Finally, put the tip of the glass pipet in the solution and apply the nitrogen stream until all the ethanol is evaporated. The final solution must have the same volume marked in **step 1**. Using this method will generally take around 1 h to remove all the ethanol from the solution (Fig. 1).
4. Add saline to the centrifuge tube until the required solution's final volume and vortex to homogenize the solution.
5. Finally, wrap the preparation tube with aluminum foil and store it at 4 ° C. This step prevents the oxidation of THC to cannabinol, a less psychoactive compound [62].

Note: For the adolescent treatment with escalating dose (*see* Subheading 3.1, (ii) protocol treatment), it is recommended to prepare the three solutions ($S_{2.5}$, S_5, and S_{10}). Each stock solution

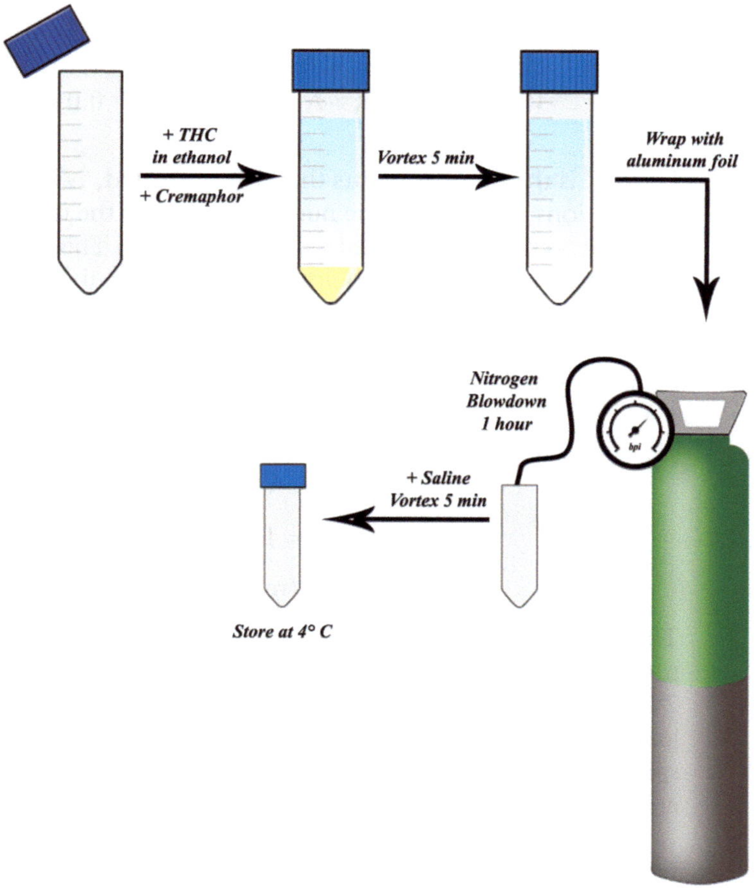

Fig. 1 Schematic summary of THC dissolution in 5% Cremaphor

wrapped in aluminum foil and refrigerated remains stable for around 20 days.

- *Calculus of Cremaphor volume and THC.*

It is necessary to prepare a stock solution at 5% (v/v) Cremaphor® in saline for THC systemic injections. In the case of adolescent escalating dose treatments (*see* Subheading 3.1, (ii) protocol treatment), it is required to prepare three different solutions for the three concentrations of THC used: 2.5 mg/ml ($S_{2.5}$), 5 mg/ml (S_5), and 10 mg/ml (S_{10}). The 2.5 mg/ml solution will be used for six injections per rat, and the 5 and 10 mg/ml solutions will be used for eight injections per rat. In addition, the increase of weights in adolescent rats for the 11 days (postnatal days 35–45) injection period should be factored into drug preparations. Then, the equation to calculate the volume stock used with a daily increase of 10 g per day in each rat is:

$$V\text{stock (ml)} = \text{R} \times N \times \left(\sum_{i=0}^{d-1} W + i \times 0.01 \right)$$

Here, R is the number of rats that will be treated, N is the number of injections per day, d is the number of days of the treatment, and W is the average of the initial weight of the rats. This equation was formulated considering a volume injection of 1 ml/kg in each rat. For example, for ten rats with an average weight initial of 250 g using our protocol (*see* Subheading 3.1, (ii)) of treatment, the equation should be:

$$V\text{stock (ml)} = 10 \times 2 \times \left(\sum_{i=0}^{10} 0.25 + i \times 0.01 \right)$$

$$V\text{stock (ml)} = 10 \times 2 \times (3.3)$$

$$V\text{stock (ml)} = 66$$

Then, the total volume for $S_{2.5}$, S_5, and S_{10} is 66 ml. Since the proportion $S_{2.5}$:S_5:S_{10} is 3:4:4, the volume for $S_{2.5}$ is 18 ml, S5 is 24 ml, and S_{10} is 24 ml. However, it is recommended to prepare at least 10% more due to the dead volume in the syringe, accidental spills, or leaks. Continuing with this example and considering a 10% error, the final volume of $S_{2.5}$ is 20 ml, S_5 is 26 ml, and S_{10} is 26 ml.

The next step is to calculate the 5% of each volume solution which will be the volume of Cremaphor required. In this example, it is necessary 1 ml of Cremaphor for $S_{2.5}$, 1.3 ml for S_5, and 1.3 ml for S_{10}.

To meet a THC concentration allowing an injection volume of 1 ml/kg, preparing the solution with a concentration equivalent to the treatment dose is recommended. For example, if the treatment dose is 2.5 mg/kg, the concentration should be 2.5 mg/ml.

Finally, using the final volume of $S_{2.5}$, S_5, and S_{10} and their respective THC concentrations, it is possible to calculate the amount of THC in each solution. The $S_{2.5}$ solution requires 50 mg, S_5 needs 130 mg, and S_{10} needs 260 mg THC. To facilitate the dissolution of THC, it is necessary to dissolve the drug in ethanol. Then, using a solution of 1 g of THC per 100 ml of ethanol (THC solubility in ethanol: 35 mg/ml), it is necessary to take 5 ml for $S_{2.5}$, 13 ml for S_5, and 26 ml for S_{10}.

(ii) THC treatment protocol.

Rats should be habituated to the injection manipulation, administration room, and the responsible researcher. The habituation period should start at least 1 week before beginning the THC treatment.

The adolescent rats receive twice-daily injections of escalating doses of THC from postnatal day (PD) 35 to PD 45 with the following schedule:

1. During PD 35, 36, and 37, rats are intraperitoneally injected with THC at a dose of 2.5 mg/kg twice per day, the first dose at 9:00 AM and the second dose at 5:00 PM.
2. During PD 38, 39, 40, and 41, rats are intraperitoneally injected with THC at 5 mg/kg twice per day, the first dose at 9:00 AM and the second dose at 5:00 PM.
3. During PD 42, 43, 44, and 45, rats are intraperitoneally injected with THC at 10 mg/kg twice per day, the first dose at 9:00 AM and the second dose at 5:00 PM.
4. Finally, start the behavioral tests after a THC washout period of 30 days (PD 75).

Note: It is strongly recommended to change the i.p. location injection site after every administration of THC to reduce the possibility of damage to the injection area and possibly inducing additional irritation or stress to the rat.

The logic behind using an escalating dose of THC is to overcome the potential desensitization of CB1 receptors after chronic exposure [63]. Furthermore, this protocol has been demonstrated to reliably modify dopamine-related behaviors and increase the mesocorticolimbic dopaminergic activity when rats are tested during adulthood [52, 53, 64, 65]. For example, this THC adolescent exposure protocol has been shown to reduce social motivation and impair social recognition [52, 53, 65], to reduce the sucrose preference in males and female rats [65, 66], and to disrupt working memory performance in the novel object recognition task [53]. Interestingly, these changes were accompanied by an increase in the tonic activity of dopamine neurons in the VTA of anesthetized rats. Importantly, these behavioral effects were not observed in rats treated with THC during adulthood, demonstrating that this THC treatment protocol shows high selectivity for inducing disruptions to adolescent neurodevelopmental processes [53].

The subsequent sections will detail the preclinical behavioral procedures to measure dopamine-related processing and dopaminergic activity in the VTA.

3.2 Adolescent Nicotine Treatment

(i) Nicotine solution preparation.

In contrast with THC, nicotine tartrate salt used in this protocol is easily dissolved in water solutions such as saline or phosphate-buffered saline (PBS). In this protocol, it is recommended to use PBS for the preparation.

Now, the preparation of nicotine solution is carried out in PBS solution, following these steps:

1. Put the nicotine bitartrate in a vial (recommended centrifuge tubes).
2. Add PBS in the vial until it reaches the volume required.
3. Vortex the solution until the drug dissolves.
4. Finally, store the final solution at 4 ° C.

Similarly to THC preparations, it is necessary to calculate the amount of nicotine and the stock solution volume. The following equation calculates the stock volume considering an injection volume of 1 ml/kg per rat and an increase in the weight of 10 g per day of treatment:

$$V\text{stock (ml)} = \text{R} \times N \times \left(\sum_{i=0}^{d-1} W + i \times 0.01 \right)$$

Here, R is the number of rats that will be treated, N is the number of injections per day, d is the number of days of the treatment, and W is the average of the initial weight of the rats. For example, for ten rats with an average weight initial of 250 g using our protocol of 10 days (*see* Subheading 3.2, (ii) protocol treatment) of treatment, the equation should be:

$$V\text{stock (ml)} = 10 \times 3 \times \left(\sum_{i=0}^{9} 0.25 + i \times 0.01 \right)$$

$$V\text{stock (ml)} = 10 \times 3 \times (2.95)$$

$$V\text{stock (ml)} = 10 \times 3 \times (2.95)$$

$$V\text{stock (ml)} = 88.5$$

In this example, the V_{stock} required is 88.5 ml. However, it is recommended to prepare at least 10% more in case of dead volume in the syringe or accidental spills or leaks. Then, taking into account a 10% error, the final volume is 97 ml, approximating 100 ml.

To meet a nicotine concentration allowing an injection volume of 1 ml/kg, preparing the solution with a concentration equivalent to the treatment dose is recommended. For example, if the treatment dose is 0.4 mg/kg, the concentration should be 0.4 mg/ml.

(ii) Protocol treatment.

It is strongly recommended that rats should be habituated to injection manipulations, homeroom, and responsible researchers. The habituation period should be started at least 1 week before beginning the nicotine injection procedures.

In contrast to adolescent THC treatments, nicotine exposure uses the same dose during the adolescent exposure period. Rats receive a subcutaneous injection (to the scruff of the neck) of 0.4 mg/kg of nicotine bitartrate three times per day, the first dose at 10:00 AM, the second dose at 1:00 PM, and the third dose at 3:00 PM for 10 consecutive days (PD 35–44). Then, the behavioral tests start after 30 days of washout until the rats reach adulthood (PD 75).

Note: It is strongly suggested to change the injection site after every administration to reduce the potential of long-term cutaneous damage to the injection sites.

The direct physiological translation of nicotine exposure between rodents and humans is not accurate due to profound differences in nicotine metabolism. Nevertheless, the dose used in this protocol produces similar plasma levels of nicotine comparable to those found in heavy smokers [67, 68].

Furthermore, rats' adolescent nicotine exposure protocol disrupts mesocorticolimbic dopaminergic transmission lasting into adulthood [57, 69]. These effects are also associated with profound behavioral abnormalities associated with anxiety and depression. For example, this protocol reduces social motivation and recognition, precisely the following adolescence exposure (but not during adulthood) [57]. In addition, the same protocol diminishes sucrose preference during rat adulthood [57, 69], suggesting that adolescent exposure to nicotine produces a depressive-like endophenotype when the subjects reach adulthood.

3.3 Preclinical Behavioral Tests

This section will describe three behavioral tests: sucrose preference, social interaction, and novel object recognition. These behavioral approaches are widely used as preclinical models of symptoms of neuropsychiatric diseases associated with dysfunctional dopaminergic transmission [70–74]. Furthermore, each section will show the long-term effects of THC and nicotine treatment during adolescence in adulthood behavior. These tests only represent a small subset of potential preclinical trials for neuropsychiatric endophenotypes. Other tests to consider for examining anxiety, depression, addiction, and schizophrenia-related behavioral phenotypes may include: paired-pulse inhibition (sensorimotor gating), the light–dark box for anxiety testing, the elevated plus-maze, associative fear conditioning, open field tests for thigmotaxis, social transmission of food preference, the Porsolt forced swim test or conditioned place preference for drug reward effects.

3.3.1 Sucrose Preference

Anhedonia has been clinically defined as a diminished interest or pleasure in one's usual activities and is a core symptom in major depressive disorder and a negative symptom of schizophrenia [75]. As a preclinical model, the sucrose preference test has been widely used in rodents to determine changes in anhedonia based on the rodent's natural preference to consume sweet solution over regular water [70, 76]. This test has multiple variations, including different sucrose concentrations, habituation periods, food and water deprivation periods, or the time of intake. This section will describe the specific protocol used in our laboratory, which has demonstrated sensitivity to detect changes in reward processing following exposure to the previously described models of adolescent THC and nicotine exposure.

For this protocol, the sucrose preference test consists of two phases: habituation and testing.

(i) Habituation Phase.

The habituation phase is carried out to reduce the novelty to the sucrose and avoid a possible aversion to the unknown solution. In this phase, the rats are exposed to two bottles filled with sucrose solution at 2% for 48 h of ad libitum access in the home cage.

It is recommended to prepare a stock solution at a concentration of 20 g/100 ml and dilute it 10 times before filling the bottles.

Note 1: The rats only have access to sucrose solution during the habituation.

Note 2: The rat must be alone in the home cage during the habituation and the session test.

After the 48 h of sucrose habituation, the rats are deprived of water and sucrose for 16 h. The deprivation is required to promote the liquid intake during the test session.

(ii) Test.

After 16 h of liquid deprivation, the test session is carried out with the following steps:

1. Fill one bottle with sucrose 2% and the second bottle with regular water per rat.
2. Mark each sucrose bottle to differentiate it from the water bottle (*see* Fig. 2a).
3. Pre-weigh and record the weights of each 2% sucrose solution and water bottle.
4. Put one bottle of sucrose and one bottle of water in each home cage, as illustrated in Fig. 2a.
5. Allow unrestricted access to the two bottles for 1 h.
6. Remove both bottles from the home cage.
7. Weigh and record each bottle filled with sucrose solution and water. The difference between the weight before and after the test session (**step 5**) will be used to assess each subject's liquid intake.
8. Return the bottle of regular water to the home cage.

Note 1: Both solutions must be prepared simultaneously and using the same water to reduce other factors influencing natural preference.

Note 2: The position of the bottles must be counterbalanced in each rat to reduce potential cage location bias.

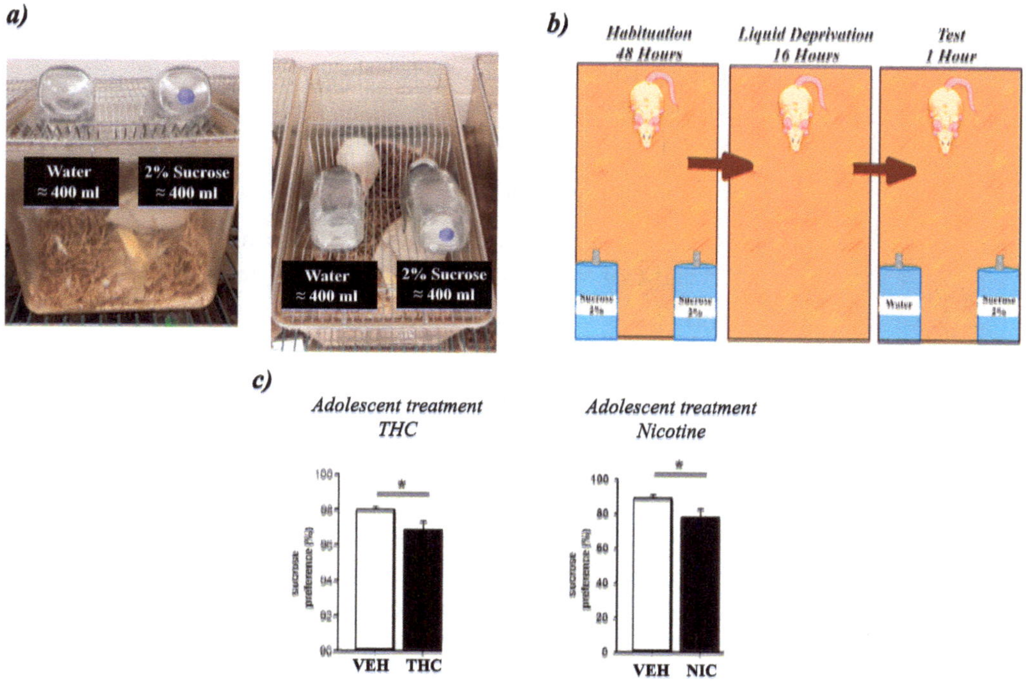

Fig. 2 Sucrose preference test. (**a**) Picture of a rat during sucrose preference test. (**b**) Schematic summary of sucrose test protocol. (**c**) Effect of THC (left) and nicotine (right) adolescent exposure in sucrose preference of adult rats. *$p < 0.05$ using unpaired *t*-test. (The results were extracted from De Felice et al. [65] and Hudson et al. [69])

(iii) Results, analysis, and interpretations.

Two main parameters are assessed to study anhedonia-like behavior: the liquid intake index and the sucrose preference index. The amount of liquid intake is the sum of the water and the sucrose solution consumed during the 1-h session normalized for the weight of the rat, which can be obtained using the following equation:

$$\text{Liquid intake index} = \frac{\text{Weight of water intake} + \text{Weight of sucrose solution intake}}{\text{Weight of the rat}} \times 100$$

The water and sucrose solution intake are obtained from the difference between the bottles' weight before and after the 1-h test (**steps 3** and 7). In addition, it is required to normalize by the rat's weight, given that the liquid intake is directly proportional to the weight [70]. The liquid intake index determines that the rats were capable of typical liquid consumption. Usually, the rats show a liquid intake index between 3 and 6 g/100 g of body weight. If any rat presents an index lower than 1 g/100 g body weight, it should be excluded from the experiment due to lack of task performance.

The sucrose preference index shows the natural preference of the rats to the sweet solution over the regular water. This index can be expressed with the following equation:

$$\text{Sucrose preference index} = \frac{\text{Weight of sucrose solution intake}}{\text{Weight of water intake} + \text{Weight of sucrose solution intake}} \times 100$$

A reduction in this index has been attributed to an increase in anhedonia in the rodents. For example, chronic mild stress exposure diminishes sucrose preferences. In addition, antidepressant treatment has been shown to restore sucrose preferences [70, 76, 77]. In both the THC and nicotine adolescent treatment protocols described previously in this chapter, our research has demonstrated that both treatments reduce sucrose preferences when the rats reach adulthood, suggesting that adolescent exposure to these drugs can cause long-term modifications to neuronal reward-processing pathways [57, 65, 69].

3.3.2 Social Interaction: Motivation and Memory

The social interaction task is used to assess the social motivation of the rats to a novel "stranger" rat and social memory through the differential exploration between a familiar rat and a stranger rat. The task takes advantage of most rodent species' natural motivation and social proclivities. Previous evidence has shown that social motivation and social memory changes are associated with dopamine transmission in mesocorticolimbic and nigrostriatal pathways [78–81]. Social motivation and social memory also play a crucial role in various neuropsychiatric diseases in humans, such as schizophrenia, major depression, and autism spectrum disorder [82–84]. This behavioral test uses a three-chambered box (*see* Subheading 2), and it has three phases: habituation, motivation, and recognition (memory).

(i) Habituation Phase.

The habituation phase is required to mitigate potential confounds to exploratory behaviors, such as neophobia to the apparatus. The habituation phase is carried out 1 day before the social interaction test and subsequent memory recognition phase. The steps in the habituation phase are:

1. Place the rat home cage in the same room as the testing apparatus for 30 min.
2. Situate the wire cages in the middle of the right and left chambers. Then, ensure that inter-chamber doorways are closed (*see* Fig. 3a).

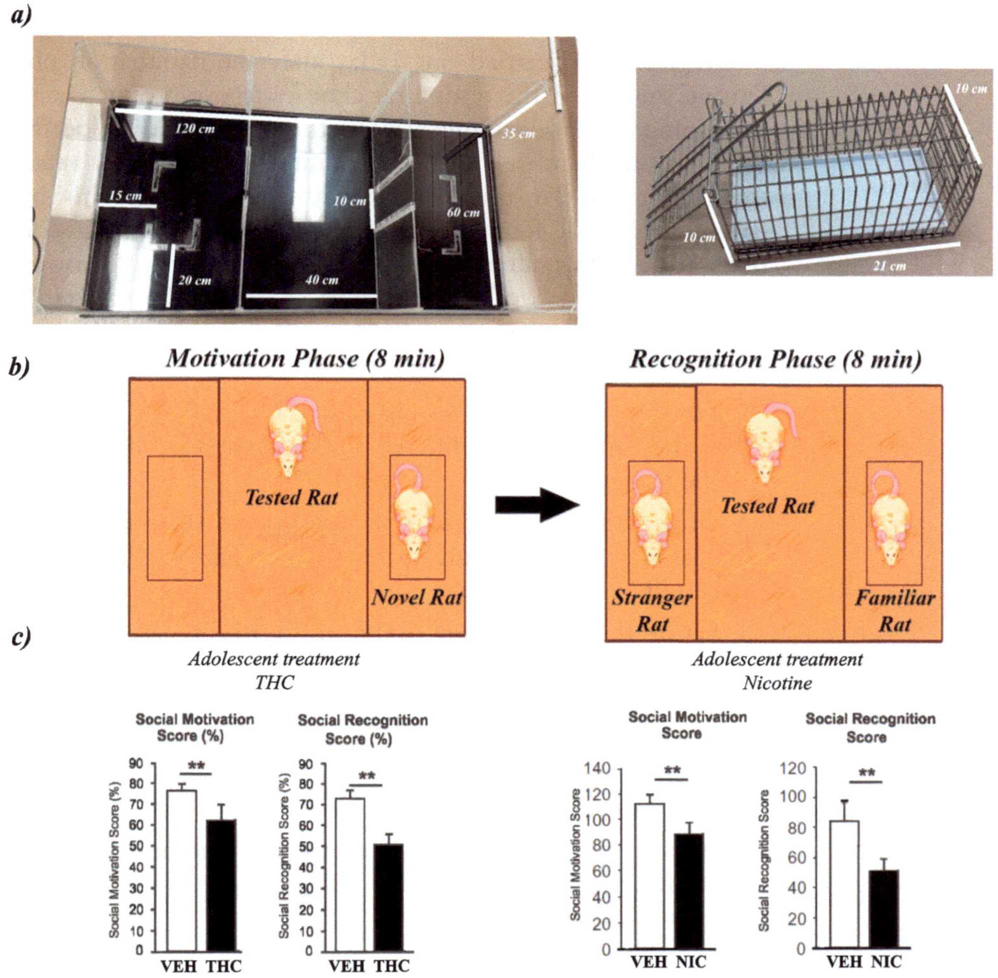

Fig. 3 Social interaction test. (**a**) Picture of three-chambered box (left) and wire cage (right) showing the measurements of each apparatus. (**b**) Schematic summary of the procedure of the social interaction test. (**c**) Effect of THC (left) and nicotine (right) adolescent exposure in social motivation and social recognition of adult rats. $^{**}p < 0.01$ using unpaired *t*-test. (Data shown is adapted from Renard et al. [53] and Jobson et al. [57])

3. Place the rat in the middle chamber of the three-chambered box with the doorways closed and let it explore the chamber for 5 min.
4. Then, open the doorways and let the rat freely explore all three chambers for 8 min.
5. Finally, return the rat to the home cage, and clean the three-chambered box (including floor and walls) and the wire cages with >70% ethanol to avoid olfactory cue biases. Wait for the ethanol to dry before proceeding with the following test rat.

(ii) Social Motivation Phase.

This phase is carried out 1 day after the habituation phase. During this phase, the testing rat chooses to explore the chamber containing the novel (stranger) rat vs. the empty rat cage (located in the opposite chamber). In naive conditions, the healthy rats will spend more time in the chamber with the novel rat, showing a natural preference for socializing. However, in pathologic conditions, for example, after chronic stress exposure [85, 86] or pharmacological models of schizophrenia [87, 88], the spontaneous exploration of the chamber with the novel rat is reduced.

The procedure of Social Motivation Phase is the following:

1. Put the cage with the testing rat and the rat used as a novel rat in the same room (in different cages) for at least 30 min before the testing.
2. Situate the wire cages in the middle of the right and left chambers. Then, ensure that the inter-chamber doorways are closed (*see* Fig. 3).
3. Set up the video camera/webcam, so it is prepared to visually record the whole area of the three-chambered box and start the video recording.
4. Place the tested rat in the middle chamber of the three-chambered box with the inter-chamber doorways closed and let it explore the chamber for 5 min.
5. After 5 min, place the novel rat in a wire cage and put the cage in the middle of a chamber (left or right, counterbalanced between subjects).
6. Open the inter-chamber doorways, and let the test rat explore the three chambers (the chamber with the empty wire cage, the middle section, and the chamber with the novel rat inside the wire cage) for 8 min.

Note: The novel rat is placed in the chamber counterbalanced between trials.

(iii) Social Recognition (Memory) Phase.

This phase is carried out immediately after the social motivation (interaction) phase. In this stage, the novel rat inside the first cage (that was just explored in the motivation phase) becomes the familiar rat, and it is necessary to place a second rat that will now serve as the new "stranger" rat. Healthy rats will choose to spend more time with the newly introduced "stranger" rat and less time with the previously experienced rat due to the loss of novelty of the previous encounter and motivation to explore a new, novel rat. Rats with social cognition/memory impairments will have difficulty remembering the last meeting and will not prefer the new stranger rat. The steps of this phase are the following:

1. After 8 min (motivation phase, **step 6**), place the stranger rat in the previously empty cage in the previously empty chamber.
2. Take the tested rat, put it in the middle chamber, and let it explore the three-chambered box for 8 min.
3. Then, return the tested, familiar, and stranger rats to their respective home cages and stop the video recording.
4. Finally, clean the three-chambered box (including floor and walls) and the wire cages with >70% ethanol to avoid olfactory cue bias.

Note 1: The familiar (novel stranger in the motivation phase) and stranger rats must be of the same sex and of a similar age to the tested rat to mitigate any confounds due to pheromonal differences, baseline activity states, aggression levels, etc. In addition, both rats must not have had any previous contact with the test rat before the test.

Note 2: Keeping the testing room as quiet as possible is recommended during the behavioral test. Any disturbing noises could modify the spontaneous exploration of the tested rat.

(iv) Results, Analysis, and Interpretations.

To determine changes induced by experimental treatments in the social motivation and recognition test, it is required to measure the time spent in each chamber in both test phases. It is recommended to use software to analyze the video offline. In Subheading 3.3.4, software analysis will be fully detailed using the free software BehaView®. It is necessary to describe in the Results (*see* Subheading 3.3.4, (i), **step 3**) the exploration in each chamber of the apparatus (motivation phase: rat chamber and empty chamber; recognition phase: familiar chamber and stranger chamber). An entry in each chamber is defined as all four paws being located in one chamber.

Once the time spent in each chamber in the motivation phase is obtained, the social motivation score is calculated using the following equation:

$$\text{Social motivation index} = \frac{\text{Time in the chamber with Novel Rat}}{\text{Time in chamber with Novel Rat} + \text{Time in empty chamber}} \times 100$$

Thus, in natural conditions, rats prefer to spend more time in the chamber with novel rat resulting in a social motivation index higher than 50%. Chronic unpredictable stress or pharmacological treatments (such as phencyclidine) reduce this natural preference to novelty. On the other hand, benzodiazepines drugs or selective serotonin reuptake inhibitors can improve social motivation indices [72, 89–91]. In the case of THC or nicotine exposure during adolescence described in this chapter, a significant reduction in the social motivation index is observed compared to vehicle controls (*see* Fig. 3c).

The following equation is used to assess the social recognition (memory) score from the recognition phase:

$$\text{Social recognition index} = \frac{\text{Time in the chamber with Stranger Rat}}{\text{Time in chamber with Strange Rat} + \text{Time in chamber with familiar Rat}} \times 100$$

The healthy rat will prefer to spend more time in the chamber with the stranger rat than the chamber with the already familiar rat, demonstrating the social recognition and the preference for novelty. However, acute stress, pharmacological manipulations, or exposure to aversive stimuli during development impair social recognition [92–95]. In particular, exposure to THC or nicotine during adolescence reduces social recognition scores when tested in adulthood, demonstrating that these treatments can induce long-term impairments in social cognition (*see* Fig. 3c).

3.3.3 Novel Object Recognition

The novel object recognition task (NOR) is a behavioral test used to assess working memory. The test is based on the rodent's ability to discriminate between the familiarity vs. novelty of encountered objects [96]. This task is mainly modulated by prefrontal cortex activity and hippocampus, and regulated by different neurotransmitters [73, 97]. In particular, it has been observed that D2 receptor activity in the prefrontal cortex is critical in the working memory of rodents and humans [73, 98, 99]. A deficit in working memory has been observed in various neuropsychiatric disorders in humans such as schizophrenia, major depression, and autism spectrum disorder [74, 100–103].

This test requires an open field box and two pairs of distinguishable objects (*see* the Subheading 2 for more details). Similar to the social interaction task, the novel object recognition task consists of three phases: habituation, acquisition, and recall.

(i) Habituation Phase.

This phase is required to reduce the novelty to the open field box that can modify the exploratory behavior during the task. The habituation phase is carried out 1 day before the acquisition and recall phases. The steps in the habituation are:

1. Place the test rat's home cage in the same test room for at least 30 min.

2. Place the rat in the middle of the empty open field box and let it explore the chamber for 10 min.
3. Finally, return the rat to the home cage, and clean the open field box (including floor and walls) using >70% ethanol to avoid olfactory cue bias. After the ethanol dries, it is possible to proceed with the next rat.

(ii) Acquisition Phase.

The acquisition phase is carried out 24 h after the habituation phase. During this phase, the rat must explore two identical objects and acquire a memory for the objects.

The procedure is the following:

1. Place the home cage with the tested rat in the same room for at least 30 min before testing.
2. Situate the two identical objects in the open field box. Each object is placed 15 cm from the sidewall in the corner, illustrated in Fig. 4.

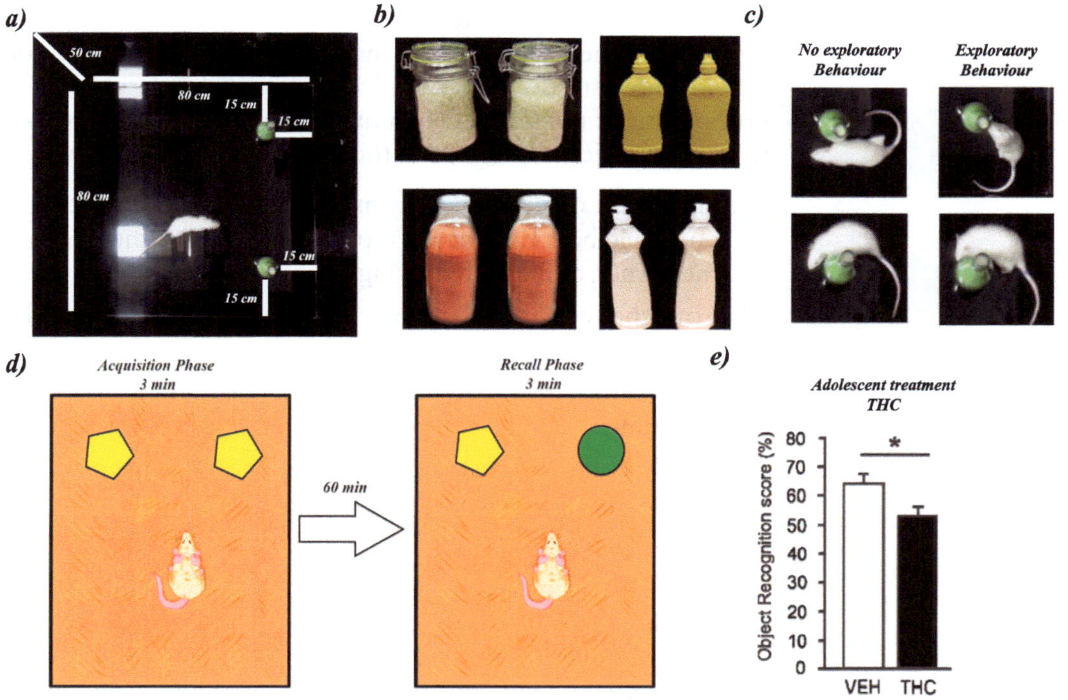

Fig. 4 Novel object recognition test. (**a**) Picture of the open field box showing the measures of the apparatus and the relative position of the object. (**b**) Examples of pairs of objects used in the test. (**c**) Examples of no exploratory behavior (left) and exploratory behavior (right) during the novel object recognition test. (**d**) Schematic summary of the protocol of novel object recognition. (**e**) Effects of adolescent THC exposure in object recognition of adult rats. *$p < 0.05$ using unpaired *t*-test. (Data is adapted from Renard et al. [53])

3. Set up the video camera/webcam such that it is allowed to record the whole open field box area and start the video recording.
4. Place the test rat in the center of the open field and let it explore the chamber for 3 min.
5. Then, return the rat to the home cage and stop the video recording.
6. Clean the open field box and the objects using a 70% ethanol solution. After the ethanol dries, it is possible to proceed with the next rat.

Note 1: The home cage must stay in the test room until finishing the recall phase.

Note 2: The objects used in this phase must be counterbalanced in each trial.

Note 4: Considering that **steps 4–6** could take around 5 min, it is recommended to carry out the acquisition phase a maximum of 10 times (different rats) before starting with the recall phase.

(iii) Recall Phase.

This phase is carried out 60 min after the acquisition phase (longer or shorter time intervals can be tested to examine shorter or longer term memory retention intervals, depending on the question of interest). In this stage, one of the objects used in the previous phase (random and counterbalanced election) is replaced with a new object. The steps of this phase are the following:

1. Situate the object previously presented and a new object in the open field box. Each object is placed 15 cm from the sidewall in the corner, as illustrated in Fig. 4a.
2. Place the test rat in the center of the open field box and let it explore both objects for 3 min.
3. Then, return the tested rat to the home cage and stop the video recording.
4. Finally, clean the open field box (including floor and walls) with >70% ethanol to avoid olfactory cue bias. After the ethanol dries, you may proceed with the next tested rat.

Note 1: The relative positions of the familiar and the novel objects must be counterbalanced in each trial.

Note 2: During the behavioral test, keep the testing room as quiet as possible as any disturbing noises could impact the spontaneous exploration of the testing rat.

(iv) Results, Analysis, and Interpretations.

If a test subject has not displayed exploration of the objects during the acquisition phase (for whatever reason), it is necessary to exclude the subject from further analyses. Several factors can cause this issue, and if it is found that most of the rats did not explore any object, it is recommended to make some protocol adjustments. For example, one explanation is that the experimental subjects are more anxious than usual. In that case, it is recommended to prolong the duration of the acquisition phase, perform more handling sessions, or reduce the light intensity of the room, all of which may mitigate anxiety issues. If the lack of exploration is specific to a particular object, the object should be replaced.

It is recommended to use software to analyze the video recordings. In Subheading 3.3.4, the procedure to analyze the behavior is fully detailed using the free software BehaView®. It is necessary to define the method (*see* Subheading 3.3.4, (i), **step 3**) for the exploration of the familiar and the novel object. The exploratory behavior of an object is considered valid when the rat positions its nose within 2 cm of the object while looking at the object, sniffing it, or touching it with its nose. However, climbing, sitting, or playing with the object is not considered exploratory behaviors per se.

Once the times spent in the exploration of the familiar and novel object in the recall phase is obtained, the recognition index is calculated using the following equation:

$$\text{Recognition index} = \frac{\text{Exploration time of Novel Object}}{\text{Exploration time of Novel Object} + \text{Exploration time of familiar Object}} \times 100$$

In the recall phase under normal conditions, the rat can discriminate between the familiarity of previously encountered objects, resulting in a high recognition index. If an experimental condition reduces this index, rats may be displaying deficits in working memory. In the case of THC exposure during adolescence, we have observed significant reductions in the recognition index when rats are tested in adulthood (*see* Fig. 4e).

3.3.4 Software Analyses

In our protocol, BehaView® software is used for accessibility and versatility, allowing the customization of multiple methods. This section will describe the procedural details necessary to analyze the behavioral protocols described previously.

The procedure will be explained in two sections: set method and video analysis.

(i) Set method.

1. Open BehaView®.
2. Click on "New."
3. Set the name of behavior (e.g., grooming, risk assessment, exploration to Chamber 1, etc.), key (the key will be used to

start and finish the behavior), color, and define if the behavior is a state or an event (e.g., time in the Chamber 1 is a state and the number of entries is an event). Then, click "OK."

4. Repeat **step 3** for each behavior that you want to analyze.
5. Save the setting. Click on file, then put the arrow on "behaviors," click on "save." Finally, name the setting with the respective behavioral task (e.g., "Phase 1 Social interaction").

(ii) Video analysis.

1. Open the desired method configured in the previous section. Click on file, put the arrow on "behavior," click on "Open." Select the method you want to use.
2. Open the video. Click on the "" icon and select the video.
3. Define the time when starting the task. Move the green cursor until the task begins and press the "Start" button.
4. There are two methods for analysis: (1) click on the "play" button. The recorded video will play, and the green cursor will indicate the specific part of the playing video. (2) Move the mouse's scroll wheel forward or rewind the video.

Note: It is recommended to use the scroll wheel in behaviors required to detect coarse movements (e.g., movements between test chambers). When analyzing more subtle behaviors (e.g., grooming or sniffing), it is recommended to use the "play" button method for a more granular overview of the recordings.

5. To measure a specific behavior, press the key configurated in the "Set method" (Subheading 3.3.4, (i), **step 3**) when the animal starts to do the behavior of interest. When the rat stops the behavior, press the same key to record the behavior.

Note: In case of mistakes, it is possible to edit the recording. Select the beginning of the behavior you want to erase in the lateral list (left) and press "DEL" or "SUPR" on the keyboard. Repeat the previous step with the end of the behavior you want to erase.

6. Repeat **step 5** each time the rat performs any behavior of interest until the task is finished.
7. Finish the behavioral recording by pressing the "END" button. Then save the recording, click on "file," move the arrow on "sequence," and click save. Finally, name the sequence of the behavior and save it.

Note: If the animal is still doing the behavior after the test, you must press the assigned key to end the behavior with the "END" button.

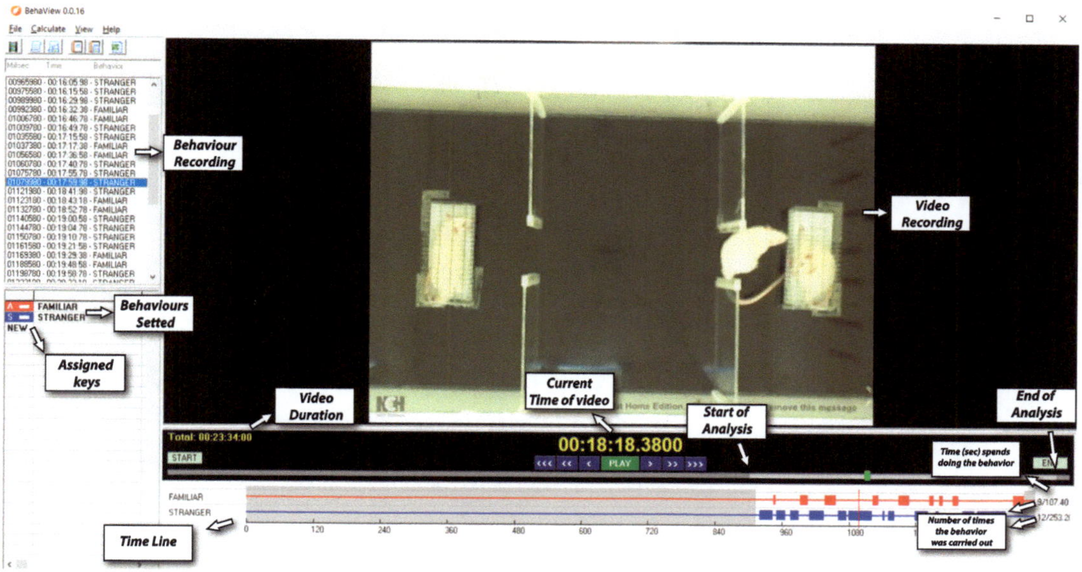

Fig. 5 Interface of BehaView® video analysis

8. The result of each behavior is shown on the right side of the timeline (*see* Fig. 5). The first number is the number of times the behavior was carried out. The second number is the total time the animal spends doing this behavior. To export the data, copy the results and paste in the statistical software (e.g., Excel, SPSS, or GraphPad Prims).

3.4 Single-Unit Recording of Dopamine Neurons in VTA

The activity of dopamine neurons from VTA is critical in several cognitive processes, which highlight its role in reward processing, working memory, decision-making, and sociability [104–106]. This section will detail the procedure used in our laboratory to measure the firing rate and the bursting patterns of dopaminergic neurons from VTA of anesthetized rats. It will be shown the effects of adolescent THC and nicotine exposure during adulthood in dopaminergic activity.

3.4.1 Set up Multiclamp® 700B Amplifier and Clampex® Acquisition Software

This section will show the configuration of the amplifier and acquisition software used in our laboratory for a successful single-unit recording. For more detail about software, it is recommended to visit the official website and check the respective manuals.

- Multiclamp 700B amplifier configuration.

Multiclamp 700B is an analog input/output instrument operated using a control panel program, Multiclamp® 700B amplifier Commander. The steps to set up the program are the following:

1. Turn on the Multiclamp® 700B amplifier and open the Multiclamp® 700B amplifier Commander.

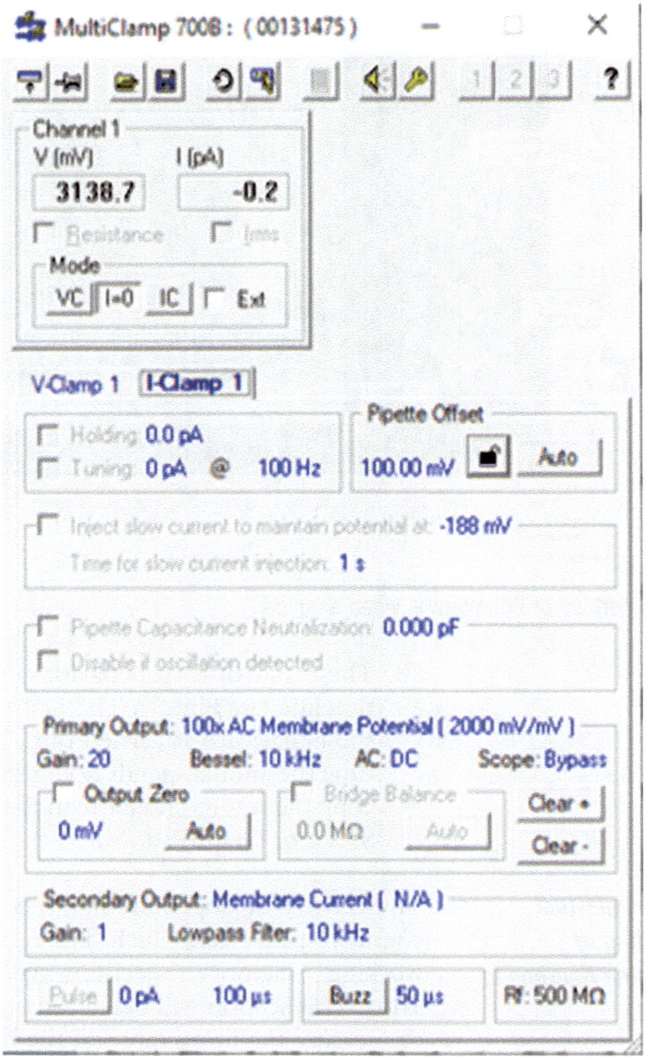

Fig. 6 Multiclamp® 700B amplifier Commander configuration

2. In commander, click on "Select Device." In the new window, select "Multiclamp Hardware" to locate "Multiclamp® 700B," then click on "Scan." After that, the serial number should be displayed, then press "OK."
3. In the commander main window, select mode "I = 0."
4. In I-Clamp 1 menu, set up primary output using "100× AC Membrane Potential (2000 mV/mV)," gain "20," Bessel "10 kHz," AC "DC," scope "Bypass" (*see* Fig. 6).
5. Finally, set the resistance reference (Rf) at 500 MΩ (*see* Fig. 6).

- Digidata® 1440A and Clampex® acquisition software configuration.

The Digidata® 1440A digitizer is a high-resolution, low-noise digitizer, particularly designed for electrophysiology experiments to send and receive signals from microelectrode amplifiers. This digitizer is supported by Clampex 10 acquisition software. This section will describe the configuration to start to use this software.

For the first use of the digitizer, it is necessary to link the digitizer with the software following the next steps:

1. Connect and switch on the digitizer, then open Clampex software.
2. Open the "configure" menu and click on "digitizer...".
3. In the new window, click on "change...," in the menu "digitizer type" selects "digidata 1440 series" and click on "Scan." Now the software will display the digitizer connected with the computer, then click on "OK" in the two windows.

Then, it is necessary to link the digitizer with the amplifier and configure the digitizer channel with the following steps:

1. Open the "configure" menu and click on "Telegraph Instrument...".
2. Select the channel connected in the digitizer and select "Axon Multiclamp 700B" in the new window. Then, select the option "Start new trial when mode, scale factor or units change."
3. In the "Amplifier Configuration" menu, select "Primary output 1." Then, click on "OK."
4. Open the "configure" menu and click on "Lab Bench."
5. Select the channel connected in the digitizer and click on "Add..." in the new window.
6. Write an appropriated signal name, for example, "VTA unit," and click on "OK."
7. Select in the signal unit "m" and "V."
8. In software RC filter, set low-pass filter at 10 kHz and high-pass filter at 300 Hz. Then click on "OK."

Before starting the first recording experiment, it is needed to set up the protocol of acquisition following these steps:

1. In the Clampex software, open the "acquisition" menu and click on "new protocol."
2. In the new window, select "Gap-free" in acquisition mode; in trial length, select "Use available disk space" and set the sampling rate per signal at 25,000 Hz.
3. Click on the "Input" menu, select the channel connected in the digitizer, open the menu, and select the previously configured signal (in our example, "VTA unit"). Then, click on "OK."

4. Open the "acquisition" menu and click on "Save protocol" to use the same protocol in your future experiment without configuring the protocol again.
5. To use the same protocol for future experiments, open the "acquisition" menu, click on "open protocol," and open the previously saved protocol.

Finally, the software is ready to begin electrophysiological experiments. If you want to set up the folder in which the recording file is saved, follow the following steps:

1. Open the "File" menu and click on "set data file names...".
2. Search the folder (or create a new one) in the new window to save the recording file.
3. Also, you can change the predefined name of your files.
4. Click on "OK."

3.4.2 Experimental Protocol

This section will show the steps to carry out a single-unit recording of putative dopaminergic neurons in the VTA of anesthetized rats:

1. Pull glass tubing electrodes on a vertical electrode puller and fill the glass electrode with filtered 2% pontamine Sky Blue in 2 M sodium acetate solution.

Note 1: Avoid the air bubbles inside of the electrode. The air bubbles may modify the impedance of the electrode and obstruct the electrophysiological recording.

Note 2: Given the fragility of the glass electrode, it is recommended to craft around six electrodes per experiment if you should replace it.

2. Test the impedance of the electrodes using a micro electrode tester. The recommended in situ range is from 6 to 15 MΩ.
3. Anesthetize the rat using a dose of 1.4 g/kg of urethane. Then, wait until the foot pinch reflex is not present.

Note: The estimated time to wait is around 45 min. If the rat is not completely anesthetized, inject 15% more of the dose.

4. Keep rat body temperature at 36 °C using a heating pad.

Note 1: Keep contact of the tail with the heating pad because the rat's tail is critical in the temperature control.

Note 2: Avoid using an electric heat pad because it may increase the noise in the baseline of the electrophysiological recording.

5. Place the anesthetized rat in a stereotaxic apparatus.
6. Remove the skin and clean the skull using saline solution and gauze.

7. Place an electrode in the holder of the stereotaxic apparatus arm. Then, cut the tip of the electrode and use it to mark bregma in the skull. Record the bregma coordinate.
8. Move the electrode using stereotaxic apparatus arm to the coordinate of VTA, in this case: AP −5.2, ML +0.8. Mark the position with the electrode in the skull.
9. Draw a circle with a marker in the skull, taking the center the mark of **step 8**.
10. Drill a hole using the circle of **step 9** as reference. Remove the bone using a tweezer.

Note: You should drill with cautions in this step because the hole is close to a blood vessel. In case of bleeding, place the gauze in the hole and pressure. After a couple of minutes, the bleeding should stop unless replacing the gauze and repeat the procedure.

11. Remove dura mater using a tweezer.
12. Place a new electrode in the holder of the stereotaxic apparatus arm. Connect the electrode with the headstage and set up the ground connections.

Note: You should set up the stereotaxic apparatus inside a Faraday cage to reduce the external electromagnetic interference. For more information about Faraday cage and how to craft it visit the following link: https://backyardbrains.com/experiments/faraday

13. Turn on the Digidata® 1440A digitizer (set up to 25 kHz sample frequency and band-pass filter between 0.3 and 3 kHz), the MultiClamp® 700B amplifier and the audio monitor. Then prepare the acquisition software. For more details about the acquisition parameter and software, *see* Subheading 3.4.1.
14. Slowly lower the electrode with the stereotaxic manipulator until it detects a baseline change. Then, record the dorsoventral coordinate and lower the electrode slowly until 6.5 mm to track dopamine cells.
15. When you find a stable spontaneous neuronal activity with a signal-to-noise ratio greater than 3:1 signal, click on the "•" symbol in Clampex software to record the signal for at least 3 min. The dopaminergic neurons are identified with the following criteria: (1) action potential width > 2.2 ms, (2) spontaneous firing rate 1–10 Hz, (3) a triphasic waveform consisting of a notch on the rising phase followed by a delayed-after potential, and (4) a single irregular or burst firing pattern. Burst firing pattern is defined as two consecutive spikes with a burst interval of 80 ms. Click on the "/" symbol in Clampex software to finish the record.

16. Sample the VTA through at least six vertical tracks separated by 200 mm at the following coordinates: AP −5.0 to −5.4, ML 0.8 to 1.0 from bregma, and −6.5 to −9.0 DV from the brain surface. It is recommended during the sampling to lower the electrode 10 μm every 3 s to assure a high probability of finding active neurons.

Note: If you do not detect any cell or the positive phase of a neuronal signal in two consecutive vertical tracks, you should replace the electrode and repeat the tracks.

17. After the last track, place the electrode in the coordinate of the most lateral cell that you detect. Then, stop and close the acquisition software, switch off the MultiClamp® 700B amplifier and the Digidata® 1440A digitizer.
18. Disconnect the electrode of the headstage and connect the iontophoresis pump to the electrode and the ground.
19. Turn on the iontophoresis pump at 20 μA constant negative current for 15 min.
20. Finally, turn off and disconnect the pump, remove the electrode, and decapitate the rat with a sharp guillotine. Take and store the brain in a vial with 4% paraformaldehyde.

3.4.3 Results, Analyses, and Interpretations

Two parameters are used to measure the electrical activity of dopaminergic neurons: average firing rate and the percentage of spikes that occurred in bursts. To calculate the firing rate, it is necessary to determine the number of spikes and the time of the record.

In our laboratory, we use Clampfit® 10.7 software to determine the firing rate of each recorded cell (*see* Fig. 7). The procedure is the following:

1. Open the file of the neuron recording that you want to analyze (use Clampfit® 10.7 software to open the file recorded with Clampex® software).
2. Set cursor 1 and cursor 2 at 0 s and 180 s, respectively.
3. Select "Event Detection" and click on "Threshold Search."
4. In the right menu, set the voltage to trigger, which can discriminate the signal from the noise (*see* Fig. 7a). Then, click on "OK."
5. Click on "" symbol, and you will get the number of events between cursor 1 and 2 (*see* Fig. 7b).
6. Divide the number of events with 180 s, and you will get the firing rate of the respective neuron.

We also calculate the percentage of spikes that are in burst to quantify the burst pattern of dopamine neurons, following these steps:

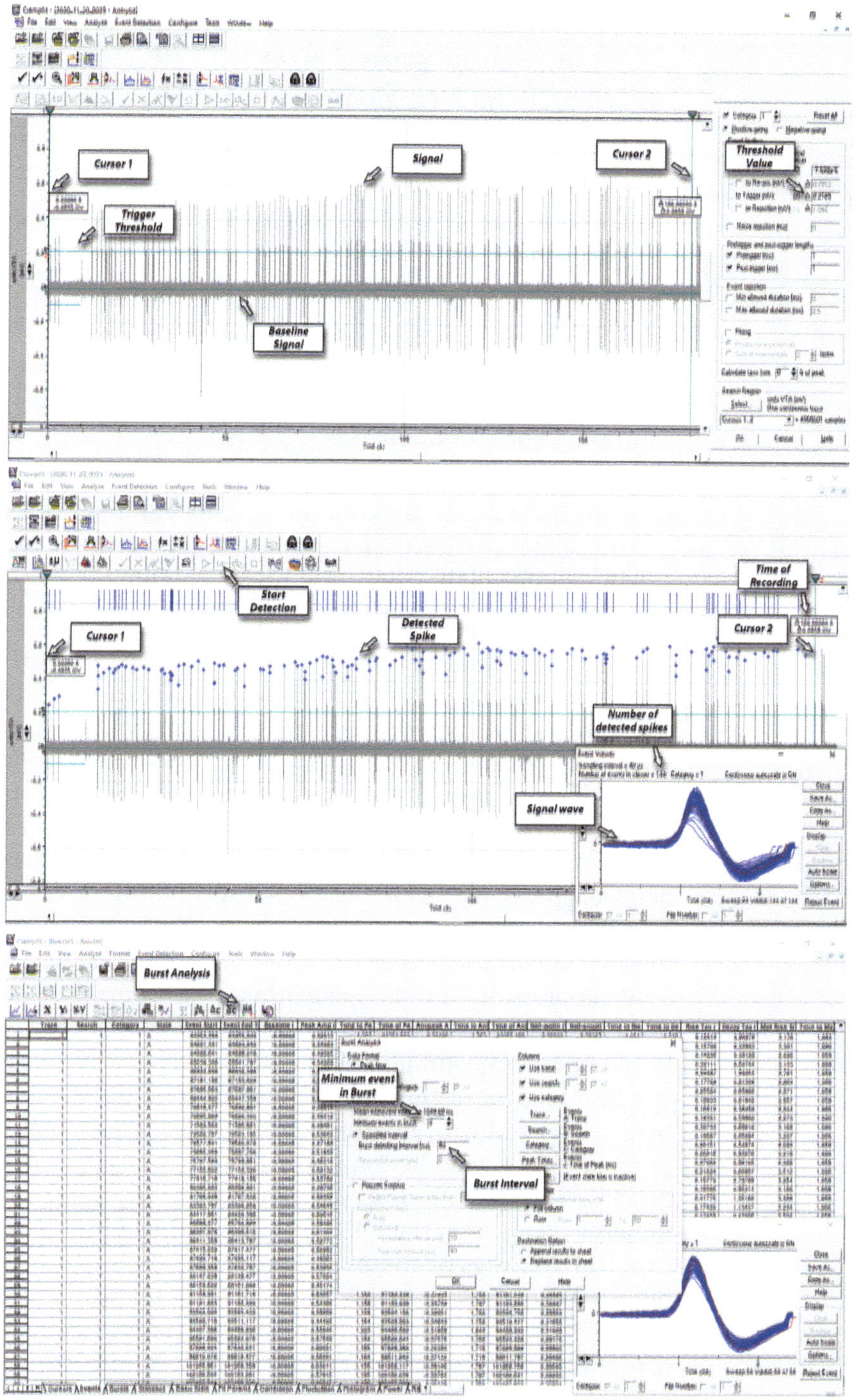

Fig. 7 Clampfit® 10.7 acquisition software. (**a**) Example of threshold value configuration in the "Event Detection" function of a dopamine activity recording. (**b**) Example of detected events and the waveform of the signal of a putative dopamine neuron recorded. (**c**) Example of "Burst Analysis" of a dopamine activity recording

1. Open the file of the neuron recording that you want to analyze.
2. Set cursor 1 and cursor 2 at 0 s and 180 s, respectively.
3. Select "Event Detection" and click on "Threshold Search."
4. In the right menu, set the voltage to trigger to discriminate the signal from the noise (*see* Fig. 7a). Then, click on "OK."
5. Click on "" symbol, and you will get the number of events between cursor 1 and 2.
6. Minimize the file window and maximize the "results1" window.
7. Click on "Burst Analysis" ("" symbol) and set minimum event per burst in "2" and burst delimiting interval (ms) at "80." Then, click on "OK" (*see* Fig. 7c).
8. Now, you can observe a table with the burst events (Fig. 7c, in "Bursts" tab). In the column "Events in Burst" you can obtain the spikes per burst. To calculate the percentage of spikes per burst, sum all the "Events in Burst" column numbers, divide by the number of events obtained in **step 5**, and multiply per 100.

Midbrain dopamine neurons exhibit two distinct modes of activity: tonic firing state and phasic firing state [107]. Tonic firing refers to spontaneous baseline activity characterized by a slow frequency of spikes between 1 and 10 Hz [108, 109]. Phasic firing state is the transition into a rapid burst-firing mode which can reach an intra-burst frequency of 30 spikes per second [110, 111]. These different states have been associated with wide behavioral and neuronal processes [112]. The burst spike firing of dopamine neurons is triggered by the presentation of unexpected rewards or sensory signals predicting such rewards [113]. Furthermore, transient suppression of tonic spike firing of dopaminergic neurons has been observed in response to an omission of expected rewards [114] or aversive stimuli [115].

In addition, disruption of dopaminergic electrical activity has been associated with neuropsychiatric pathologies such as schizophrenia and drug addiction [116, 117]. For example, in preclinical models, depression is observed a prolonged hypofunction of dopaminergic neurons in the VTA [118, 119]. In addition, the repeated exposure to drugs of abuses produces an augmentation of dopaminergic mesolimbic transmission [120, 121], which is associated with a higher probability of psychosis episode [122, 123] and compulsive drug-seeking behavior [124]. Adolescent exposure to THC or nicotine increases the mesolimbic dopaminergic transmission in adulthood (*see* Fig. 8c). The evidence in this chapter suggests that THC and nicotine adolescent exposure induces long-term dopaminergic changes, which are associated with impairment in sociability, anhedonia, and working memory.

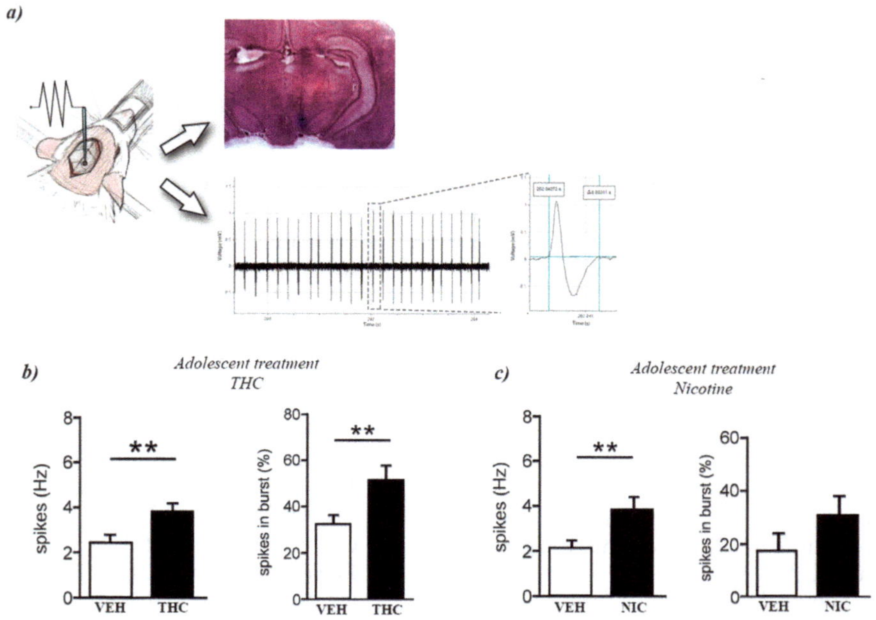

Fig. 8 Extracellular single-unit recording of dopamine neurons from VTA. (**a**) Representation of anesthetized rat in the stereotaxic apparatus (left), example of electrode placement in the VTA (top, right), and example of dopaminergic neuron recording and waveform of action potential (bottom, right). (**b**) Effects of adolescent THC exposure in firing rate and spikes in burst of dopaminergic neurons of adult rats. $^{**}p < 0.01$ using unpaired *t*-test. (Data were adapted from De Felice et al. [65]). (**c**) Effects of adolescent Nicotine treatment in firing rate and spikes in burst of dopaminergic neurons of adult rats. $^{**}p < 0.01$ using unpaired *t*-test. (Data were adapted from Jobson et al. [57])

4 Conclusions

In the current global context, with an increase in the consumption of cannabis during adolescence and the diminished perception of health risks associated with the use of e-cigarettes, it is critical to determine the long-term neurobiological and neuropsychiatric consequences of THC and nicotine adolescent exposure. Despite the extensive clinical evidence demonstrating associative links between adolescent consumption of tobacco and cannabis with higher neuropsychiatric risk, it is difficult to prove causal relationships due to inherent experimental limitations found in most clinical research approaches. With this in mind, preclinical animal models are indispensable tools to determine the causal and neurobiological mechanisms underlying the relationship between early neurodevelopmental drug exposure and psychiatric disease risk.

The methodological tools described in this chapter aim to expand the possibilities in neurodevelopmental research and allow for a more detailed investigation into the pathophysiological dopaminergic mechanisms associated with adolescent exposure to cannabinoids and nicotine-based drug products.

References

1. Spear LP (2000) The adolescent brain and age-related behavioral manifestations. 24(4). https://doi.org/10.1016/S0149-7634(00)00014-2
2. World Health Organization (2019) WHO global report on trends in prevalence of tobacco use 2000–2025, 2nd ed. https://apps.who.int/iris/bitstream/handle/10665/272694/9789241514170-eng.pdf
3. Saddleson ML, Kozlowski LT, Giovino GA, Homish GG, Mahoney MC, Goniewicz ML (2016) Assessing 30-day quantity-frequency of U.S. adolescent cigarette smoking as a predictor of adult smoking 14 years later. Drug Alcohol Depend 162:92–98. https://doi.org/10.1016/j.drugalcdep.2016.02.043
4. Ambrose BK et al (2014) Perceptions of the relative harm of cigarettes and E-cigarettes among U.S. youth. *Am J Prev Med* 47(2 Suppl. 1):S53–S60. https://doi.org/10.1016/j.amepre.2014.04.016
5. Jamal A (2000) Tobacco use among middle and high school students-United States, 1999. Conn Med 64(4):216–218. https://doi.org/10.15585/mmwr.mm6623a1
6. Perikleous EP, Steiropoulos P, Paraskakis E, Constantinidis TC, Nena E (2018) E-cigarette use among adolescents: an overview of the literature and future perspectives. *Front Public Health* 6:1. https://doi.org/10.3389/fpubh.2018.00086. Frontiers Media S.A.
7. United Nations Office on Drugs and Crime (2020) World drug report 2020: drug use and health consequences. https://wdr.unodc.org/wdr2020/field/WDR20_Booklet_2.pdf
8. Paus T, Keshavan M, Giedd JN (2008) Why do many psychiatric disorders emerge during adolescence? Nat Rev Neurosci 9(12):947–957. https://doi.org/10.1038/nrn2513
9. Merikangas KR, Nakamura EF, Kessler RC (2009) Epidemiology of mental disorders in children and adolescents. *Dialogues Clin Neurosci* 11(1):7–20. https://doi.org/10.31887/dcns.2009.11.1/krmerikangas
10. Schneider M (2008) Puberty as a highly vulnerable developmental period for the consequences of cannabis exposure. Addict Biol 13(2):253–263. https://doi.org/10.1111/j.1369-1600.2008.00110.x
11. Schneider M (2013) Adolescence as a vulnerable period to alter rodent behavior. *Cell Tissue Res* 354(1):99–106. https://doi.org/10.1007/s00441-013-1581-2. Springer.
12. Brook JS, Schuster E, Zhang C (2004) Cigarette smoking and depressive symptoms: a longitudinal study of adolescents and young adults. Psychol Rep 95(1):159–166. https://doi.org/10.2466/pr0.95.1.159-166
13. Ranjit A et al (2019) Testing the reciprocal association between smoking and depressive symptoms from adolescence to adulthood: a longitudinal twin study. Drug Alcohol Depend 200:64–70. https://doi.org/10.1016/j.drugalcdep.2019.03.012
14. Ranjit A, Latvala A, Kinnunen TH, Kaprio J, Korhonen T (2020) Depressive symptoms predict smoking cessation in a 20-year longitudinal study of adult twins. Addict Behav 108:106427. https://doi.org/10.1016/j.addbeh.2020.106427
15. Choi WS, Patten CA, Christian Gillin J, Kaplan RM, Pierce JP (1997) Cigarette smoking predicts development of depressive symptoms among U.S. adolescents. Ann Behav Med 19(1):42–50. https://doi.org/10.1007/BF02883426
16. Dalack GW, Meador-Woodruff JH (1996) Smoking, smoking withdrawal and schizophrenia: case reports and a review of the literature. Schizophr Res 22(2):133–141. https://doi.org/10.1016/S0920-9964(96)80441-5
17. Hughes JR, Hatsukami DK, Mitchell JE, Dahlgren LA (1986) Prevalence of smoking among psychiatric outpatients. Am J Psychiatry 143(8):993–997. https://doi.org/10.1176/ajp.143.8.993
18. Mustonen A et al (2018) Smokin‘ hot: adolescent smoking and the risk of psychosis. Acta Psychiatr Scand 138(1):5–14. https://doi.org/10.1111/acps.12863
19. Weiser M et al (2004) Higher rates of cigarette smoking in male adolescents before the onset of schizophrenia: a historical-prospective cohort study. Am J Psychiatry 161(7):1219–1223. https://doi.org/10.1176/appi.ajp.161.7.1219
20. Fergusson DM, Horwood LJ, Swain-Campbell N (2002) Cannabis use and psychosocial adjustment in adolescence and young adulthood. Addiction 97(9):1123–1135. https://doi.org/10.1046/j.1360-0443.2002.00103.x
21. Andréasson S, Engström A, Allebeck P, Rydberg U (1987) Cannabis and schizophrenia a longitudinal study of Swedish conscripts. Lancet 330(8574):1483–1486. https://doi.org/10.1016/S0140-6736(87)92620-1

22. Arseneault L, Cannon M, Poulton R, Murray R, Caspi A, Moffitt TE (2002) Cannabis use in adolescence and risk for adult psychosis: longitudinal prospective study. BMJ 325(7374):1212–1213. https://doi.org/10.1136/bmj.325.7374.1212
23. Stefanis NC, Delespaul P, Henquet C, Bakoula C, Stefanis CN, Van Os J (2004) Early adolescent cannabis exposure and positive and negative dimensions of psychosis. Addiction 99(10):1333–1341. https://doi.org/10.1111/j.1360-0443.2004.00806.x
24. Laviolette SR (2021) Exploring the impact of adolescent exposure to cannabinoids and nicotine on psychiatric risk: insights from translational animal models. *Psychol Med* 51(6): 940–947. https://doi.org/10.1017/S0033291719003325. Cambridge University Press.
25. Doremus-Fitzwater TL, Spear LP (2016) Reward-centricity and attenuated aversions: an adolescent phenotype emerging from studies in laboratory animals. Neurosci Biobehav Rev 70:121–134. https://doi.org/10.1016/j.neubiorev.2016.08.015
26. Gardner M, Steinberg L (2005) Peer influence on risk taking, risk preference, and risky decision making in adolescence and adulthood: an experimental study. Dev Psychol 41(4):625–635. https://doi.org/10.1037/0012-1649.41.4.625
27. Douglas LA, Varlinskaya EI, Spear LP (2004) Rewarding properties of social interactions in adolescent and adult male and female rats: impact of social versus isolate housing of subjects and partners. Dev Psychobiol 45(3): 153–162. https://doi.org/10.1002/dev.20025
28. Foulkes L, Blakemore SJ (2016) Is there heightened sensitivity to social reward in adolescence? *Curr Opin Neurobiol* 40:81–85. https://doi.org/10.1016/j.conb.2016.06.016. Elsevier Ltd.
29. Desor JA, Beauchamp GK (1987) Longitudinal changes in sweet preferences in humans. Physiol Behav 39(5):639–641. https://doi.org/10.1016/0031-9384(87)90166-1
30. Wilmouth CE, Spear LP (2009) Hedonic sensitivity in adolescent and adult rats: taste reactivity and voluntary sucrose consumption. Pharmacol Biochem Behav 92(4):566–573. https://doi.org/10.1016/j.pbb.2009.02.009
31. Steinberg L, Cauffman E, Woolard J, Graham S, Banich M (2009) Are adolescents less mature than adults? Minors' access to abortion, the juvenile death penalty, and the alleged APA 'flip-flop. Am Psychol 64(7): 583–594. https://doi.org/10.1037/a0014763
32. Doremus-Fitzwater TL, Barreto M, Spear LP (2012) Age-related differences in impulsivity among adolescent and adult Sprague-Dawley rats. Behav Neurosci 126(5):735–741. https://doi.org/10.1037/a0029697
33. Gogtay N et al (2004) Dynamic mapping of human cortical development during childhood through early adulthood. Proc Natl Acad Sci 101(21):8174–8179. https://doi.org/10.1073/pnas.0402680101
34. Simmonds DJ, Hallquist MN, Asato M, Luna B (2014) Developmental stages and sex differences of white matter and behavioral development through adolescence: a longitudinal diffusion tensor imaging (DTI) study. Neuroimage 92:356–368. https://doi.org/10.1016/j.neuroimage.2013.12.044
35. Weickert CS et al (2007) Postnatal alterations in dopaminergic markers in the human prefrontal cortex. Neuroscience 144(3): 1109–1119. https://doi.org/10.1016/j.neuroscience.2006.10.009
36. Larsen B, Luna B (2015) In vivo evidence of neurophysiological maturation of the human adolescent striatum. Dev Cogn Neurosci 12: 74–85. https://doi.org/10.1016/j.dcn.2014.12.003
37. Larsen B, Luna B (2018) Adolescence as a neurobiological critical period for the development of higher-order cognition. *Neurosci Biobehav Rev* 94:179–195. https://doi.org/10.1016/j.neubiorev.2018.09.005
38. Sowell ER, Thompson PM, Holmes CJ, Jernigan TL, Toga AW (1999) In vivo evidence for post-adolescent brain maturation in frontal and striatal regions [1]. Nat Neurosci 2(10):859–861. https://doi.org/10.1038/13154. Nature Publishing Group.
39. Ernst M, Luciana M (2015) Neuroimaging of the dopamine/reward system in adolescent drug use. CNS Spectr 20(04):427–441. https://doi.org/10.1017/s1092852915000395
40. Galvan A, Hare T, Voss H, Glover G, Casey BJ (2007) Risk-taking and the adolescent brain: who is at risk? Dev Sci 10(2):F8–F14. https://doi.org/10.1111/j.1467-7687.2006.00579.x
41. Geier CF, Terwilliger R, Teslovich T, Velanova K, Luna B (2010) Immaturities in reward processing and its influence on inhibitory control in adolescence. Cereb Cortex 20(7):1613–1629. https://doi.org/10.1093/cercor/bhp225

42. Wahlstrom D, Collins P, White T, Luciana M (2010) Developmental changes in dopamine neurotransmission in adolescence: behavioral implications and issues in assessment. Brain Cogn 72(1):146–159. https://doi.org/10.1016/j.bandc.2009.10.013. Academic Press.
43. Pérez-Valenzuela E, Castillo-Faúndez R, Fuentealba JA (2019) Comparing dopaminergic dynamics in the dorsolateral striatum between adolescent and adult rats: effect of an acute dose of WIN55212-2. Brain Res 1719: 235–242. https://doi.org/10.1016/j.brainres.2019.06.005
44. McCutcheon JE, Marinelli M (2009) Age matters. Eur J Neurosci 29(5):997–1014. https://doi.org/10.1111/j.1460-9568.2009.06648.x
45. Naneix F, Marchand AR, Di Scala G, Pape J-R, Coutureau E (2012) Parallel maturation of goal-directed behavior and dopaminergic systems during adolescence. J Neurosci 32(46):16223–16232. https://doi.org/10.1523/JNEUROSCI.3080-12.2012
46. Padmanabhan A, Luna B (2014) Developmental imaging genetics: linking dopamine function to adolescent behavior. *Brain Cogn* 89:27–38. https://doi.org/10.1016/j.bandc.2013.09.011. Academic Press Inc.
47. Badanich KA, Adler KJ, Kirstein CL (2006) Adolescents differ from adults in cocaine conditioned place preference and cocaine-induced dopamine in the nucleus accumbens septi. Eur J Pharmacol 550(1–3):95–106. https://doi.org/10.1016/j.ejphar.2006.08.034
48. Spear LP (2013) Adolescent neurodevelopment. *J Adolesc Health* 52(2 Suppl. 2):S7–S13. https://doi.org/10.1016/j.jadohealth.2012.05.006
49. Pérez-Valenzuela EJ, Andrés Coke ME, Grace AA, Fuentealba Evans JA (2021) Adolescent exposure to WIN 55212-2 render the nigrostriatal dopaminergic pathway activated during adulthood. Int J Neuropsychopharmacol 23(9):626–637. https://doi.org/10.1093/ijnp/pyaa053
50. Gomes FV, Guimarães FS, Grace AA (2015) Effects of pubertal cannabinoid administration on attentional set-shifting and dopaminergic hyper- responsivity in a developmental disruption model of schizophrenia. Int J Neuropsychopharmacol:1–10. https://doi.org/10.1093/ijnp/pyu018
51. Pistis M, Perra S, Pillolla G, Melis M, Muntoni AL, Gessa GL (2004) Adolescent exposure to cannabinoids induces long-lasting changes in the response to drugs of abuse of rat midbrain dopamine neurons. Biol Psychiatry 56(2):86–94. https://doi.org/10.1016/j.biopsych.2004.05.006
52. Renard J et al (2017) Adolescent cannabinoid exposure induces a persistent sub-cortical hyper-dopaminergic state and associated molecular adaptations in the prefrontal cortex. Cereb Cortex 27(2):1297–1310. https://doi.org/10.1093/cercor/bhv335
53. Renard J et al (2017) Adolescent THC exposure causes enduring prefrontal cortical disruption of GABAergic inhibition and dysregulation of sub-cortical dopamine function. Sci Rep 7(1):1–14. https://doi.org/10.1038/s41598-017-11645-8
54. Higuera-Matas A et al (2010) Periadolescent exposure to cannabinoids alters the striatal and hippocampal dopaminergic system in the adult rat brain. Eur Neuropsychopharmacol 20(12):895–906. https://doi.org/10.1016/j.euroneuro.2010.06.017
55. Bortolato M et al (2014) Juvenile cannabinoid treatment induces frontostriatal gliogenesis in Lewis rats. Eur Neuropsychopharmacol 24(6):974–985. https://doi.org/10.1016/j.euroneuro.2013.12.011
56. Zamberletti E et al (2012) Gender-dependent behavioral and biochemical effects of adolescent delta-9-tetrahydrocannabinol in adult maternally deprived rats. Neuroscience 204: 245–257. https://doi.org/10.1016/J.NEUROSCIENCE.2011.11.038
57. Jobson CLM et al (2019) Adolescent nicotine exposure induces dysregulation of mesocorticolimbic activity states and depressive and anxiety-like prefrontal cortical molecular phenotypes persisting into adulthood. Cereb Cortex 29(7):3140–3153. https://doi.org/10.1093/cercor/bhy179
58. Counotte DS et al (2009) Long-lasting cognitive deficits resulting from adolescent nicotine exposure in rats. Neuropsychopharmacology 34(2):299–306. https://doi.org/10.1038/npp.2008.96
59. Gumbleton M, Benet LZ (1991) Drug metabolism and laboratory anesthetic protocols in the rat: examination of antipyrine pharmacokinetics. Pharm Res 8(4):544–546. https://doi.org/10.1023/A:1015827917684
60. Howard SG, Feigenbaum JJ (1997) Effect of γ-hydroxybutyrate on central dopamine release in vivo: a microdialysis study in awake and anesthetized animals. Biochem Pharmacol 53(1):103–110. https://doi.org/10.1016/S0006-2952(96)00664-8
61. Tepper JM, Creese I, Schwartz DH (1991) Stimulus-evoked changes in neostriatal

dopamine levels in awake and anesthetized rats as measured by microdialysis. Brain Res 559(2):283–292. https://doi.org/10.1016/0006-8993(91)90013-L

62. Carbone M et al (2010) Chemical characterisation of oxidative degradation products of Δ9-THC. Tetrahedron 66(49):9497–9501. https://doi.org/10.1016/j.tet.2010.10.025
63. González S, Cebeira M, Fernández-Ruiz J (2005) Cannabinoid tolerance and dependence: a review of studies in laboratory animals. Pharmacol Biochem Behav 81(2 Spec. Iss):300–318. https://doi.org/10.1016/j.pbb.2005.01.028
64. Rubino T et al (2009) The depressive phenotype induced in adult female rats by adolescent exposure to THC is associated with cognitive impairment and altered neuroplasticity in the prefrontal cortex. Neurotox Res 15(4):291–302. https://doi.org/10.1007/s12640-009-9031-3
65. De Felice M et al (2021) L-theanine prevents long-term affective and cognitive side effects of adolescent Δ-9-tetrahydrocannabinol exposure and blocks associated molecular and neuronal abnormalities in the mesocorticolimbic circuitry. J Neurosci 41(4):739–750. https://doi.org/10.1523/JNEUROSCI.1050-20.2020
66. Rubino T et al (2008) Chronic Δ9-tetrahydrocannabinol during adolescence provokes sex-dependent changes in the emotional profile in adult rats: behavioral and biochemical correlates. Neuropsychopharmacology 33(11):2760–2771. https://doi.org/10.1038/sj.npp.1301664
67. Hill P, Haley NJ, Wynder EL (1983) Cigarette smoking: carboxyhemoglobin, plasma nicotine, cotinine and thiocyanate vs self-reported smoking data and cardiovascular disease. J Chronic Dis 36(6):439–449. https://doi.org/10.1016/0021-9681(83)90136-4
68. Murrin LC, Ferrer JR, Wanyun Z, Haley NJ (1987) Nicotine administration to rats: methodological considerations. Life Sci 40(17):1699–1708. https://doi.org/10.1016/0024-3205(87)90020-8
69. Hudson R et al (2021) Adolescent nicotine induces depressive and anxiogenic effects through ERK 1–2 and Akt-GSK-3 pathways and neuronal dysregulation in the nucleus accumbens. *Addict Biol* 26(2):e12891. https://doi.org/10.1111/adb.12891
70. Liu MY et al (2018) Sucrose preference test for measurement of stress-induced anhedonia in mice. Nat Protoc 13(7):1686–1698. https://doi.org/10.1038/s41596-018-0011-z
71. Meyerolbersleben L, Winter C, Bernhardt N (2020) Dissociation of wanting and liking in the sucrose preference test in dopamine transporter overexpressing rats. Behav Brain Res 378:112244. https://doi.org/10.1016/j.bbr.2019.112244
72. File SE, Seth P (2003) A review of 25 years of the social interaction test. *Eur J Pharmacol* 463(1–3):35–53. https://doi.org/10.1016/S0014-2999(03)01273-1. Elsevier.
73. D'Esposito M, Postle BR (2015) The cognitive neuroscience of working memory. Annu Rev Psychol 66:115–142. https://doi.org/10.1146/annurev-psych-010814-015031
74. Eich TS, Nee DE, Insel C, Malapani C, Smith EE (2014) Neural correlates of impaired cognitive control over working memory in schizophrenia. Biol Psychiatry 76(2):146–153. https://doi.org/10.1016/j.biopsych.2013.09.032
75. American Psychiatric Association (2013) Diagnostic and statistical manual of mental disorders. https://doi.org/10.1176/appi.books.9780890425596.744053
76. Papp M, Willner P, Muscat R (1991) An animal model of anhedonia: attenuation of sucrose consumption and place preference conditioning by chronic unpredictable mild stress. Psychopharmacology (Berl) 104(2):255–259. https://doi.org/10.1007/BF02244188
77. Wiborg O (2013) Chronic mild stress for modeling anhedonia. *Cell Tissue Res* 354(1):155–169. https://doi.org/10.1007/s00441-013-1664-0. Springer.
78. Lee Y et al (2018) Excessive D1 dopamine receptor activation in the dorsal striatum promotes autistic-like behaviors. Mol Neurobiol 55(7):5658–5671. https://doi.org/10.1007/s12035-017-0770-5
79. Espejo EF (2003) Prefrontocortical dopamine loss in rats delays long-term extinction of contextual conditioned fear, and reduces social interaction without affecting short-term social interaction memory. Neuropsychopharmacology 28(3):490–498. https://doi.org/10.1038/sj.npp.1300066
80. Rodriguiz RM, Chu R, Caron MG, Wetsel WC (2004) Aberrant responses in social interaction of dopamine transporter knockout mice. Behav Brain Res 148(1–2):185–198. https://doi.org/10.1016/S0166-4328(03)00187-6
81. Liu Q, Shi J, Lin R, Wen T (2017) Dopamine and dopamine receptor D1 associated with decreased social interaction. Behav Brain Res

324:51–57. https://doi.org/10.1016/j.bbr.2017.01.045

82. Wilson CA, Koenig JI (2014) Social interaction and social withdrawal in rodents as readouts for investigating the negative symptoms of schizophrenia. Eur Neuropsychopharmacol 24(5):759–773. https://doi.org/10.1016/j.euroneuro.2013.11.008
83. Iwata K (2019) Characterization of an animal model of autism and social interaction. *Methods Mol Biol* 1916:149–155. https://doi.org/10.1007/978-1-4939-8994-2_14. Humana Press Inc.
84. Kupferberg A, Bicks L, Hasler G (2016) Social functioning in major depressive disorder. *Neurosci Biobehav Rev* 69:313–332. https://doi.org/10.1016/j.neubiorev.2016.07.002. Elsevier Ltd.
85. Lu Q et al (2019) Chronic unpredictable mild stress-induced behavioral changes are coupled with dopaminergic hyperfunction and serotonergic hypofunction in mouse models of depression. Behav Brain Res 372:112053. https://doi.org/10.1016/j.bbr.2019.112053
86. Harro J, Tõnissaar M, Eller M, Kask A, Oreland L (2001) Chronic variable stress and partial 5-HT denervation by parachloroamphetamine treatment in the rat: effects on behavior and monoamine neurochemistry. Brain Res 899(1–2):227–239. https://doi.org/10.1016/S0006-8993(01)02256-9
87. Osborne AL, Solowij N, Babic I, Huang XF, Weston-Green K (2017) Improved social interaction, recognition and working memory with cannabidiol treatment in a prenatal infection (poly I:C) rat model. Neuropsychopharmacology 42(7):1447–1457. https://doi.org/10.1038/npp.2017.40
88. Sams-Dodd F (1999) Phencyclidine in the social interaction test: an animal model of schizophrenia with face and predictive validity. *Rev Neurosci* 10(1):59–90. https://doi.org/10.1515/REVNEURO.1999.10.1.59. Freund Publishing House Ltd.
89. Sams-Dodd F (2014) Phencyclidine in the social interaction test: an animal model of schizophrenia with face and predictive validity PhD project view project. https://doi.org/10.1515/REVNEURO.1999.10.1.59
90. Sams-Dodd F (1996) Phencyclidine-induced stereotyped behaviour and social isolation in rats: a possible animal model of schizophrenia. Behav Pharmacol 7(1):3–23. https://doi.org/10.1097/00008877-199601000-00001
91. Pandey DK, Pati D, Joshi A (2010) Chronic unpredictable stress: possible animal model of comorbid depression. Int J Preclin Pharm Res 1(1):54–63. Accessed: 7 May 2021. [Online]. Available: https://www.researchgate.net/publication/266142540.
92. Eagle AL, Fitzpatrick CJ, Perrine SA (2013) Single prolonged stress impairs social and object novelty recognition in rats. Behav Brain Res 256:591–597. https://doi.org/10.1016/j.bbr.2013.09.014
93. Schneider M, Schömig E, Leweke FM (2008) Acute and chronic cannabinoid treatment differentially affects recognition memory and social behavior in pubertal and adult rats. Addict Biol 13(3–4):345–357. https://doi.org/10.1111/j.1369-1600.2008.00117.x
94. Hodges TE, Baumbach JL, Marcolin ML, Bredewold R, Veenema AH, McCormick CM (2017) Social instability stress in adolescent male rats reduces social interaction and social recognition performance and increases oxytocin receptor binding. Neuroscience 359:172–182. https://doi.org/10.1016/j.neuroscience.2017.07.032
95. Holman PJ et al (2021) Altered social recognition memory and hypothalamic neuropeptide expression in adolescent male and female rats following prenatal alcohol exposure and/or early-life adversity. Psychoneuroendocrinology 126:105146. https://doi.org/10.1016/j.psyneuen.2021.105146
96. Ennaceur A, Delacour J (1988) A new one-trial test for neurobiological studies of memory in rats. 1: behavioral data. Behav Brain Res 31(1):47–59. https://doi.org/10.1016/0166-4328(88)90157-X
97. Cohen SJ, Stackman RW (2015) Assessing rodent hippocampal involvement in the novel object recognition task. A review. Behav Brain Res 285:105–117. https://doi.org/10.1016/J.BBR.2014.08.002
98. Naef M, Müller U, Linssen A, Clark L, Robbins TW, Eisenegger C (2017) Effects of dopamine D2/D3 receptor antagonism on human planning and spatial working memory. Transl Psychiatry 7(4):e1107. https://doi.org/10.1038/tp.2017.56
99. Ott T, Nieder A (2019. . Elsevier Ltd.) Dopamine and cognitive control in prefrontal cortex. *Trends Cogn Sci* 23(3):213–234. https://doi.org/10.1016/j.tics.2018.12.006
100. Chatterjee I et al (2019) Identification of brain regions associated with working memory deficit in schizophrenia [version 1; peer review: 2 approved]. F1000Research 8:124. https://doi.org/10.12688/f1000research.17731.1

101. Hill EL (2004) Executive dysfunction in autism. *Trends Cogn Sci* 8(1):26–32. https://doi.org/10.1016/j.tics.2003.11.003. Elsevier Ltd.
102. Lai MC, Lombardo MV, Baron-Cohen S (2014) Autism. The Lancet 383(9920): 896–910. https://doi.org/10.1016/S0140-6736(13)61539-1
103. Knight MJ, Baune BT (2018) Cognitive dysfunction in major depressive disorder. Curr Opin Psychiatry 31(1):26–31. https://doi.org/10.1097/YCO.0000000000000378
104. Floresco SB, Magyar O (2006) Mesocortical dopamine modulation of executive functions: beyond working memory. Psychopharmacology (Berl) 188(4):567–585. https://doi.org/10.1007/s00213-006-0404-5
105. Salamone JD, Pardo M, Yohn SE, López-Cruz L, SanMiguel N, Correa M (2015) Mesolimbic dopamine and the regulation of motivated behavior. Curr Top Behav Neurosci 27:231–257. https://doi.org/10.1007/7854_2015_383
106. Baik J-H (2013) Dopamine signaling in reward-related behaviors. Front Neural Circuits 7:152. https://doi.org/10.3389/fncir.2013.00152
107. Gomes FV, Rincón-Cortés M, Grace AA (2016) Adolescence as a period of vulnerability and intervention in schizophrenia: insights from the MAM model. Neurosci Biobehav Rev:1–11. https://doi.org/10.1016/j.neubiorev.2016.05.030
108. Bunney BS, Walters JR, Roth RH, Aghajanian GK (1973) Dopaminergic neurons: effect of antipsychotic drugs and amphetamine on single cell activity. J Pharmacol Exp Ther 185(3): 560–571
109. Aghajanian GK, Bunney BS (1973) Central dopaminergic neurons: neurophysiological identification and responses to drugs. Front Catech Res:643–648. https://doi.org/10.1016/B978-0-08-017922-3.50120-9
110. Grace AA, Bunney BS (1984) The control of firing pattern in nigral dopamine neurons: single spike firing. J Neurosci 4(11): 2866–2876. https://doi.org/10.1523/JNEUROSCI.04-11-02866.1984
111. Hyland BI, Reynolds JNJ, Hay J, Perk CG, Miller R (2002) Firing modes of midbrain dopamine cells in the freely moving rat. Neuroscience 114(2):475–492. https://doi.org/10.1016/S0306-4522(02)00267-1
112. Goto Y, Otani S, Grace AA (2007) The Yin and Yang of dopamine release: a new perspective. Neuropharmacology 53(5):583–587. https://doi.org/10.1016/j.neuropharm.2007.07.007
113. Schultz W, Apicella P, Ljungberg T (1993) Responses of monkey dopamine neurons to reward and conditioned stimuli during successive steps of learning a delayed response task. J Neurosci 13(3):900–913. https://doi.org/10.1523/JNEUROSCI.13-03-00900.1993
114. Tobler PN, Dickinson A, Schultz W (2003) Coding of predicted reward omission by dopamine neurons in a conditioned inhibition paradigm. J Neurosci 23(32): 10402–10410. https://doi.org/10.1523/JNEUROSCI.23-32-10402.2003
115. Ungless MA, Magill PJ, Bolam JP (2004) Uniform inhibition of dopamine neurons in the ventral tegmental area by aversive stimuli. Science (80-) 303(5666):2040–2042. https://doi.org/10.1126/SCIENCE.1093360/SUPPL_FILE/UNGLESS.SOM.PDF
116. Robinson MJF, Fischer AM, Ahuja A, Lesser EN, Maniates H (2016) Roles of 'wanting'; and 'liking' in motivating behavior: gambling, food, and drug addictions. Curr Top Behav Neurosci 27:105–136. https://doi.org/10.1007/7854_2015_387
117. Everitt BJ, Robbins TW (2016) Drug addiction: updating actions to habits to compulsions ten years on. Annu Rev Psychol 67(1): 23–50. https://doi.org/10.1146/annurev-psych-122414-033457
118. Chang CH, Grace AA (2014) Amygdala-ventral pallidum pathway decreases dopamine activity after chronic mild stress in rats. Biol Psychiatry 76(3):223–230. https://doi.org/10.1016/j.biopsych.2013.09.020
119. Palacios RD y, Verhoye M, Henningsen K, Wiborg O, Van der Linden A (2014) Diffusion kurtosis imaging and high-resolution MRI demonstrate structural aberrations of caudate putamen and amygdala after chronic mild stress. *PLoS One* 9(4):e95077. https://doi.org/10.1371/JOURNAL.PONE.0095077
120. Henry DJ, Greene MA, White FJ (1989) Electrophysiological effects of cocaine in the mesoaccumbens dopamine system: repeated administration. *J Pharmacol Exp Ther* 251(3):833–839
121. Wanat MJ, Willuhn I, Clark JJ, Phillips PEM (2009) Phasic dopamine release in appetitive behaviors and drug addiction. Curr Drug Abuse Rev 2:0–000
122. Tost H, Alam T, Meyer-Lindenberg A (2010) Dopamine and psychosis: theory,

pathomechanisms and intermediate phenotypes. Neurosci Biobehav Rev 34(5): 689–700. https://doi.org/10.1016/J.NEUBIOREV.2009.06.005

123. Grace AA (2010) Dopamine system dysregulation by the ventral subiculum as the common pathophysiological basis for schizophrenia psychosis, psychostimulant abuse, and stress. Neurotox Res 18(3–4): 367–376. https://doi.org/10.1007/S12640-010-9154-6/FIGURES/3

124. Marinelli M, Cooper DC, Baker LK, White FJ (2003) Impulse activity of midbrain dopamine neurons modulates drug-seeking behavior. Psychopharmacology (Berl) 168(1–2): 84–98. https://doi.org/10.1007/S00213-003-1491-1/FIGURES/10

Chapter 10

The MAM Model to Study the Role of Dopamine in Schizophrenia

X. Zhu, D. L. Uliana, and A. A. Grace

Abstract

Offspring derived from prenatal administration of methylazoxymethanol acetate (MAM) represent a well-validated animal model to study neuropathological states, many of which are analogous to those observed in schizophrenia. MAM is a DNA-methylating agent that, when injected at gestational day 17, produces in the adult offspring a constellation of alterations such as anxiety, increased locomotor response to amphetamine injection, increased number of spontaneous active dopaminergic neurons in the ventral tegmental area (VTA), and anatomical changes including loss of parvalbumin interneurons and cortical thinning. The excessive spontaneous dopamine neuron firing represents a construct validity of this model that could be related to increased mesolimbic dopaminergic activity observed in schizophrenia patients. Therefore, the MAM model is a reliable and feasible system helpful for the advance of the understanding of psychiatric conditions. However, a consistent protocol must be followed to assess the desired constructs and also to provide the necessary safety for the experimenters. Here we describe in a point-by-point manner the procedures of injection and the material management involved in the MAM model.

Key words schizophrenia, dopamine, in vivo electrophysiology, rat, methylazoxymethanol acetate (MAM)

1 Introduction

There are two ways in which animals can be used to study human disorders: (1) studying selective systems (e.g., genetic, neuropharmacological, neurophysiological) that have been shown to be impacted in a human condition, and (2) using an animal model of the disorders themselves with exceedingly strong construct, face, and predictive validity. While some have suggested that animal models of diseases may be less effective in replicating certain human conditions, we posit that the study of disrupted systems based on construct validity approaches are essential for advancing

X. Zhu and D. L. Uliana contributed equally with all other contributors.

José Antonio Fuentealba-Evans and Pablo Henny (eds.), *Dopaminergic System Function and Dysfunction: Experimental Approaches*, Neuromethods, vol. 193, https://doi.org/10.1007/978-1-0716-2799-0_10,

our understanding of disease risks, systems alterations, treatment, and prevention of such uniquely human disorders. Therefore, while the model system may not replicate all of the symptoms characteristic of humans with diseases, they can provide unparalleled insight into systems-level dysfunction that underlie human disorders. Thus, valid, reliable, and predictive animal models are essential to study complex neuropsychiatric disorders, such as schizophrenia. In schizophrenia research, animal models are typically categorized based on induction methods, yielding endpoints resembling diverse symptom domains. Although the precise etiology of schizophrenia is still elusive, it is widely considered as a neurodevelopmental disorder. Thus, compelling evidence points to a neurodevelopmental origin of schizophrenia, as the pathological neurodevelopmental processes in schizophrenia may begin as early as the first trimester [1], and are modified by a complex interaction between genetic and perinatal environmental risk factors throughout one's lifespan. Accordingly, researchers have devised animal models based on manipulations of perinatal environments during early sensitive periods to produce long-lasting neurodevelopmental changes that recapitulate distinct features of human schizophrenia.

The methylazoxymethanol acetate (MAM) rodent model has been demonstrated to be exceptionally useful in schizophrenia research. Specifically, prenatal exposure to MAM on gestational day (GD) 17 can lead to anatomical [2], neurophysiological [3–5], and behavioral abnormalities [2, 6–8], resembling distinct symptomatology of schizophrenia (for review, *see* [9]). Moreover, the predictive validity of the MAM model has been demonstrated repeatedly, as several current antipsychotic treatments were shown to reverse selective schizophrenia-related deficits [10]. Furthermore, while originally designed to target perinatal development, recent evidence has suggested that MAM animals also present developmental abnormalities analogous to selective prodromal alterations in the disease trajectory. For example, peri-adolescent MAM rats display blunted HPA axis response to repeated stress [7], consistent with the dysfunctional stress responsivity observed in humans [11]. Moreover, prepubertal anxiety has also been found in the MAM rats, which if alleviated can circumvent the dopamine dysfunctions in adults [6]. These novel findings have inspired emerging studies to explore the pathophysiological mechanisms in schizophrenia, and to screen innovative preventative and therapeutic approaches [3, 6, 12]. Therefore, the MAM model provides a unique opportunity to study both the functional deficits of and the abnormal neurodevelopment in schizophrenia, both of which are important elements in the NIMH's Research Domain Criteria (RDoC) framework [13].

While extensively used in schizophrenia research, gestational injection of MAM at a higher dose on GD15 has also been utilized to model broad abnormal neurodevelopment, leading to conditions broadly termed as malformations of cortical development [14]. MAM is a rapid-acting, neuro-specific mitotoxin believed to act primarily on dividing cells to inhibit cell proliferation [15]. Therefore, the resultant neurodevelopmental deficits from MAM injection critically depend on the dose and the timing of injection. Moreover, increasing evidence shows that MAM offspring may be especially sensitive to postnatal environments and stress such that the adult schizophrenia-related phenotypes can be largely modulated by the environments during several critical periods in peri-puberty. Thus, in this chapter we describe critical steps and procedures to effectively and safely generate MAM GD17 model rats that closely recapitulate the core symptomology of schizophrenia in adults, namely the dopamine dysregulation and hippocampal pathology. In addition, we also describe important experiments to validate the effectiveness of the MAM model, and extensively review current literature on behavioral and histological endpoints relevant to schizophrenia symptoms.

2 Material

2.1 MAM Compound

MAM is commercially available through MRIGlobal Chemical Carcinogen Repository (catalog #213). The handling of MAM should always be done with gloves. MAM is shipped under room temperature. Upon the arrival, MAM should be stored at 4 °C, with aliquoting best done within 24 h after arrival. Each vial of MAM stock contains 100 mg/0.1 mL, which can be diluted to generate approximately 4–5 aliquots of 1 mL, each at a final concentration of 20 mg/mL. Each aliquot is sufficient for two injections for pregnant dams (body weight: 300–400 g; dose: 20 mg/kg), but the exact usage depends on the weight of the pregnant rats. The aliquots are stored at −20 °C until the injection day.

2.2 Timed Pregnant Rats

The MAM injection takes place at GD17 of female Sprague-Dawley (SD) rats. The pregnant rats should be ordered such that they arrive around GD13–15 and GD17 would fall on a Friday. In our experience, the pregnant rat arrives on Wednesday at GD15. This will decrease the risk of hazardous exposure to the experimenter during the following 2 days after injection (i.e., weekends). All experiments were conducted in accordance with the USPHS Guide for the Care and Use of Laboratory Animals and approved by the University of Pittsburgh Animal Care and Use Committee.

2.3 MAM Aliquoting Materials

Aliquoting should be performed under a chemical fume hood. The materials are needed as given in Table 1.

Table 1
Materials for aliquot MAM

Personnel protection equipment	Materials		Cleaning products
Nitrile gloves	Pipettes (volume: 980 μL and 20 μL)	Saline	Bleach, 10%
Latex gloves	Pipette tips	MAM	Cavicide™
Tyvek suits	Scissors	Autoclave bags for hazardous waste	Paper towel
Face mask	Eppendorf 1 mL (4 for each MAM stock vial)	Zip ties	
Goggles	Surgical absorbent pad	Ziplock bags	

Table 2
Materials for MAM injection in pregnant rats

Personnel protection equipment	Materials		Cleaning products
Nitrile gloves	Syringes (1 mL)	Saline	Bleach, 10%
Latex gloves	26-gauge needles	MAM aliquots	Cavicide™
Tyvek suits	Sharp container	Autoclave bag for hazardous waste	Paper towel
Face mask	Chemical hazard sticker	Zip ties	
Goggles	Surgical absorbent pad	Ziplock bag	
	Hazard sign – Door		

2.4 Injection Materials

On the day of MAM injection, prior to entering the injection site the materials given in Table 2 should be made available.

2.5 Bedding Changing Materials (Table 3)

2.6 Safety Concerns

All the mentioned protective materials should be used to avoid direct exposure. When handling MAM always use double gloves for protection. As MAM is a reproductive hazard, do not proceed with the method if there is a possibility of pregnancy.

Table 3
Materials for bedding change of MAM-injected dams

Personnel protection equipment	Materials	Cleaning products
Nitrile gloves	Surgical absorbent pad	Bleach, 10%
Latex gloves	Autoclave bag for hazardous waste	Cavicide™
Tyvek suits	Zip ties	Paper towel
Face mask	Ziplock bag	
Goggles		

3 Methods

3.1 Aliquoting MAM

Before aliquoting MAM, check if all necessary materials are ready for use. Place hazard signage outside the door of the room and plan accordingly to avoid other personal exposure. In the fume hood, place the surgical absorbent pad on the workstation surface and the hazard bag close to the manipulation area. Label the Ziplock bag (storage for MAM aliquots) and Eppendorf tubes with substance name, concentration, date, and experimenters' initials. First, fill all the Eppendorf tubes with 980 μL of saline, and proceed by adding 20 μL of MAM stock solution to each vial using a different pipette tip. Rinse the pipette tip a few times in saline to assure all MAM is transferred to the aliquot. Next, close all vials and transfer them into the Ziplock bag which is then stored in the original MAM stock shipping container. After aliquoting MAM, bleach and cavicide™ are needed to clean the area and the materials used, before disposing them into the designated bags for biohazard wastes. First, spray bleach on the MAM vials, pipettes, pipette tips, and all the other materials used in the procedure slowly to avoid splash. Following bleach treatment, spray the affected area and materials with cavicide™, and thoroughly wipe them with paper towels. The MAM vial (stock solution), pipette tips, absorbent pads, used protective gears, and the cleaning supplies should then be disposed into a designated autoclave bag for hazardous waste according to institution's research safety guideline. Use a Zip tie to tightly close the opening of the autoclave bag and store it in a safe place until institution's hazard trash pickup. Store MAM aliquots in −20 °C until further use.

3.2 Injection Procedure

All materials and supplies need to be checked before going to the injection room. The procedure is best done with two people, one holding the dam and one performs injections. This allows a designated person to hold the dam to decrease the stress during animal handling, as well as to provide sufficient immobilization. This step is critical to ensure appropriate injection location and for the maximal safety of the experimenters. On the door of the injection room, place the chemical hazard signage and coordinate with the other users to minimize the possibility of a nonaffiliated individual entering the room. The hazard sign should remain on the door for the next 3 days and is removed after the cage cleaning and the bedding change.

Place the absorbent pads on the work surface under a laboratory hood, and organize the needles, syringes, saline, MAM aliquots, and a scale. The experimenters should both wear Tyvek suits, double gloves, face mask, and goggles to proceed. Weigh the dams and handle them for approximately 1 min to reduce stress. The dams might not be pregnant if their weight is near 200 g, because the average weight for a pregnant dam in our experience is between 290 and 350 g. While the dams are being handled, under a chemical hood the person performing the injection can prepare the saline (0.1 mL/100 g of body weight; 1 mL/kg) and MAM (0.1 mL/ 100 g of body weight; 20 mg/kg, final dose) syringes. This step should be carefully carried out to avoid splashing. After syringes are prepared, the second experimenter should grasp the pregnant rat just below the forelimbs and use the other hand to control both hindlimbs. This will allow maximal animal restraint and avoid lower body movements upon injection. For injection, aim at the third pair of nipples laterally (counting from the chest to the abdomen), as represented in Fig. 1a, and use an upward angle to avoid penetrating the womb (Fig. 1b). After injection, dispose the syringes in a sharp container, mark the animal cage with the respective

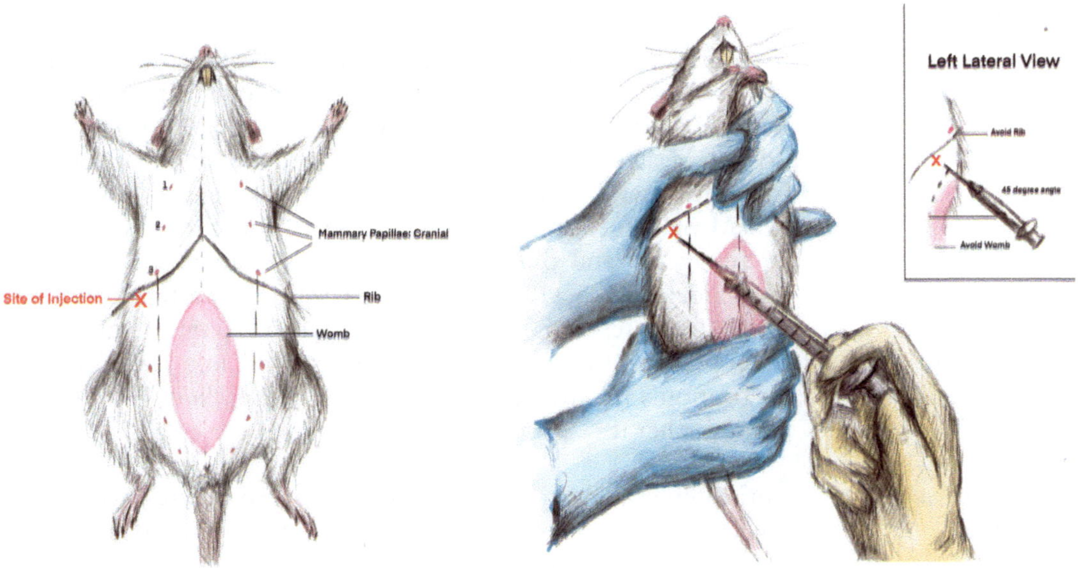

Fig. 1 MAM injection procedure. The injection site is done below and lateral to the third nipple (**a**). The pregnant dam immobilization procedure is used to avoid abrupt movements and accidents (**b**). (Illustrations courtesy of Ms. Loretta Liu)

treatment, and place a chemical hazard sticker and "Do not change bedding" sign on cages containing the injected dams. Similar to aliquoting, spray and clean all materials with bleach followed by cavicide™ and place all the wastes in an autoclave hazard bag. Next, remove the double gloves and Tyvek suit and dispose them in the waste bag. Close the bag with a Zip tie and place it in a safe area until institution's hazardous trash pickup.

3.3 Hazardous Period and Bedding Disposal

The hazardous period is considered 3 days for the animal room affected by MAM injection. Until then, the hazard sign on the door should remain in place, and for safety reason the use of the room should be minimal. During this period, experimenters may check the injected dams to ensure the health of rats.

After 3 days of hazardous period (Monday), the bedding from MAM-treated dams needs to be changed. Similar to previous steps, all personal protective equipment needs to be worn to ensure safety, including double gloves, Tyvek suits, goggles, and face masks. The MAM-injected dams are transferred to a clean cage with fresh new rodent chow, and the remaining food and the bedding must be disposed of into an autoclavable hazard bag. The empty cage and lids should be cleaned with 10% bleach followed by cavicide™. After this step, all the used cleaning materials and protective equipment should be disposed of to the same hazard bags. The empty cages and lids need to be appropriately indicated for later cleaning by the animal care staff. After this procedure, the animal room and the dams are no longer considered hazardous.

3.4 Disposal of MAM Materials

All the hazardous wastes created by aliquoting, injecting, or bedding change should be stored in a safe place, with minimal people traffic until the appropriate collection is available. Autoclave hazard bags should be disposed of by the Institutional Environmental Health and Safety Office.

3.5 Postnatal Period

The date of birth (DOB) will take place usually on Wednesday following the cage change, i.e., on gestational day 22. The pregnant rats need to be monitored to determine the exact DOB. After birth (i.e., postnatal day, PD0), the mom and pups remain undisturbed until weaning. The weaning happens on PD21 and it is recommended to not mix the offspring from saline or MAM moms to trace the litter in case of an unexpected outcome. The animals remain housed in groups of 2–3 depending on housing conditions. The adult period testing usually takes place between PD65 and 80.

4 Discussion

The prenatal MAM exposure at GD17 has been extensively used to model neuropathological states present in schizophrenia and other developmental disorders. Here we described materials and point-by-point methods to successfully perform and reproduce the MAM phenotype. In the discussion, we will describe the science behind MAM exposure and speculate regarding some important issues that may contribute to unexpected outcomes in the MAM protocol.

How Systemic MAM Injection Inhibits Embryogenesis
MAM can pass through the placenta to inhibit cell proliferation and migration via DNA methylation. It is considered a rapid and short-acting DNA methylation agent, as demonstrated by radioactive [^{3}H]MAM injection studies showing the peak fetal DNA methylation between 2 and 22 h after systemic injection in dams [16]. MAM-induced cellular deficits during embryogenesis are highly neural-specific, and given its confined time window of action, MAM appears to be an ideal tool to induce specific cellular ablation in the developing brains [15]. The selectivity of MAM-induced cellular deficits is multifaceted, and previous studies have shown different prenatal MAM exposure paradigms, varied by dose and injection time, can induce region-specific, cortical layer-specific, and cell type-specific neuronal ablation in the postnatal brains [2, 15]. Thus, different MAM exposure paradigms could induce distinct histological, physiological, and behavioral alterations. For this reason, when using MAM to induce adult pathology that closely mimics schizophrenia-related changes, it is important to consider appropriate doses and injection timeline.

Exposure Timing and Adult Pathology Differences, Overview
It is known that MAM outcomes are highly dependent on the gestational exposure day. While MAM exposure at GD9–17 has been used to model selective pathologies related to schizophrenia, it is now known that earlier MAM exposure tends to induce gross morphological changes in the brain that are not closely parallel with clinical observations in patients. Current literature shows that earlier time points such as GD9–15 are more suitable to model microcephaly, as exposed offspring demonstrate robust disarranged cytoarchitecture not only in the neocortex [2, 17] but also in hindbrain structures [18]. Behaviorally, it is described that the injections of MAM at GD12–15 failed to induce significant changes in working memory or behavioral flexibility [19], which is usually observed in animal models of schizophrenia including the MAM GD17 models [9]. Moreover, MAM injections at early time points between GD9 and 12 failed to impair social behaviors and locomotion in young rats but induced rather nonspecific changes in nociception [20]. Thus, these inadequate effects of MAM GD9–15 models with respect to schizophrenia-relevant behavioral and morphological endophenotypes are linked to an early exposure time, highlighting an essential role for gestational injection timeline for

consistent adult outcomes. Importantly, although MAM GD15 models do recapitulate characteristics of abnormal cortical activity [21], striatal hyperdopaminergic state [22, 23], and cognitive deficits [24, 25] relevant to schizophrenia, these models also tend to produce nonspecific morphological changes related to microcephaly and gross motor disruption [2, 26–29]. Therefore, MAM administration at GD17 appears to produce a more significant conjunct of changes relevant to schizophrenia, such as sensorimotor gating deficits, cognitive deficits, striatal dopaminergic dysfunctions, reduced cortical thickness and hippocampal structural alterations, and increased sensitivity to stimulants [2, 30, 31]. Such advantages of the MAM GD17 model are potentially linked to the specific gestational neurodevelopmental events occurring in late gestation, a stage when interneuron development peaks and approaches the completion in the proliferative stage [32]. Therefore, the injection of MAM at GD17 is effective in modeling the neuropathological states observed in schizophrenia. For this reason, other factors interfering with the MAM outcomes will focus specifically on the MAM GD17 model. Given that dosage and rat strain [33] are also important factors deciding outcomes in MAM-exposed offspring, we also included dose information in the models discussed below.

Dopaminergic Activity, Comparison of Different MAM Doses Electrophysiological studies have focused primarily on rats injected with MAM on GD17 and using doses of 20–25 mg/kg. In our experience, one of the primary and consistent electrophysiological changes observed using the dose of 20 mg/kg in the MAM-offspring is the increased number of spontaneous active dopamine neurons in the ventral tegmental area (VTA) [4, 6, 10, 12, 34–40], specifically in the lateral portion of VTA that projects to associative striatum [5, 41]. These alterations are consistent with the increased dopaminergic activity in terms of amphetamine-induced raclopride displacement and increased f-DOPA uptake observed in schizophrenia patients [42, 43]. Indeed, other doses of MAM were found to also consistently replicate the dopaminergic electrophysiological endpoints, as shown in studies using dose of 22 mg/kg [44–48] and the dose of 25 mg/kg in males and females [49, 50]. Although the dopaminergic changes represent a pivotal alteration related to behavioral and anatomical changes in this model, it has been proposed that the dopaminergic dysfunction observed is due to alterations in the afferents areas, such as the basolateral amygdala and ventral hippocampus [51]. The hippocampus hyperactivity is a core alteration in clinical schizophrenia and the MAM-offspring also present increased hippocampal activity [4, 12, 40]. In contrast, increased activity of the basolateral amygdala is observed earlier during neurodevelopment, in which the circuit seems to be more vulnerable to stressors [3, 12]. In fact, the hyperdopaminergic state observed in MAM litters has been demonstrated to be present before puberty, which is consistent with the idea of perinatal MAM injection conferring neurodevelopmental vulnerability [52]. These electrophysiological changes are consistently found in the MAM model that uses a dose range of 20–25 mg/kg administered at GD17.

Parvalbumin Deficits in MAM Rats

Loss of parvalbumin (PV) is a robust feature in schizophrenia, which has been successfully recapitulated by various animal models, including the MAM rats [53]. In particular, hippocampal PV loss has been observed in patients in postmortem studies and has been postulated to contribute to functional deficits and perhaps pathogenic processes in schizophrenia (*see* review [54]). The MAM model also consistently produces PV deficits [55], which is analogous to the observed human PV deficits in the limbic, anterior hippocampus [56]. Immunohistochemistry (IHC) studies reported that adult (>P60) MAM rats have reduced numbers of PV cells in the ventral hippocampus (vHipp), with doses ranging from 20 mg to 25 mg/kg [35, 52]. This effect is chronic, with the latest PV reduction detected at >P240 in the 20 mg/kg model [35]. Weaning age did not seem to affect this adult endpoint, as previously IHC studies with a weaning age ranging from P21 to P30 all showed significant PV reductions, albeit at different effect sizes [35, 57–59]. The PV reduction in IHC studies has been confirmed by quantitative western blot in adult (>P56) MAM rats exposed to 25 mg/kg [52]. Altogether, these findings indicate that adult vHipp PV decrease is a robust endophenotype of MAM G17 models.

Emerging evidence posit a potential pathogenic role of developmental PV deficits in schizophrenia [53]. With regard to the effects of MAM exposure on PV development, this has been examined directly in only two studies. Specifically, we reported a vHipp PV decrease in P27 juvenile rats previously exposed to 20 mg/kg MAM [35], which is consistent with a study by Chen et al., 2014, reporting decreased PV expression using western blot in rats previously exposed to 25 mg/kg MAM [52]. Given the dearth of studies currently, it is unclear whether the extent of developmental PV deficits in MAM rats are sensitive to different dosages. However, since a low dose of 20 mg/kg can already induce adolescent PV loss, vHipp PV is presumably vulnerable to a higher dose as well.

PV loss in PFC is another important feature of PV deficiency in schizophrenia [60]. To our knowledge, only three studies have examined mPFC PV expression in adult MAM rats, and the results are mixed. Using IHC, Penschuck et al., 2006 [59] reported no alteration in mPFC PV number in MAM (22 mg/kg) Wistar offspring between P90 and 180, whereas Lodge et al. 2009 reported decreased PV density in the mPFC of MAM (20 mg/kg) SD offspring at >P84 [55]. Using WB and qPCR, Maćkowiak et al., 2018 [61] confirmed the latter finding, reporting MAM (22 mg/kg) exposure significantly reduced mPFC PV protein and mRNA expression in early adulthood (P70–71), although in Wistar Han rats. The variability may have to do with what is being measured, in that in human postmortem studies [62] showed that there is not a loss of PV neurons in the PFC, but instead a decrease in PV content per neuron. Thus, given the variability in rodent findings, it is important in future studies to ensure that cell counts and PV density are both correlated. Noteworthy, age might be a critical factor in PV expression change in the mPFC of MAM rats. Although the exact age was not reported by the authors but rather an age range, the age discrepancy in these studies might be as far apart as more than 100 days, and therefore MAM exposure may have a more pronounced effect on PV development specifically during early adulthood.

Behavioral Deficits of MAM Models Generated at Different Doses

Adult MAM GD17 rats can recapitulate schizophrenia-related behavioral deficits across symptom domains [63]. Overall, positive- and negative-symptom-related behaviors are consistently observed in MAM GD17 models generated in different rat strains and at various doses. The present literature on cognitive deficits and anxiety-like behaviors are mixed.

Spontaneous and Stimulant-Induced Locomotion

Positive symptoms, which are typically based on a hyperdopaminergic state and include behaviors such as psychomotor agitation and hypersensitivity to psychotomimetic drugs, are currently modeled in rodent research by measuring the spontaneous locomotor activity and locomotor response to psychostimulants (e.g., amphetamine, MK801, PCP, etc.), respectively [64]. In MAM GD17 models, deficits in these positive symptom-related behaviors are among the most consistent findings, observable in adult MAM GD17 offspring generated in both Wistar [59, 65] and SD rats [2, 66], and at 20 [6], 22 [45], and 25 mg/kg [67] dosages.

Although adult deficits are commonly observed, relatively few studies have directly examined the developmental trajectory of hyperlocomotion and hypersensitivity in MAM GD17 models. The available studies overall reported a postpubertal onset of hyperlocomotion and hypersensitivity to stimulants [2, 65, 66]. This developmental onset of positive symptom-relevant behavioral abnormalities agrees with the clinical course of schizophrenia [68], in which positive symptoms mainly emerge after adolescence. However, recently the relevance of amphetamine-induced hyperlocomotion as an indices of DA overactivity related to schizophrenia has come into question [51, 69, 70]. Specifically, hyperlocomotion is generally believed to be a ventral striatal-related behavior, whereas in schizophrenia and in the MAM rat, it is the associative striatal-projecting DA neurons that are hyperactive [5, 71]. Indeed, in prepubertal stress models of schizophrenia, the amphetamine hyperlocomotion is nominal along with medial VTA DA neuron activity, while lateral VTA-associative striatal activity remains elevated [72].

Altogether, given the numerous studies demonstrating consistent locomotor dysfunction in MAM rats, we have proposed to use low-dose amphetamine-induced hyperlocomotion as a control experiment to correlate changes in dopamine system functions and to verify adult "MAM phenotypes" [73].

Prepulse Inhibition and Latent Inhibition

Prepulse inhibition (PPI) of the startle response is an operational behavioral measure of sensorimotor gating. In a clinical study, PPI has been successfully used to detect attention deficits in schizophrenia, a core feature of the disease [74]. In MAM GD17 models, adult auditory PPI deficits appeared to be a robust behavioral abnormality. Thus, since the first report on PPI deficits in MAM (22 mg/kg) GD17 SD rats in 2006 [2], adult PPI deficits have been replicated by two independent groups [65, 66]. Similar to locomotor hyperactivity and hypersensitivity to stimulants, one consistent finding of PPI deficits in MAM GD17 models is the postpubertal emergence. This was first demonstrated by Le Pen et al., 2006 [66] in SD rats, and recently confirmed in Wistar rats by Kállai et al., 2017 [65]. These results indicate that adult-onset PPI deficits are perhaps independent of animal strains and/or MAM doses (22 vs. 25 mg/kg). Interestingly, while all available MAM GD17 studies reported consistent adult PPI deficits, results from MAM GD15 models and earlier models are mixed [27, 75]. Given the lack of consistent effect in PPI in MAM models with earlier gestational injections, Jongen-Rêlo et al., 2004 [27] concluded that prenatal G15 and earlier MAM models are not valid behavioral models for schizophrenia. Therefore, MAM GD17 rats are comparatively a more useful model for attentional deficits in schizophrenia.

Besides PPI deficits, latent inhibition is another useful index for attention deficits in schizophrenia [76], and MAM GD17 SD (20 mg/kg, [55]) and Wistar rats (22 mg/kg, [77]) both displayed adult latent inhibition deficits. Furthermore, a more recent study utilizing touchscreen-based tasks [78] also confirmed attentional deficits in this model. Altogether, these results corroborate the PPI deficits, indicating prominent attention deficits in MAM rats.

Social Interaction

Social withdrawal is a key feature of the negative symptoms of schizophrenia [3]. MAM GD17 rats display chronic social deficits, as measured by the rat social interaction test. This effect so far has been demonstrated by several independent laboratories using various rat strains, MAM dosages, and weaning paradigms [30, 66, 79–81], and hence appears to be another robust behavioral endophenotype. Noteworthy, as reported by Le Pen et al. 2006 [66], unlike positive symptom-relevant deficits that emerged in adulthood, social deficits in MAM rats can occur as early as P27. Such early onset of social deficits further highlights the face validity of the MAM model, as it recapitulates the clinical course of negative symptoms in prodromal phases [82]. Given that social interaction can possibly be tested in a nonintrusive manner (e.g., home-caging camera), social behaviors could be measured as early validation for the overall MAM effect.

Cognitive Impairment

Cognitive deficits in schizophrenia involve multiple domains [83] and previous studies in sum indicate MAM rats might only behaviorally recapitulate certain deficits, such as working memory and behavioral flexibility. Working memory impairment is a core feature in schizophrenia. In rodent research, working memory is currently operationally measured by maze tasks, such as radial maze, Y or T maze, and Morris water maze, etc. [84]. MAM GD17 models overall display working/short-term memory-related deficits in these tasks. For example, Gourevitch et al., 2004 [31] assessed spatial working memory deficits in MAM GD15 and GD17 models using a radial maze task, concluding that whereas GD15 MAM rats display gross learning disability, the GD17 model displays putative working memory-related deficits at set delays. Using Y-maze or 4-arm radial maze paradigms, while Le Pen et al., 2006 [66] and Gastambide et al., 2015 [85] reported reduced novelty seeking, Moore et al., 2006 [2], Hazane et al., 2009 [86], and Brown et al., 2013 [87] reported alternation deficits in MAM GD17 rats, all of which have been interpreted as working-memory impairments. Three studies examined the short-term memory of the MAM rats using novel object recognition paradigms, consistently reporting that adult MAM offspring has worse object recognition [77, 88, 89]. Accordingly, poorer performance in the Morris water maze has also been demonstrated in some MAM GD17 models (e.g., [77, 86]), which has been construed as working/short-term memory deficits [63, 86]. Moreover, cognitive flexibility impairments constitute another important aspect of cognitive deficits in schizophrenia [90], which has been demonstrated by MAM GD17 models using several behavioral paradigms [36, 88, 91]. Thus, given those numerous studies reporting working/short-term memory and cognitive flexibility impairments, we tentatively conclude that MAM GD17 models can recapitulate certain cognitive deficits.

Noteworthy, rat strain appeared to be a critical caveat of the above-mentioned cognitive performance tests in adult MAM GD17 models. For example, as also speculated by Gastambide et al., 2015 [85], the lack of effect of MAM on Morris water maze performance in a previous study [77] could be due to the use of Wistar instead of SD rats. Interestingly, O'Reilly et al., 2016 reported intact cognitive flexibility in a two-frame active place avoidance task in MAM GD17 (26 mg/kg) Long Evans rats [92], in contrast to the poorer performance observed using the same paradigm in SD rats [93]. Intriguingly, these results of MAM GD17 models are similar to those of the GD15 models, in which SD rats were characterized as "considerably more susceptible to MAM-induced learning impairments and neurological damage [33]." Altogether, given the possible selective susceptibility of SD rats to neurotoxicity effects of MAM and the comparatively more consistent schizophrenia-like cognitive endophenotypes, we recommend using this rat strain when possible to study schizophrenia-relevant cognitive deficits. Alternately, it would be necessary to confirm dose/timing effects of MAM administration in different rat strains, as susceptibility may differ among them.

Anxiety

Stress and anxiety are common in schizophrenia [94] and are thought to be important risk factors for psychosis onset [95, 96]. In particular, abnormal cortisol levels [97] and blunted HPA axis response [98, 99] are found in individuals at risk or in early phases of schizophrenia. In agreement with those clinical findings, in adult MAM GD17 (20 mg/kg) SD rats we reported increased anxiety-like behaviors in the elevated plus maze (EPM) [6, 12], as well as blunted HPA axis activation by acute stressors [7]. Importantly, these results have been independently replicated by another study, in which Yang et al., 2019 [100] reported MAM offspring displayed chronic anxiety-like behaviors in the open field test (OFT) starting at prepuberty (P32–39). Therefore, given the consistent anxiety-like behaviors in two different assays (i.e., EPM and OFT), we concluded that MAM (20 mg/kg) GD17 SD rats have increased anxiety.

We are aware that some studies report decreased or unchanged anxiety-like behaviors in other MAM GD17 models. However, we do not believe these results disprove our conclusion, as MAM offspring in those studies were generated in different strains and/or using different paradigms. To our knowledge, four studies have reported inconsistent anxiety-like behaviors compared to our results, of which three studies used Wistar or Wistar Han rat strains. Specifically, Kállai et al., 2017 [65] conducted a repeated-testing EPM experiment and reported decreased anxiety-like behaviors in prepubertal (P28–42) MAM (25 mg/kg) GD17 Wistar rats. Moreover, Flagstad et al., 2004 [30] reported unchanged anxiety-like behavior in OFT in young adult MAM GD17 Wistar rats (22 mg/kg, P58–90). Using the light/dark box anxiety test, Bilecki et al., 2021 [101] found no anxiety-like behaviors in adolescent and adult Wistar Han rats. These negative results in Wistar or Wistar Han rats might suggest a bona fide strain effect on MAM-induced anxiety-like behaviors, or a different sensitivity to the dose of MAM administered, as robust schizophrenia-relevant pathologies (i.e., elevated dopamine functions and medial temporal lobe/hippocampal histopathology [9]) has not been performed in these strains. Noteworthy, besides the strain difference, these studies weaned MAM offspring at a considerably older (P28–30) age compared to our settings (P21), and/or the MAM-treated weanlings were raised in relatively nonstandard animal housing environments (single house [65]) and social house in four in large boxes [101]. Thus, such manipulations of early postweaning social environments might have an unexpected impact on postnatal anxiety-like behaviors. In MAM-GD17 SD rats, however, only one study reported decreased anxiety. Specifically, Gastambide et al., 2015 [85] used MAM GD17 (22 mg/kg) model weaned at P28 and reported decreased anxiety-like behaviors in EPM tested in >P120, which again could be confounded by the late weaning age. Given juvenile social isolation or enrichment, which are directly linked to the weaning condition and age, can significantly modulate adult anxiety in rats [102], it might be necessary to confirm diminished anxiety-like behaviors of these models using comparable weaning age (i.e., P21) and postweaning social environments (i.e., 2–3 rats).

Early Postnatal Environment as an Emerging Factor in MAM Models

Most studies have used MAM models to study the *pathophysiology* of schizophrenia, mainly focusing on electrophysiological, behavioral, and morphological endpoints. Comparatively, the molecular impacts of MAM exposure had been relatively understudied until recently. Although MAM-induced inhibition to cell proliferation itself is short-acting, emerging evidence suggests such immediate MAM effects might have broad, long-lasting molecular impacts on postnatal brains. For example, within 18 h of MAM (22 mg/kg) exposure, 33 genes were already differentially regulated in the developing hippocampus and neocortex [103], many of which point to early transcriptional deficits in interneuron development.

This novel finding indicates that the well-validated adult deficits in GABAergic interneurons of MAM GD17 model might have a developmental origin, thus providing further construct validity. Moreover, one previous proteomics study in MAM GD17 (22 mg/kg) SD rats indicates that glutamatergic makers are altered by MAM treatment, with expressional differences localized to the hippocampus rather than the frontal cortex [104]. Based on pathway analysis and slice physiology validation, the authors further concluded altered hippocampal glutamatergic neurotransmission, linking this model to the glutamatergic hypothesis of schizophrenia pathogenesis. Intriguingly, those null effects of MAM-treatment on frontal cortical protein expression seemed opposite to results from another proteomics study using MAM GD17 Wistar Han rats, which found 23 differentially expressed proteins in the mPFC [101]. This finding implies that, in addition to the described distinct behavioral phenotypes (e.g., anxiety and working memory), some molecular changes associated with MAM treatment might be unique to different strains of rats. Moreover, these null effects of MAM GD17 (22 mg/kg) exposure on PFC glutamatergic neurotransmission conflict with another study using MAM at a higher dosage (25 mg/kg) [105]. Therefore, the hippocampal glutamatergic network may be more susceptible to a lower MAM concentration, as opposed to that of the PFC. Overall, although recent findings on transcriptional and translational effects of MAM treatment are promising, comprehensive molecular characterizations of MAM models are still needed to explore other developmental time windows and to confirm current data. Given that dosage and rat strain could be the critical factor determining the molecular effect of MAM exposure, future studies should comprehensively examine these potentially confounding effects.

Findings in MAM rats also lend support to the "two hit" hypothesis of schizophrenia pathogenesis. Briefly, the two hit model predicts that individuals exposed to perinatal risk factors (i.e., the "first hit") are on altered developmental/maturational trajectories, and thereby establish increased vulnerability to other risk factors later in life (i.e., the "second hit") [106], probably during multiple sensitive periods around adolescence [107]. Accordingly, late gestational exposure to MAM may represent a "first hit," characterized by apparent, albeit somewhat nonspecific, DNA hyper-methylation, which is presumed to mainly affect proliferation of interneurons (e.g., PV interneurons) considering the fetal neurogenesis timeline [108]. During postnatal development and especially during prepuberty (P21–40), MAM GD17 rats display increased anxiety and abnormal stress responsivity [3], which precedes the adult manifestation of dopamine-related behavioral alterations [2]. Intriguingly, pharmacological treatment of anxiety can prevent dopamine dysfunction in adults [6], and in normal rats, sufficiently strong stressors specifically during prepuberty have been shown to recapitulate multiple MAM-related phenotypes [109]. Moreover, rearing animals in stressful (e.g., social isolation) vs. enriched peripubertal environments can exacerbate [61, 89] or prevent [12, 81] schizophrenia-relevant phenotypes in MAM rats, respectively. All those findings point that the schizophrenia-like phenotypes in MAM models are NOT directly caused by prenatal MAM exposure per se, but rather primed by complex interaction among MAM exposure, developmental stress exposure, and postnatal rearing environments, occurring during a susceptible window to "second hits." Indeed, this is consistent with the notion that MAM, or in humans genetic susceptibility, may not "cause" schizophrenia, but instead render the individual more susceptible to prepubertal stress-induced damage [107]. This would be consistent with the findings that concordance rate for schizophrenia in monozygotic twins is only around 50%, and that at least one-third of schizophrenia is nonfamilial [110]. Given such heightened environmental sensitivity when conducting research in MAM models, one has to ensure consistent postweaning rearing environments. Since atypical social environment (single-house- vs. social enrichment) can putatively alter adult schizophrenia-related phenotypes such as PV expression [58], dopamine dysfunction [12], hippocampal function [12], we recommend housing 2–3 rat weanlings in each cage for MAM GD17 model-related studies. This would also impact studies in which MAM-treated juvenile rats are shipped during this critical period, often resulting in excessive tissue loss in the hippocampus and neocortex and disrupted behavioral phenotypes [111].

Summary

In this chapter, we have described in our experience the methodology to effectively model schizophrenia-related deficits in rats using perinatal MAM exposure at GD17. We also highlighted in the discussion the main outcomes observed in adult rats that correlate with schizophrenia conditions and speculated some factors that may be leading to divergent results. The time of MAM injection and the environmental conditions during neurodevelopment are critical for the later outcome, and controlling these factors is essential in order to achieve the neuropathological features in the MAM model. MAM exposure during pregnancy has been continually demonstrated to recapitulate change in rats that correlate with clinical schizophrenia which demonstrates to be a valid and reliable approach to study and advance in our understanding of the neurobiology, risk factors, early markers, treatment, and prevention of schizophrenia.

Conflicts of Interest

AAG has received consulting fees from Alkermes, Lundbeck, Takeda, Roche, Lyra, Concert, and SynAgile and research funding from Lundbeck. DLU and XZ declare no conflict of interest.

References

1. Ursini G, Punzi G, Chen Q, Marenco S, Robinson JF, Porcelli A et al (2018) Convergence of placenta biology and genetic risk for schizophrenia. Nat Med 24:792–801
2. Moore H, Jentsch JD, Ghajarnia M, Geyer MA, Grace AA (2006) A neurobehavioral systems analysis of adult rats exposed to methylazoxymethanol acetate on E17: implications for the neuropathology of schizophrenia. Biol Psychiatry 60:253–264
3. Du Y, Grace AA (2016) Amygdala hyperactivity in MAM model of Schizophrenia is normalized by Peripubertal Diazepam administration. Neuropsychopharmacology 41:2455–2462
4. Lodge DJ, Grace AA (2007) Aberrant hippocampal activity underlies the dopamine dysregulation in an animal model of schizophrenia. J Neurosci 27:11424–11430
5. Lodge DJ, Grace AA (2012) Divergent activation of ventromedial and ventrolateral dopamine systems in animal models of amphetamine sensitization and schizophrenia. Int J Neuropsychopharmacol 15:69–76
6. Du Y, Grace AA (2013) Peripubertal diazepam administration prevents the emergence of dopamine system hyperresponsivity in the MAM developmental disruption model of schizophrenia. Neuropsychopharmacology 38:1881–1888
7. Zimmerman EC, Bellaire M, Ewing SG, Grace AA (2013) Abnormal stress responsivity in a rodent developmental disruption model of schizophrenia. Neuropsychopharmacology 38:2131–2139
8. Gill KM, Miller SA, Grace AA (2018) Impaired contextual fear conditioning in MAM rodent model of schizophrenia. Schizophr Res 195:343–352
9. Modinos G, Allen P, Grace AA, McGuire P (2015) Translating the MAM model of psychosis to humans. Trends Neurosci 38:129–138
10. Valenti O, Cifelli P, Gill KM, Grace AA (2011) Antipsychotic drugs rapidly induce dopamine neuron depolarization block in a developmental rat model of schizophrenia. J Neurosci 31:12330–12338
11. Borges S, Gayer-Anderson C, Mondelli V (2013) A systematic review of the activity of the hypothalamic-pituitary-adrenal axis in first episode psychosis. Psychoneuroendocrinology 38:603–611
12. Zhu X, Grace AA (2021) Prepubertal environmental enrichment prevents Dopamine dysregulation and Hippocampal hyperactivity in MAM Schizophrenia model rats. Biol Psychiatry 89:298–307
13. Insel T, Cuthbert B, Garvey M, Heinssen R, Pine DS, Quinn K et al (2010) Research domain criteria (RDoC): toward a new classification framework for research on mental disorders. Am J Psychiatry 167:748–751
14. Jacobs KM, Kharazia VN, Prince DA (1999) Mechanisms underlying epileptogenesis in cortical malformations. Epilepsy Res 36:165–188
15. Cattabeni F, Di Luca M (1997) Developmental models of brain dysfunctions induced by targeted cellular ablations with methylazoxymethanol. Physiol Rev 77:199–215
16. Matsumoto H, Spatz M, Laqueur GL (1972) Quantitative changes with age in the DNA content of methylazoxymethanol-induced microencephalic rat brain. J Neurochem 19:297–306
17. Funahashi A, Fujimiya M, Kimura H, Maeda T (1998) Abnormal development of serotonin nerve fibers in the visual cortex in rats with methylazoxymethanol-induced microcephaly. Acta Neuropathol (Berl). 95:5–14
18. Loupe P, Schroeder S, Tessel R (1997) The behavior and neurochemistry of the Methylazoxymethanol-induced microencephalic Rat. Int. Rev. Res. Ment. Retard, vol 21. Elsevier Inc, pp 187–220
19. Leng A, Jongen-Rêlo AL, Pothuizen HHJ, Feldon J (2005) Effects of prenatal methylazoxymethanol acetate (MAM) treatment in rats on water maze performance. Behav Brain Res 161:291–298

20. Fiore M, Talamini L, Angelucci F, Koch T, Aloe L, Korf J (1999) Prenatal methylazoxymethanol acetate alters behavior and brain NGF levels in young rats: a possible correlation with the development of schizophrenia-like deficits. Neuropharmacology 38: 857–869
21. Chevassus-Au-Louis N, Congar P, Represa A, Ben-Ari Y, Gaïarsa JL (1998) Neuronal migration disorders: heterotopic neocortical neurons in CA1 provide a bridge between the hippocampus and the neocortex. Proc Natl Acad Sci U S A 95:10263–10268
22. Archer T, Fredriksson A, Sundström E, Luthman J, Lewander T, Söderberg U et al (1988) Prenatal methylazoxymethanol treatment potentiates d-amphetamine- and methylphenidate-induced motor activity in male and female rats. Pharmacol Toxicol 63: 233–239
23. Watanabe M, Shimizu K, Kodama Y, Takishima K, Mamiya G, Ichinowatari N (1995) Potentiating effects of methamphetamine on the hyperactivity of microencephalic rats treated prenatally with methylazoxymethanol: possible implication of hyperdopaminergia. Brain Res 670:173–180
24. Talamini LM, Ellenbroek B, Koch T, Korf J (2000) Impaired sensory gating and attention in rats with developmental abnormalities of the mesocortex. Implications for schizophrenia. Ann N Y Acad Sci 911:486–494
25. Mohammed AK, Jonsson G, Sundström E, Minor BG, Söderberg U, Archer T (1986) Selective attention and place navigation in rats treated prenatally with methylazoxymethanol. Brain Res 395:145–155
26. Dambska M, Haddad R, Kozlowski PB, Lee MH, Shek J (1982) Telencephalic cytoarchitectonics in the brains of rats with graded degrees of microencephaly. Acta Neuropathol (Berl). 58:203–209
27. Jongen-Rêlo AL, Leng A, Lüber M, Pothuizen HHJ, Weber L, Feldon J (2004) The prenatal methylazoxymethanol acetate treatment: a neurodevelopmental animal model for schizophrenia? Behav Brain Res 149: 159–181
28. Kabat K, Buterbaugh GG, Eccles CU (1985) Methylazoxymethanol as a developmental model of neurotoxicity. Neurobehav Toxicol Teratol 7:519–525
29. Singh SC (1980) Deformed dendrites and reduced spine numbers on ectopic neurones in the hippocampus of rats exposed to methylazoxymethanol-acetate. A Golgi-Cox study. Acta Neuropathol (Berl) 49:193–198
30. Flagstad P, Mørk A, Glenthøj BY, van Beek J, Michael-Titus AT, Didriksen M (2004) Disruption of neurogenesis on gestational day 17 in the rat causes behavioral changes relevant to positive and negative schizophrenia symptoms and alters amphetamine-induced dopamine release in nucleus accumbens. Neuropsychopharmacology 29:2052–2064
31. Gourevitch R, Rocher C, Le Pen G, Krebs M-O, Jay TM (2004) Working memory deficits in adult rats after prenatal disruption of neurogenesis. Behav Pharmacol 15:287–292
32. Bayer S, Altman J (1995) Neurogenesis and neuronal migration. Rat Nerv. Syst, 2nd edn. Academic Press, San Diego, pp 1041–1078
33. Zmarowski A, Beekhuijzen M, Lensen J, Emmen H (2012) Differential performance of Wistar Han and Sprague Dawley rats in behavioral tests: differences in baseline behavior and reactivity to positive control agents. Reprod Toxicol Elmsford N 34:192–203
34. Gill KM, Miller SA, Grace AA (2018) Impaired contextual fear-conditioning in MAM rodent model of schizophrenia. Schizophr Res 195:343–352
35. Gill KM, Grace AA (2014) Corresponding decrease in neuronal markers signals progressive parvalbumin neuron loss in MAM schizophrenia model. Int J Neuropsychopharmacol 17:1609–1619
36. Gomes FV, Guimarães FS, Grace AA (2014) Effects of pubertal cannabinoid administration on attentional set-shifting and dopaminergic hyper-responsivity in a developmental disruption model of schizophrenia. Int J Neuropsychopharmacol 18:pyu018
37. Neves GA, Grace AA (2018) α7 Nicotinic receptor-modulating agents reverse the hyperdopaminergic tone in the MAM model of schizophrenia. Neuropsychopharmacology 43:1712–1720
38. Bortz DM, Grace AA (2018) Medial septum activation produces opposite effects on dopamine neuron activity in the ventral tegmental area and substantia nigra in MAM vs. normal rats. NPJ Schizophr 4:17
39. Sonnenschein SF, Gill KM, Grace AA (2019) State-dependent effects of the D2 partial agonist aripiprazole on dopamine neuron activity in the MAM neurodevelopmental model of schizophrenia. Neuropsychopharmacology 44:572–580
40. Sonnenschein SF, Grace AA (2020) The mGluR2/3 agonist pomaglumetad methionil normalizes aberrant dopamine neuron activity via action in the ventral hippocampus. Neuropsychopharmacology 45:2106–2113

41. Ikemoto S (2007) Dopamine reward circuitry: two projection systems from the ventral midbrain to the nucleus accumbens-olfactory tubercle complex. Brain Res Rev 56:27–78
42. Kegeles LS, Abi-Dargham A, Frankle WG, Gil R, Cooper TB, Slifstein M et al (2010) Increased synaptic dopamine function in associative regions of the striatum in schizophrenia. Arch Gen Psychiatry 67:231–239
43. Egerton A, Chaddock CA, Winton-Brown TT, Bloomfield MAP, Bhattacharyya S, Allen P et al (2013) Presynaptic striatal dopamine dysfunction in people at ultra-high risk for psychosis: findings in a second cohort. Biol Psychiatry 74:106–112
44. Perez SM, Lodge DJ (2021) Orexin modulation of VTA Dopamine neuron activity: relevance to Schizophrenia. Int J Neuropsychopharmacol 24:344–353
45. Perez SM, Shah A, Asher A, Lodge DJ (2013) Hippocampal deep brain stimulation reverses physiological and behavioural deficits in a rodent model of schizophrenia. Int J Neuropsychopharmacol 16:1331–1339
46. Donegan JJ, Boley AM, Yamaguchi J, Toney GM, Lodge DJ (2019) Modulation of extrasynaptic GABAA alpha 5 receptors in the ventral hippocampus normalizes physiological and behavioral deficits in a circuit specific manner. Nat Commun 10:2819
47. Perez SM, Carreno FR, Frazer A, Lodge DJ (2014) Vagal nerve stimulation reverses aberrant dopamine system function in the methylazoxymethanol acetate rodent model of schizophrenia. J Neurosci 34:9261–9267
48. Perez SM, Lodge DJ (2018) Convergent inputs from the Hippocampus and Thalamus to the nucleus Accumbens regulate Dopamine neuron activity. J Neurosci 38:10607–10618
49. Perez SM, Lodge DJ (2012) Aberrant dopamine D2-like receptor function in a rodent model of schizophrenia. J Pharmacol Exp Ther 343:288–295
50. Perez SM, Chen L, Lodge DJ (2014) Alterations in dopamine system function across the estrous cycle of the MAM rodent model of schizophrenia. Psychoneuroendocrinology 47:88–97
51. Grace AA (2016) Dysregulation of the dopamine system in the pathophysiology of schizophrenia and depression. Nat Rev Neurosci 17: 524–532
52. Chen L, Perez SM, Lodge DJ (2014) An augmented dopamine system function is present prior to puberty in the methylazoxymethanol acetate rodent model of schizophrenia. Dev Neurobiol 74:907–917
53. Steullet P, Cabungcal J-H, Coyle J, Didriksen M, Gill K, Grace AA et al (2017) Oxidative stress-driven parvalbumin interneuron impairment as a common mechanism in models of schizophrenia. Mol Psychiatry 22: 936–943
54. Lieberman JA, Girgis RR, Brucato G, Moore H, Provenzano F, Kegeles L et al (2018) Hippocampal dysfunction in the pathophysiology of schizophrenia: a selective review and hypothesis for early detection and intervention. Mol Psychiatry 23:1764–1772
55. Lodge DJ, Behrens MM, Grace AA (2009) A loss of parvalbumin-containing interneurons is associated with diminished oscillatory activity in an animal model of schizophrenia. J Neurosci 29:2344–2354
56. Fanselow MS, Dong H-W (2010) Are the dorsal and ventral hippocampus functionally distinct structures? Neuron 65:7–19
57. Du Y, Grace AA (2016) Loss of Parvalbumin in the Hippocampus of MAM Schizophrenia model rats is attenuated by Peripubertal Diazepam. Int J Neuropsychopharmacol 19: pyw065
58. Hernan AE, Mahoney JM, Curry W, Richard G, Lucas MM, Massey A et al (2018) Environmental enrichment normalizes hippocampal timing coding in a malformed hippocampus. PLoS One 13: e0191488
59. Penschuck S, Flagstad P, Didriksen M, Leist M, Michael-Titus AT (2006) Decrease in parvalbumin-expressing neurons in the hippocampus and increased phencyclidine-induced locomotor activity in the rat methylazoxymethanol (MAM) model of schizophrenia. Eur J Neurosci 23:279–284
60. Lewis DA, Curley AA, Glausier JR, Volk DW (2012) Cortical parvalbumin interneurons and cognitive dysfunction in schizophrenia. Trends Neurosci 35:57–67
61. Maćkowiak M, Latusz J, Głowacka U, Bator E, Bilecki W (2019) Adolescent social isolation affects parvalbumin expression in the medial prefrontal cortex in the MAM-E17 model of schizophrenia. Metab Brain Dis 34: 341–352
62. Chung DW, Fish KN, Lewis DA (2016) Pathological basis for deficient excitatory drive to cortical Parvalbumin interneurons in Schizophrenia. Am J Psychiatry 173:1131–1139
63. Lodge DJ, Grace AA (2009) Gestational methylazoxymethanol acetate administration:

a developmental disruption model of schizophrenia. Behav Brain Res 204:306–312

64. Powell CM, Miyakawa T (2006) Schizophrenia-relevant behavioral testing in rodent models: a uniquely human disorder? Biol Psychiatry 59:1198–1207
65. Kállai V, Tóth A, Gálosi R, Péczely L, Ollmann T, Petykó Z et al (2017) The MAM-E17 schizophrenia rat model: comprehensive behavioral analysis of pre-pubertal, pubertal and adult rats. Behav Brain Res 332:75–83
66. Le Pen G, Gourevitch R, Hazane F, Hoareau C, Jay TM, Krebs M-O (2006) Peri-pubertal maturation after developmental disturbance: a model for psychosis onset in the rat. Neuroscience 143:395–405
67. Le Pen G, Jay TM, Krebs M-O (2011) Effect of antipsychotics on spontaneous hyperactivity and hypersensitivity to MK-801-induced hyperactivity in rats prenatally exposed to methylazoxymethanol. J Psychopharmacol Oxf Engl 25:822–835
68. Lewis DA, Lieberman JA (2000) Catching up on schizophrenia: natural history and neurobiology. Neuron 28:325–334
69. Sonnenschein SF, Gomes FV, Grace AA (2020) Dysregulation of midbrain Dopamine system and the pathophysiology of Schizophrenia. Front Psych 11:613
70. Grace AA, Gomes FV (2019) The circuitry of Dopamine system regulation and its disruption in Schizophrenia: insights into treatment and prevention. Schizophr Bull 45:148–157
71. Gill KM, Lodge DJ, Cook JM, Aras S, Grace AA (2011) A novel α5GABA(A)R-positive allosteric modulator reverses hyperactivation of the dopamine system in the MAM model of schizophrenia. Neuropsychopharmacology 36:1903–1911
72. Gomes FV, Grace AA (2017) Prefrontal cortex dysfunction increases susceptibility to Schizophrenia-like changes induced by adolescent stress exposure. Schizophr Bull 43: 592–600
73. Lodge DJ (2013) The MAM rodent model of schizophrenia. Curr Protoc Neurosci Chapter 9:Unit9.43
74. Braff DL (1993) Information processing and attention dysfunctions in schizophrenia. Schizophr Bull 19:233–259
75. Talamini LM, Ellenbroek B, Koch T, Korf J (1999) Acoustic startle responses of rats with cerebral developmental abnormalities: implications for schizophrenia. Acta Neuropsychiatr 11:110–113
76. Moser PC, Hitchcock JM, Lister S, Moran PM (2000) The pharmacology of latent inhibition as an animal model of schizophrenia. Brain Res Brain Res Rev 33:275–307
77. Flagstad P, Glenthøj BY, Didriksen M (2005) Cognitive deficits caused by late gestational disruption of neurogenesis in rats: a preclinical model of schizophrenia. Neuropsychopharmacology 30:250–260
78. Mar AC, Nilsson SRO, Gamallo-Lana B, Lei M, Dourado T, Alsiö J et al (2017) MAM-E17 rat model impairments on a novel continuous performance task: effects of potential cognitive enhancing drugs. Psychopharmacology 234:2837–2857
79. Stark T, Ruda-Kucerova J, Iannotti FA, D'Addario C, Di Marco R, Pekarik V et al (2019) Peripubertal cannabidiol treatment rescues behavioral and neurochemical abnormalities in the MAM model of schizophrenia. Neuropharmacology 146:212–221
80. Drazanova E, Ruda-Kucerova J, Kratka L, Stark T, Kuchar M, Maryska M et al (2019) Different effects of prenatal MAM vs. perinatal THC exposure on regional cerebral blood perfusion detected by Arterial Spin Labelling MRI in rats. Sci Rep 9:6062
81. Bator E, Latusz J, Wędzony K, Maćkowiak M (2018) Adolescent environmental enrichment prevents the emergence of schizophrenia-like abnormalities in a neurodevelopmental model of schizophrenia. Eur Neuropsychopharmacol J Eur Coll Neuropsychopharmacol 28: 97–108
82. Piskulic D, Addington J, Cadenhead KS, Cannon TD, Cornblatt BA, Heinssen R et al (2012) Negative symptoms in individuals at clinical high risk of psychosis. Psychiatry Res 196:220–224
83. Lewis DA (2012) Cortical circuit dysfunction and cognitive deficits in schizophrenia--implications for preemptive interventions. Eur J Neurosci 35:1871–1878
84. Dudchenko PA (2004) An overview of the tasks used to test working memory in rodents. Neurosci Biobehav Rev 28:699–709
85. Gastambide F, Taylor AM, Palmer C, Svard H, Karjalainen M, Janhunen SK et al (2015) Alterations in spatial memory and anxiety in the MAM E17 rat model of hippocampal pathology in schizophrenia. Psychopharmacology 232:4099–4112
86. Hazane F, Krebs M-O, Jay TM, Le Pen G (2009) Behavioral perturbations after prenatal neurogenesis disturbance in female rat. Neurotox Res 15:311–320

87. Brown JW, Whitehead CA, Basso AM, Rueter LE, Zhang M (2013) Preclinical evaluation of non-imidazole histamine H3 receptor antagonists in comparison to atypical antipsychotics for the treatment of cognitive deficits associated with schizophrenia. Int J Neuropsychopharmacol 16:889–904
88. Potasiewicz A, Holuj M, Litwa E, Gzielo K, Socha L, Popik P et al (2020) Social dysfunction in the neurodevelopmental model of schizophrenia in male and female rats: behavioural and biochemical studies. Neuropharmacology 170:108040
89. Bator E, Latusz J, Głowacka U, Radaszkiewicz A, Mudlaff K, Maćkowiak M (2018) Adolescent social isolation affects Schizophrenia-like behavior in the MAM-E17 model of Schizophrenia. Neurotox Res 34:305–323
90. Waltz JA (2017) The neural underpinnings of cognitive flexibility and their disruption in psychotic illness. Neuroscience 345:203–217
91. Featherstone RE, Rizos Z, Nobrega JN, Kapur S, Fletcher PJ (2007) Gestational methylazoxymethanol acetate treatment impairs select cognitive functions: parallels to schizophrenia. Neuropsychopharmacology 32:483–492
92. O'Reilly KC, Perica MI, Fenton AA (2016) Memory deficits with intact cognitive control in the methylazoxymethanol acetate (MAM) exposure model of neurodevelopmental insult. Neurobiol Learn Mem 134 Pt B:294–303
93. Jenks KR, Lucas MM, Duffy BA, Robbins AA, Gimi B, Barry JM et al (2013) Enrichment and training improve cognition in rats with cortical malformations. PLoS One 8:e84492
94. Temmingh H, Stein DJ (2015) Anxiety in patients with Schizophrenia: epidemiology and management. CNS Drugs 29:819–832
95. Owens DGC, Miller P, Lawrie SM, Johnstone EC (2005) Pathogenesis of schizophrenia: a psychopathological perspective. Br J Psychiatry J Ment Sci 186:386–393
96. Holtzman CW, Shapiro DI, Trotman HD, Walker EF (2012) Stress and the prodromal phase of psychosis. Curr Pharm Des 18: 527–533
97. Walker EF, Brennan PA, Esterberg M, Brasfield J, Pearce B, Compton MT (2010) Longitudinal changes in cortisol secretion and conversion to psychosis in at-risk youth. J Abnorm Psychol 119:401–408
98. Goldman MB, Gnerlich J, Hussain N (2007) Neuroendocrine responses to a cold pressor stimulus in polydipsic hyponatremic and in matched schizophrenic patients. Neuropsychopharmacology 32:1611–1621
99. Jansen LM, Gispen-de Wied CC, Gademan PJ, De Jonge RC, van der Linden JA, Kahn RS (1998) Blunted cortisol response to a psychosocial stressor in schizophrenia. Schizophr Res 33:87–94
100. Yang J, Guo H, Sun D, Duan J, Rao X, Xu F et al (2019) Elevated glutamate, glutamine and GABA levels and reduced taurine level in a schizophrenia model using an in vitro proton nuclear magnetic resonance method. Am J Transl Res 11:5919–5931
101. Bilecki W, Latusz J, Gawlińska K, Chmelova M, Maćkowiak M (2021) Prenatal MAM treatment altered fear conditioning following social isolation: relevance to schizophrenia. Behav Brain Res 406:113231
102. Lukkes JL, Watt MJ, Lowry CA, Forster GL (2009) Consequences of post-weaning social isolation on anxiety behavior and related neural circuits in rodents. Front Behav Neurosci 3:18
103. Donegan JJ, Boley AM, Glenn JP, Carless MA, Lodge DJ (2020) Developmental alterations in the transcriptome of three distinct rodent models of schizophrenia. PLoS One 15:e0232200
104. Hradetzky E, Sanderson TM, Tsang TM, Sherwood JL, Fitzjohn SM, Lakics V et al (2012) The methylazoxymethanol acetate (MAM-E17) rat model: molecular and functional effects in the hippocampus. Neuropsychopharmacology 37:364–377
105. Gulchina Y, Xu S-J, Snyder MA, Elefant F, Gao W-J (2017) Epigenetic mechanisms underlying NMDA receptor hypofunction in the prefrontal cortex of juvenile animals in the MAM model for schizophrenia. J Neurochem 143:320–333
106. Davis J, Eyre H, Jacka FN, Dodd S, Dean O, McEwen S et al (2016) A review of vulnerability and risks for schizophrenia: beyond the two hit hypothesis. Neurosci Biobehav Rev 65:185–194
107. Gomes FV, Zhu X, Grace AA (2019) Stress during critical periods of development and risk for schizophrenia. Schizophr Res 213: 107–113
108. Lim L, Mi D, Llorca A, Marín O (2018) Development and functional diversification of cortical interneurons. Neuron 100: 294–313
109. Gomes FV, Zhu X, Grace AA (2020) The pathophysiological impact of stress on the dopamine system is dependent on the state

of the critical period of vulnerability. Mol Psychiatry 25(12):3278–3291

110. Kahn RS, Sommer IE, Murray RM, Meyer-Lindenberg A, Weinberger DR, Cannon TD et al (2015) Schizophrenia. Nat Rev Dis Primer 1:15067

111. Howe WM, Tierney PL, Young DA, Oomen C, Kozak R (2015) MAM (E17) rodent developmental model of neuropsychiatric disease: disruptions in learning and dysregulation of nucleus accumbens dopamine release, but spared executive function. Psychopharmacology 232:4113–4127

Part IV

Circuit Level and Behavior

Chapter 11

Fast Scan Cyclic Voltammetry to Assess Dopamine Function: From Circuits to Behavior

Philip J. Clark and Rodrigo A. España

Abstract

Electrochemical techniques have allowed investigators to measure analytes of interest for decades. Of these techniques, fast scan cyclic voltammetry (FSCV) offers advantages that make it ideal for applications in neuroscience research. FSCV is particularly valuable due to its high temporal resolution and specificity for analytes such as dopamine (DA), serotonin, and norepinephrine. Finally, the development of FSCV for use in anesthetized, ex vivo, and freely moving preparations has broadened experimental options—allowing investigators to measure DA across a wide range of contexts. In this chapter, we describe FSCV, its principles and methodologies, popular implementations, and provide examples of discoveries made with this powerful technique.

Key words Electrochemistry, Neurotransmission, Addiction, Cocaine, Neurochemistry, Catecholamines

1 Introduction

Voltammetry is a term that applies to a group of methods from analytical chemistry wherein the concentration of analytes is estimated by their reduction and oxidation (redox) in response to applied voltages. Technological advances in the past century have enabled the development of new variations of voltammetry, including fast scan cyclic voltammetry (FSCV). FSCV was first implemented for neuroscience research in the 1980s and has grown to become the de facto technique for measuring fast dopamine (DA) transmission. Although it should be noted that FSCV is also used to measure norepinephrine, serotonin, and other analytes, by far its most common application in neuroscience is to measure DA. The popularity of measuring DA with FSCV is due in part to long-standing interests in DA's role in learning and motivation, with special emphasis on the contribution of DA to these processes on a sub-second timescale. The relatively slow temporal resolution of

José Antonio Fuentealba-Evans and Pablo Henny (eds.), *Dopaminergic System Function and Dysfunction: Experimental Approaches*, Neuromethods, vol. 193, https://doi.org/10.1007/978-1-0716-2799-0_11,

other methods, such as microdialysis, limits examination of fast DA dynamics; thus, the advent of FSCV has led to a number of fascinating discoveries [1–5].

In this chapter, we compare FSCV to other methods for measuring DA system activity, including electrophysiology, microdialysis, and fluorescent imaging approaches, while emphasizing the strengths and limitations of FSCV. We then provide a background describing basic FSCV principles followed by discussions on methodological approaches and significant discoveries in neuroscience resulting from FSCV studies. Finally, we end by providing insight as to how one can leverage FSCV in combination with other popular laboratory techniques to enhance our understanding of the link between functional and biochemical aspects of DA transmission.

2 Measuring Dopamine Activity in the Brain

The mesolimbic DA system has been implicated in a variety of behavioral processes including the regulation of motor activity, learning, cognition, motivated behavior, sleep/wake activity, and substance abuse [6–9]. Although numerous observations highlight the importance of DA neuron activity in regulating DA levels throughout the brain [10], accumulating evidence also indicates that the intensity and duration of DA actions are also governed by local processes in DA terminal regions [11–17]. As such, several tools have been employed to measure both DA neuron activity at the cell body and changes in DA concentration in terminal regions. Here, we will briefly describe common methods for measuring DA system activity with an emphasis on their strengths and weaknesses.

2.1 Electrophysiology

Electrophysiological approaches have been used to make innumerable discoveries about the molecular and physiological properties of neurons in the mesolimbic system. *Ex vivo* patch-clamp electrophysiology has been used to examine midbrain DA neuron activity [18–21], and has revealed considerable heterogeneity in the properties of these cells [22]. In vivo electrophysiology has been used to demonstrate that the firing of putative DA neurons may contribute to learning by conveying a reward prediction error [23, 24], and to show that drugs of abuse tend to increase putative DA neuron activity [25–27]. However, the electrophysiological properties of DA neurons share considerable overlap with that of other neuronal types within the VTA, making it difficult to confirm the identity of DA neurons using electrophysiological criteria [22]. Furthermore, while electrophysiological methods have provided invaluable discoveries about DA neuron properties and contributions to behavior, such studies do not directly capture information about DA transmission itself [15].

While electrophysiology is ideal for measuring neuronal activity, the technique has also been used to estimate changes in neurotransmitter release by examining responses in postsynaptic neurons. For example, the probability of glutamate release can be estimated from the paired pulse ratio (PPR) of postsynaptic currents following electrical stimulation [28, 29]. PPR values greater than 1 indicates a low presynaptic release probability, whereas values less than 1 suggest the opposite [30, 31]. However, such measures are only possible for neurotransmitters that induce an immediate postsynaptic current, as is the case for ionotropic glutamate and GABA receptors. In contrast, DA receptors are metabotropic, and thus engage signaling cascades that alter neuronal excitability via more diffuse mechanisms, particularly in the nucleus accumbens (NAc) [32, 33]. Thus, DA receptor activation does not yield a single, rapid postsynaptic current that can be used to compute PPR or other measures of neurotransmission. Furthermore, neuronal responses to DA receptor stimulation may vary according to the cell type being recorded and whether cells are receiving coordinated input from other neurotransmitter systems [33].

Despite these limitations, methods have been developed for assessing DA transmission via D2 receptor responses using patch-clamp electrophysiology. In a transgenic mouse line, G-coupled inward-rectifying potassium channels (GIRK) were overexpressed in medium spiny neurons of the striatum [34]. Because D2 receptors readily couple to GIRK channels, overexpression of GIRK allows for measurement of large outward currents that track D2 receptor activation. The investigators use this tool to demonstrate that stimulation of DA release evokes a D2-mediated outward current in medium spiny neurons. Although promising, GIRK overexpression has not yet been widely adopted for estimating extracellular DA levels.

2.2 Microdialysis

The development of microdialysis in the early to mid-1970s led to thousands of articles investigating alterations in neurotransmitter changes in the brain. This technique uses semipermeable probes to measure changes in extracellular levels of analytes in brain tissue—including DA, other monoamines, and their metabolites [35, 36]. Among the greatest benefits of microdialysis are the high sensitivity and excellent chemical resolution obtained when paired with high-performance liquid chromatography or mass spectroscopy [37, 38]. In this manner, microdialysis allows investigators to measure absolute concentration of analytes and the ability to distinguish multiple analytes in the same sample. Furthermore, microdialysis is not as limited as FSCV by the electrochemical properties of target molecules, and thus can be applied to a broader range of analytes. Despite these strengths, microdialysis generally has relatively low temporal resolution and under typical conditions allows for detection of DA changes over the course of minutes. Consequently, microdialysis is limited in terms of resolving rapid release and uptake dynamics.

2.3 Fluorescent Biosensors

In 2014, the development of calcium imaging with optical fibers (i.e., fiber photometry) significantly enhanced our ability to detect changes in activity within select neuronal populations [39]. In fiber photometry, an optical fiber emits a wavelength of light to excite fluorophores on engineered proteins expressed in neurons of interest. In recent years, many proteins have been developed, including sensors to detect DA, such as dLight [40] and GRAB-DA [41]. Briefly, these sensors are modified DA receptors, that upon DA binding, undergo a conformational change that results in fluorescent emission. Because the concentration of DA is determined by how many and for how long DA receptors are occupied, the magnitude of the fluorescent response is an index of extracellular DA levels. Like FSCV, this method has sub-second temporal resolution and has been demonstrated to track DA concentration in vivo and *ex vivo* [40, 41]. While powerful, fiber photometry has limitations; first it is only able to measure *relative* changes in fluorescence—similar to FSCV—and second, photometry data cannot be readily calibrated to estimate changes in DA concentrations. Despite these issues, continued improvement of engineered proteins promises to produce increasingly sophisticated biological sensors, which are already beginning to revolutionize the field.

3 Fundamentals of Fast Scan Cyclic Voltammetry

The development of FSCV occurred over a period of decades, beginning with early electrochemical assessments of catecholamines and other neurochemicals by Ralph Adams and colleagues [42, 43]. As more electrode materials were tested and computing power continued to increase, new iterations of voltammetry were developed. It was not until 1985 that background-subtracted FSCV—as we know it today—was developed by Mark Wightman and Julian Millar [44].

Despite important differences among the various forms of voltammetry, they all follow the same basic principles; they use electrodes to apply voltages to elicit the oxidation and reduction of analytes of interest, which are then recorded as a current. Although setups can vary significantly with respect to software, amplification, and other hardware, the modern FSCV rig typically involves software control of a potentiostat and connections to working and reference electrodes, which can be employed using in vivo and *ex vivo* preparations (Fig. 1). The working electrode is where voltages are applied and where analyte measurements take place. Although working electrodes can be constructed from

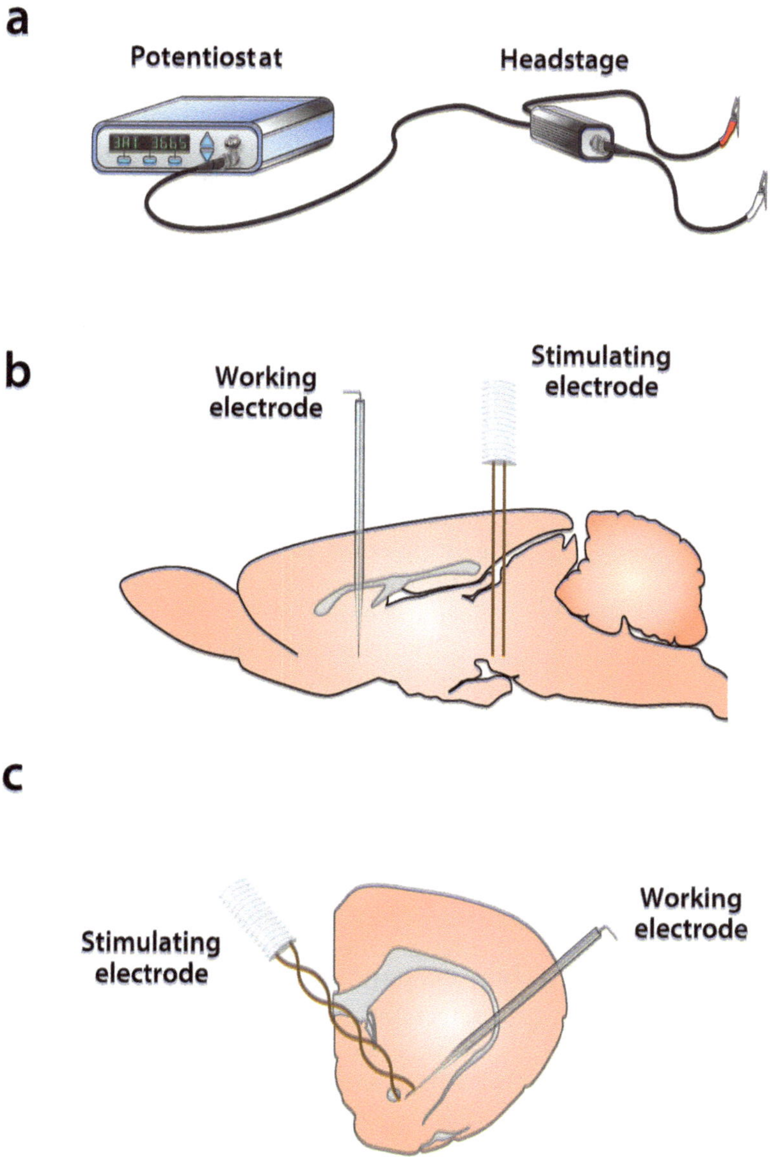

Fig. 1 Schematic showing basic components of FSCV recordings. (**a**) A potentiostat, headstage, and recording leads for reference (red) and working (white) electrodes. (**b**) Sagittal brain section with a stimulating electrode positioned in the VTA and a working electrode in the NAc. (**c**) A coronal brain slice showing configuration of electrodes with both stimulating and working electrodes in the NAc. Note: Reference electrodes not shown for either (**b**) or (**c**)

different materials, by far the most common is the carbon fiber electrode (*see* Subheading 3.1). In addition to being an inexpensive, highly conductive, and nontoxic material, the outer surface of the carbon fiber contains a layer of oxides that promote adsorption of DA to the electrode surface, thereby increasing sensitivity [45].

3.1 Working and Reference Electrode Fabrication

Working electrodes in modern FSCV are carbon fibers encased in either borosilicate glass capillaries or silicate tubing (Fig. 2). Glass capillary electrodes are typically single-use electrodes that can be used in ex vivo, anesthetized, and behaving animal preparations, whereas electrodes made using silicate tubing are typically used for chronic implants in FSCV experiments in awake, freely moving animals. For glass working electrodes, carbon fibers (Goodfellow Corporation, Coraopolis, PA; # 624503) are aspirated into a glass capillary (A-M Systems, Sequim, WA; # C 005722) and then pulled in a pipette puller (David Kopf Instruments, Tujunga, CA; model 730 or similar) such that the puller melts the glass around the carbon fiber, forming a seal. The exposed fiber is then cut under a microscope to 100–150 μM beyond the glass seal and the open end of the glass capillary is backfilled with potassium chloride or low melting point metal alloy, and a stainless steel wire is inserted with ~5 mm left protruding from the back of the electrode. If the low melting point alloy is used, it must be briefly melted after insertion of the stainless steel wire to ensure conductivity between the carbon fiber and wire.

The chronically implanted working electrodes were a significant advance by the Phillips laboratory, and have allowed for stable, long-term recordings in behaving animals [46]. Chronic electrodes are housed in a semiflexible silicate tubing (Polymicro Technologies, Phoenix, AZ; # 1068150381) which, like glass, insulates the carbon fiber, but is sturdier and better able to withstand long-term implantation in the brain [46]. To fabricate chronic working electrodes, one must first thread carbon fibers through the fused silica, such that the carbon fiber extends out of each end of the silica tubing. One end is sealed with a small amount of two-part epoxy leaving 100–150 μM of carbon fiber exposed for subsequent DA detection [46]. On the other end, a 3 mm long, 26 g stainless steel sleeve is filled with silver print (GC Electronics, Rockford, IL; 22-023) and then slid over the silica tubing such that it is in contact with the carbon fiber. After at least 12 h of drying, a gold-plated pin (Newark, Chicago, IL; # M80-0130005) is crimped to the stainless steel sleeve, which will allow the electrode to be connected to the headcap implant during surgery [47, 48].

The reference electrodes used in FSCV are typically Ag/AgCl electrodes as this material tends to have low noise and reliable performance in biological tissues [49]. Unlike the working electrode, which must rapidly fluctuate potentials, the reference electrode must maintain a relatively stable potential during the experiment so that it can be a point of comparison for the voltage applied to the working electrode. Various methods have been employed to manufacture reference electrodes [50–52] or they can be purchased to achieve excellent consistency (In Vivo Metric, Healdsburg, CA; #E255). While the reference electrode can be placed anywhere within a solution or brain tissue, the working electrode must be located at the site of expected neurotransmitter release.

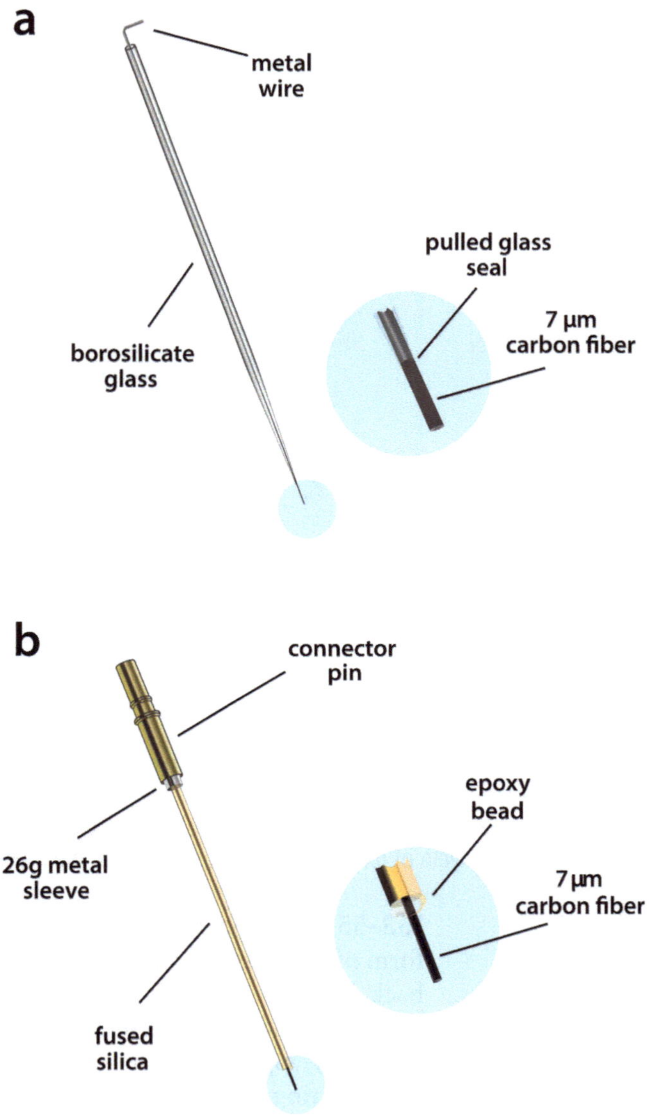

Fig. 2 Elements of acute and chronic working electrodes. (**a**) Acute working electrodes consist of a carbon fiber housed within a borosilicate glass pipette that has been pulled to form a seal at the tip. A metal wire is inserted in the rear of the electrode. (**b**) Chronic working electrodes include a carbon fiber housed in a semiflexible fused silica tube. The electrode tip is sealed with an epoxy bead and the other end is covered by a metal sleeve filled with conductive silver print and a connector pin is then crimped into place. Insets for (**a**) and (**b**) depict carbon fiber extensions from the glass or epoxy seals

3.2 Data Acquisition

The voltage parameters applied to the working electrode, including the speed at which voltages are scanned, the shape of the voltage waveform, and the range of voltages included in the waveform are key features that distinguish the various versions of voltammetry

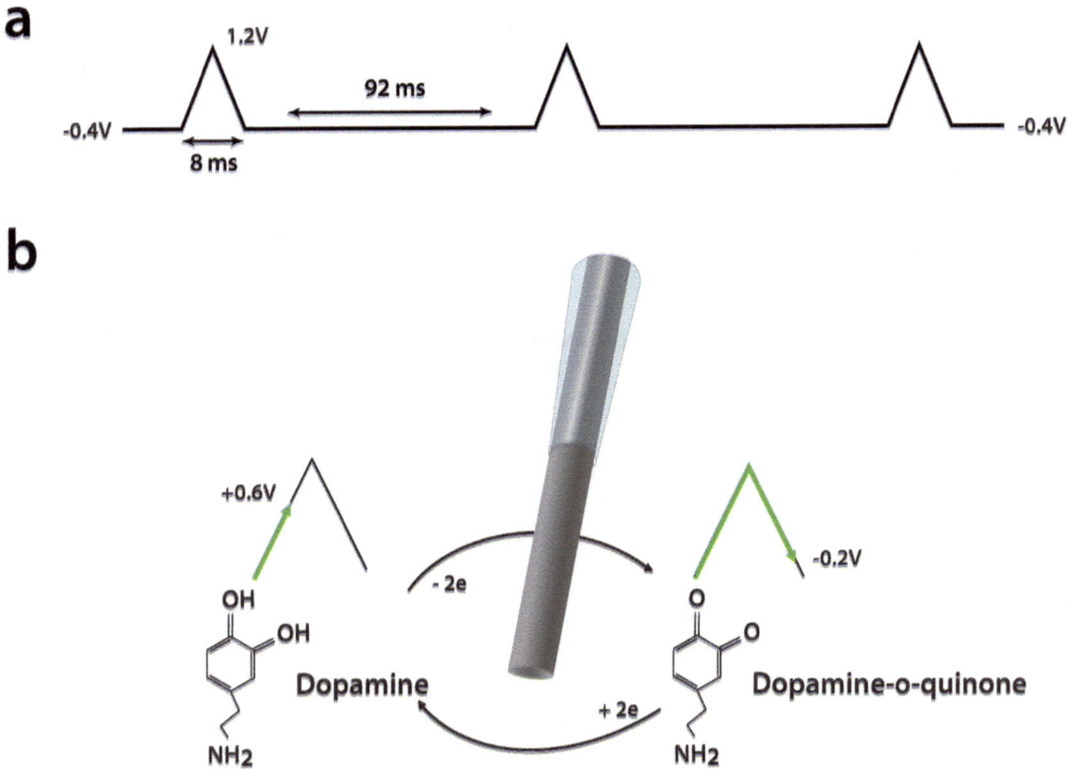

Fig. 3 Applied voltage and redox reaction in FSCV. (**a**) A series of voltage scans at 10 Hz (100 ms interval). (**b**) Schematic showing the cycle of oxidation and reduction at the electrode tip during each voltage scan

[53–55]. In modern FSCV for examining DA, a triangular waveform of voltages is typically scanned from −0.4 V up to +1.2 V, then back down to −0.4 V at a rate of 400 V/s and a frequency of 10 Hz (Fig. 3). These parameters may vary slightly depending on the particular application, but the values listed above are widely used for detecting DA [6, 56–59].

During the anodic phase of each scan, DA molecules oxidize at approximately +0.6 V, releasing two electrons and creating dopamine-O-quinone. At approximately −0.2 V on the cathodic phase of each scan, dopamine-O-quinone will reduce to DA, taking back two electrons per molecule (Fig. 3). Thus, when plotting the recorded current against the applied voltage, we should expect to see a positive deflection at +0.6 V and a negative deflection at −0.2 V. Researchers often visualize FSCV data with a pseudocolor plot (Fig. 4a), in which the *y*-axis represents the current recorded during each scan and the *x*-axis represents current measured across time. However, these processes can also be viewed in separate plots as current versus voltage, to confirm analyte identity (Fig. 4b), and current versus time, to track changes in analyte concentration (Fig. 4c).

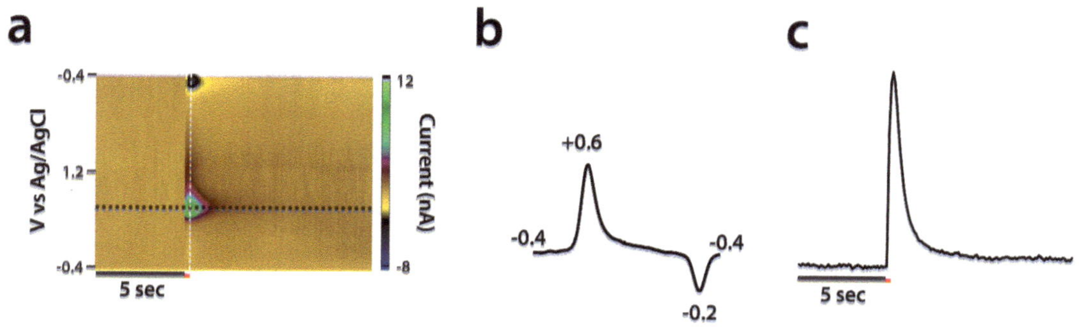

Fig. 4 Graphical depiction of FSCV data. (**a**) A colorplot demonstrating DA release resulting from a single pulse (~400 μA, 4 ms) electrical stimulation (red bar) at 5 s. Applied voltage is shown on the *y*-axis, time is on the *x*-axis, and current is on the *z*-axis. The vertical line (white) depicts the voltage vs current data shown in panel (**b**). The horizontal line (black) depicts the current vs. time data plotted in panel (**c**). (**b**) Cyclic voltammogram from panel (**a**) is used to identify DA. (**c**) Current vs. time plot from panel (**a**) is used to analyze release and uptake kinetics

3.2.1 Background Subtraction

It is important to note that the FSCV voltage scan also induces oxidation and reduction of molecules that are not DA. The faradaic current generated by the reaction of these molecules in combination with a capacitive current generated by movement of ions at the surface of the electrode during the voltage scan constitute large portions of the measured signal. In fact, the total background current is 10–100 times greater than the current generated by redox of DA [59]. Fortunately, the factors that contribute to this background current tend to be relatively stable from scan to scan and can therefore be eliminated via background subtraction. For example, in FSCV for *ex vivo* studies, DA release is commonly elicited by an electrical stimulation. Stimulation does not meaningfully alter the amount of background current in the signal, but does increase the concentration of DA. Thus, subtracting pre-stimulation scans from post-stimulation scans will cancel out the background current component, leaving behind the DA current (Fig. 5). Background subtraction is essential to FSCV, but results in relative, rather than absolute, measurements of analyte concentration, as whatever DA current may have been present in the subtracted scan is eliminated.

3.2.2 Cycling of Electrodes

Prolonged exposure to brain tissue can cause the carbon fiber to foul with polymers such as melanin. This process reduces the surface area available for electron exchange and lowers sensitivity to DA [60, 61] and is of particular concern for chronically implanted electrodes used in FSCV in freely moving animal preparations. To restore the carbon fiber surface and enhance DA detection, it is recommended that researchers cycle electrodes before experiments, which can be achieved by increasing the voltage scanning frequency [62, 63]. Electrodes are cycled by increasing the scanning

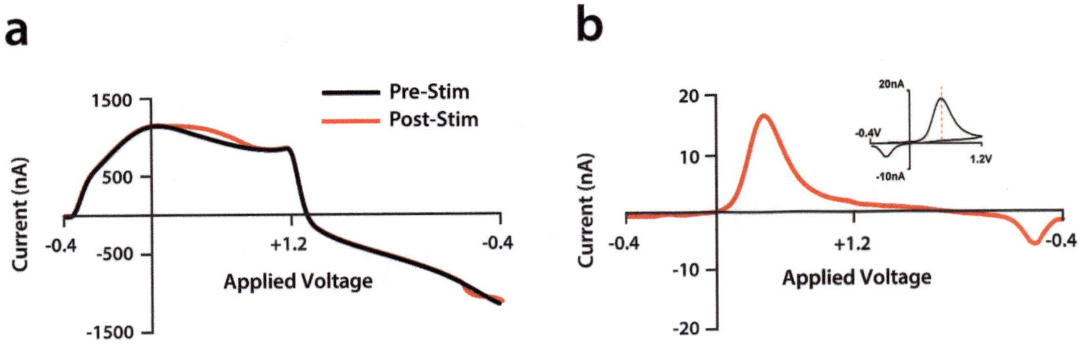

Fig. 5 Background subtraction allows for DA detection. (**a**) Total background current from scans taken prior to (Pre-Stim) and after (Post-Stim) electrical stimulation of DA release in an FSCV experiment in brain slices. (**b**) Subtracting the Pre-Stim scan from the Post-Stim scan in (**a**) yields the DA cyclic voltammogram, with oxidation at +0.6 V and reduction at −0.2 V. Inset shows a typical representation of a cyclic voltammogram in which the applied voltage is double plotted on the *x*-axis

frequency to 60 Hz for 30 min then returning it to 10 Hz for 30 min prior to each recording session [47, 48, 64].

3.3 Data Analysis

3.3.1 Electrode Calibration

Raw FSCV data depicts DA as a current. To transform the current into an estimate of DA concentration, a calibration factor is required for the working electrode. Electrode calibration often occurs using an in vitro system where the electrode is lowered into artificial cerebrospinal fluid (aCSF; NaCl (126 mM), KCl (2.5 mM), NaH_2PO_4 (1.2 mM), $CaCl_2$ (2.4 mM), $MgCl_2$ (1.2 mM), $NaHCO_3$ (25 mM), L-ascorbic acid (0.4 mM), pH adjusted to 7.4) and known concentrations of DA (1–3 μM) dissolved in aCSF are briefly washed over the electrode. As in an experiment, this produces a signal measured in current, but because the concentration of DA is known, the experimenter can calculate a ratio of current to concentration (i.e., nA per μM DA), which can be retroactively applied to the recordings from experiments to calculate DA concentrations. Once in vitro calibrations have been generated, they can be used to create an in situ calibration regression model, which uses the background current of an electrode to predict its calibration factor [65, 66]. The in situ model is particularly valuable when electrodes unexpectedly break or lose seal patency and for FSCV in freely moving animals where post-experiment electrode calibration is challenging [47, 67].

3.3.2 Analysis of Dopamine Release and Uptake

There are numerous metrics one can extract from FSCV data. Which measures are selected for interpreting data will depend on the experimental context, variation of FSCV used, data quality, and available software. When DA signals are large enough to saturate the DA transporter (DAT), as can be readily achieved with FSCV in *ex vivo* and anesthetized preparations, DAT-mediated uptake kinetics can be modeled using Michaelis-Menten kinetics [68]. When those conditions are not met, other metrics such as tau, slope, area under the curve, T20, T80, and full width at half height can be employed for examining uptake, whereas peak height is valuable for assessing release [69–72].

Unstimulated, endogenous DA transients recorded with FSCV in awake, behaving animals are typically too small to permit analysis of uptake measures, so researchers are limited to measuring the peak height and frequency of transients instead. In all above cases, observed effects can be due to a number of variables, including changes in release-regulating proteins, DAT number, DAT phosphorylation state, DAT trafficking, or other posttranslational modifications to dopaminergic proteins.

3.3.3 Chemometric Analysis

DA signals obtained from awake, behaving animals tend to be relatively small in amplitude and sometimes difficult to separate from noise. Therefore, researchers often use chemometric analysis on data obtained from freely moving animals to isolate DA signals from interferents. Chemometrics is a multistep analysis that requires obtaining example signals and interferents to train a statistical model. The model can then be deployed on experimental data to extract the signal components (DA) and eliminate interferents (pH, ascorbic acid, etc.). This process has been described in detail elsewhere [73, 74], but researchers should generally be aware that they may need to generate a training set to develop their chemometric model to estimate DA concentrations obtained from FSCV experiments in freely moving animals.

4 FSCV Variations and Applications

4.1 FSCV in Anesthetized Animals

Among the first implementations of FSCV was in the anesthetized rodent brain [5, 44, 75]. In this approach, exogenous electrical stimulation is used to elicit DA release in head-fixed, anesthetized rats and mice. One advantage of this approach is that stimulation parameters can be varied to measure different components of DA transmission within an intact circuit. For example, a train of pulses delivered to the ventral tegmental area (VTA) or the medial forebrain bundle will elicit robust DA release at terminals in the striatum. The magnitude of DA release under these conditions is sufficient to evaluate uptake rates via the DAT, which requires saturating DA levels [76, 77]. While eliciting relatively high concentrations of DA is possible in awake rodents [48, 52, 78–81], behavioral disruptions associated with stimulated release make it challenging to assess DA uptake changes with respect to behavior. Another benefit of conducting FSCV in an anesthetized preparation is the ability to examine the effects of pharmacological agents delivered systemically or locally into regions of interest while maintaining intact circuits. Effects can be examined over the course of minutes to hours which affords the opportunity to monitor onset and offset effects of various agents. Together, with the benefit of controlling for potential behavioral alterations in DA neurotransmission, these features make FSCV in anesthetized animals an invaluable technique.

Early investigation using FSCV in anesthetized animals focused on addressing fundamental questions in DA neurotransmission including DAT function [82], autoreceptor sensitivity [83], regional heterogeneity within the striatum [84], and sex differences [85]. Further, given the extensive evidence indicating that most drugs of abuse increase DA concentrations in the striatum, FSCV in anesthetized rodents has also been used to reveal the neurobiological underpinnings of how drugs of abuse influence DA neurotransmission (*see* Subheading 5).

4.1.1 Methods and Materials for FSCV in Anesthetized Animals

To examine the effects of altering the number of electrical stimulation pulses in the VTA on evoked DA release in the NAc, we performed FSCV in anesthetized male Sprague-Dawley rats (Envigo, Frederick, MD). Briefly, rats weighing 325–350 g on arrival were maintained in a vivarium with controlled temperature and humidity under a 12:12 h light:dark cycle and were provided ad libitum access to food and water. All experiments and procedures were conducted during the dark cycle and in accordance with the National Institutes of Health (NIH) Guide for Care and Use of Laboratory Animals (2011) under the supervision of the Institutional Animal Care and Use Committee (IACUC) at Drexel University.

During the experiment, rats were placed into a stereotaxic apparatus under isoflurane anesthesia and a glass working electrode was inserted into the dorsomedial striatum (DMS; +1.3 A, +1.3 L, −4.5 to 5.0 V), an Ag/AgCl reference electrode was placed into the contralateral cortex (approximately +1.3 A, +1.3 L, −2.0 V), and a bipolar stimulating electrode (Plastics One, Roanoke, VA; # MS303/1-A/SPC) was placed in the VTA (−5.2 P, +1.1 L, −7.8 V) as previously described [1, 86]. Electrodes were lowered in 100–200 μm increments until a 30 pulse, 60 Hz monophasic (4 ms; 750 μA) stimulation train produced a robust DA response. The working electrode was then lowered into the NAc (+1.3 A, +1.3 L, −6.5 to −7.5 V). Electrode dorsoventral positioning was again maximized in 100 μm increments until robust DA release was achieved in the NAc. Once stable release was obtained (less than 10% variation in release amplitude), at least three DA recordings were collected for each of the stimulation parameters (30, 60, and 120 pulses) with 5 min between stimulations.

Figure 6 shows that altering the number of electrical stimulation pulses influences evoked DA release concentrations in the NAc. DA concentrations were calculated by comparing currents at the peak oxidation potential for DA in consecutive voltammograms with electrode calibrations determined using an in situ calibration method as described previously [47, 66, 67]. DA release

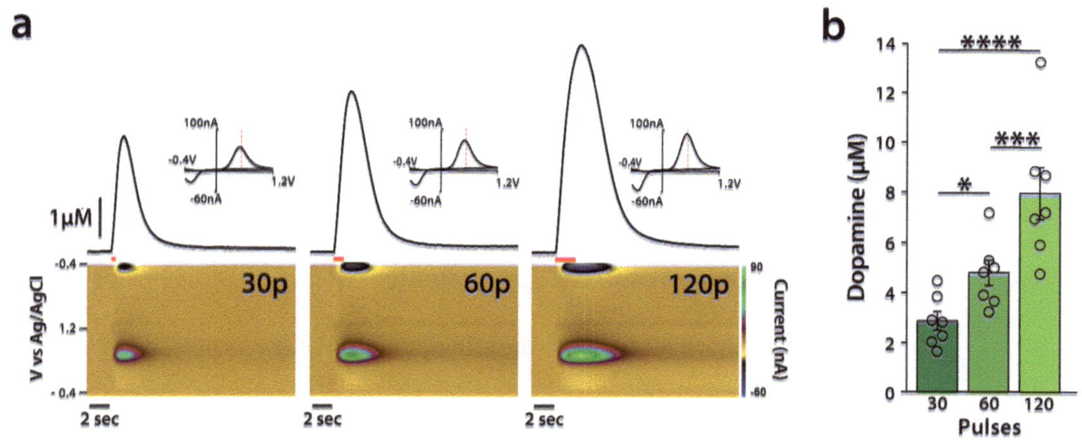

Fig. 6 Stimulation pulse number influences DA release: FSCV in anesthetized rats. (**a**) Colorplots and current vs time plots depicting DA release in the NAc following 30, 60, or 120 pulse electrical stimulation (4 ms, ~700 µA). Red bars indicate timing and length of electrical stimulation. Insets show cyclic voltammograms. (**b**) Mean ± SEM DA release across pulse number. $N = 7$ per group

across stimulation parameters was statistically compared using a repeated measures ANOVA (IBM SPSS Statistics 24, Chicago, IL). As shown in Fig. 6, increasing the number of stimulation pulses significantly increased DA release ($F(2,20) = 35.46$, $p < 0.0001$).

4.2 FSCV in Brain Slices

Shortly after the implementation of FSCV in anesthetized animals, the technique was quickly adapted to the *ex vivo* preparation [87]. Like other brain slice methods, FSCV for *ex vivo* experiments allows for a high degree of control over the recording environment but sacrifices some generalizability of findings to the intact brain. Similar to experiments in anesthetized rats, evoked DA release in brain slices results in large amplitude DA release with excellent signal-to-noise ratio that provides ideal conditions for examining release and uptake kinetics. Furthermore, applying stimulation to target regions such as the NAc allows for isolation of DA terminals from the soma [88–90]. This allows investigators to query underlying release and uptake parameters that are restricted to local mechanisms, including microcircuits that interact with heteroreceptors on DA terminals [91]. One final strength of this approach is the ability to record from multiple brain slices per animal. Thus, one can perform concentration–response experiments of several pharmacological agents independently, by testing a different agent in each brain slice [90, 92, 93]. As such, FSCV in brain slices is generally higher throughput than what is achievable with other variations of FSCV.

FSCV in *ex vivo* preparations has been commonly used to examine heterogeneity of DA transmission [87, 94], D2 autoreceptor function [95–97], DAT conformation [92], and intraspecies differences in DA transmission [98]. Furthermore, it has yielded numerous insights into the actions of addictive drugs on local DA transmission (*see* Subheading 5).

4.2.1 Methods and Materials for FSCV in Brain Slices

To test the effect of increasing cocaine concentrations on DA uptake rate, we performed FSCV in Long-Evans rats (Envigo) brain slices. Rats were maintained as described above in Subheading 4.1.1 and in accordance with the NIH Guide for Care and Use of Laboratory Animals (2011) under the supervision of the Drexel IACUC.

Briefly, rats were anesthetized with 2.5% isoflurane for 5 min and then decapitated. These anesthesia procedures have been shown previously to have no effect on DA uptake [56]. Brains were rapidly dissected and transferred to oxygenated, ice-cold aCSF (as above +11 mM glucose). A vibrating microtome was used to produce 400 μm-thick coronal sections containing the NAc. Slices rested at room temperature for 45 min before being transferred into a recording chamber flushed with aCSF (32 °C). Once in the recording chamber, a twisted bipolar stimulating electrode (Plastics One, Roanoke, VA; # MS303/3-B) was placed on the surface of the tissue and an acute working electrode was placed between stimulating electrode leads. DA release was evoked every 3 min using a single electrical pulse (~400 μA, 4 ms, monophasic) and measured using Demon Voltammetry and Analysis software. Once stable baselines were obtained (three stimulations with <10% variation), slices were exposed to five concentrations of cocaine (0.3–30 μM) as previously described [6, 57].

Figure 7 shows inhibition of DA uptake in the NAc following increasing concentrations of cocaine. DA concentrations were analyzed as in Subheading 4.1.1. Cocaine-induced DA uptake

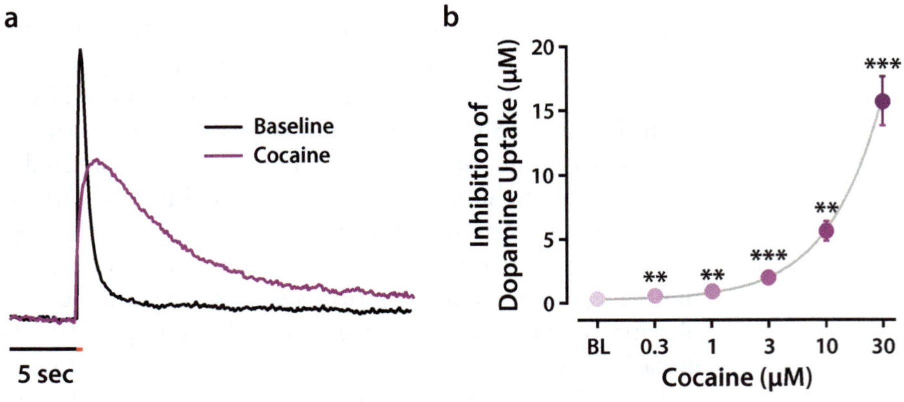

Fig. 7 Cocaine-induced inhibition of DA uptake: FSCV in brain slices. (**a**) Example current vs time plots depicting DA release and uptake at baseline and following 30 μM cocaine. Red bar depicts a 1 pulse (4 ms, ~400 μA) stimulation. (**b**) Mean ± SEM inhibition of DA uptake (apparent K_m) following cumulative cocaine concentrations (0.3–30 μM). $N = 6$ (3 females, 3 males)

inhibition (app K_m) was analyzed by initially setting K_m values to 0.18 μM to calculate DA uptake rate (V_{max}; data not shown), and then attributing cocaine-induced inhibition of DA uptake to changes in app K_m while holding V_{max} constant. Cocaine-induced DA uptake inhibition was statistically compared using a repeated measures ANOVA. As shown in Fig. 7a, b, cumulative increases in cocaine significantly increased inhibition of DA uptake ($F(4,20) = 55.99$, $p < 0.001$).

4.3 FSCV in Freely Moving Animals

FSCV in anesthetized animals or in brain slices allow for assessment of DA release and uptake dynamics under controlled conditions that eliminate behavioral influences on DA neurotransmission. While these methods have generated critical insights into DA neurobiology, lingering questions about DA neurotransmission during behaviors of interest prompted the development of FSCV in awake, behaving animals. As mentioned above, studies using electrophysiology and microdialysis have long suggested involvement of DA in goal-directed behavior, but it was challenging to examine rapid DA dynamics in awake rats with these tools. The development of FSCV in behaving rodents opened the door for interrogating sub-second DA transmission in reward and reinforcement and has been vital to recent discoveries regarding the function of this neurotransmitter in a wide range of behaviors.

Central to the use of FSCV in freely moving animals is the ability to match DA transients to behaviorally relevant events such as reward delivery and cue presentation. Observing how DA signals develop across learning has contributed to theories about the kind of information DA signals convey. Studies employing FSCV and electrophysiology in behaving animals are often consistent, but occasionally FSCV data has challenged conclusions drawn from electrophysiological studies [15] and opened the door for new interpretations about DA's role in behavior [99]. Just as in the anesthetized preparation, pharmacological agents that modulate the DA system can be administered systemically or into discrete brain regions prior to or during recordings in freely moving animals and the resulting effects on behavior and DA transmission can be observed in real time.

FSCV in freely moving animals is primarily used to observe spontaneous and cue-evoked DA signals that are of relatively small amplitude. These signals, commonly referred to as DA transients, are often too small to be analyzed for uptake kinetics as is possible for the much larger signals readily achieved in anesthetized and *ex vivo* studies. As such, information about DA transmission in the freely moving preparation is often focused on the amplitude, duration, and frequency of transients. As mentioned above, behavioral disruptions associated with stimulating DA release in freely moving animals can complicate interpretation of data from such experiments, although this approach is sometimes valuable for testing causality of DA transmission in various behavioral outcomes [52].

Early FSCV studies in freely moving animals identified DA release during conditioned and unconditioned behavior [4, 100, 101]. This approach has also been used to test for the causal influence of DA transmission in drug seeking [52] and to probe the effect of different stimulation parameters on DA transmission [102]. Together, these studies underscore the versatility of FSCV in freely moving animals to examine DA transmission across behavioral contexts.

4.3.1 Methods and Materials for FSCV in Awake, Freely Moving Animals

To test the effect of VTA stimulation on DA release and uptake in the NAc of awake, freely moving animals, rats were anesthetized with 2.5% isoflurane and placed in a stereotaxic apparatus. An internally threaded guide cannula for the working electrode (Faircloth Machine Shop, Winston-Salem, NC; custom design) was implanted above the striatum (+1.3 A, ±1.3 L), a reference electrode (In Vivo Metric; E255) was implanted in the posterior cortex (−2.5 A, −2.5 L, −2.0 V), and a bipolar stimulating electrode (Plastics One; # MS303/1-A) was implanted in the VTA (−5.2 P, +1.1 L, −7.8 V).

Prior to securing implants, a temporary working electrode was lowered into the dorsal-most aspects of the DMS (+1.3 A, ±1.3 L, −4.0 V). Working and stimulating electrodes were positioned such that a 30-pulse (60 Hz, 2 ms biphasic, 250 μA) stimulation train elicited DA release. At that point, the temporary working electrode was removed, and remaining implants were secured into place using dental acrylic cement and five screws driven into the skull. Rats received postsurgical antibiotic (Neo-Predef, Pharmacia & Upjohn Company, and 5 mg/kg Baytril, Bayer HealthCare LLC) and analgesic (5 mg/kg; Ketoprofen, Patterson Veterinary) and were allowed to recover for 5–7 days before FSCV experiments began.

On experimental days, glass working electrodes were inserted into microdrives (Faircloth Machine Shop; custom design), as detailed previously [103]. Briefly, the working electrodes were filled with a low melting point metal alloy and a stainless steel wire was inserted to achieve electrical connectivity with the carbon fiber. The glass electrode was carefully placed into the microdrive and attached to the microdrive shaft. The microdrive assembly was connected to the implanted guide cannulae and a headstage cable (Electronics and Materials Engineering, Seattle, WA; custom design) was attached to the working and stimulating electrodes. The other end of the cable was connected to FSCV equipment (Electronics and Materials Engineering; custom design) via an electrical swivel commutator (Christ Instruments, Hagerstown, MD; # 4-TBC-25-LT).

Next, the working electrode was slowly lowered into the DMS (+1.3 A, ±1.3 L, −4.5 V) and then cycled as described in

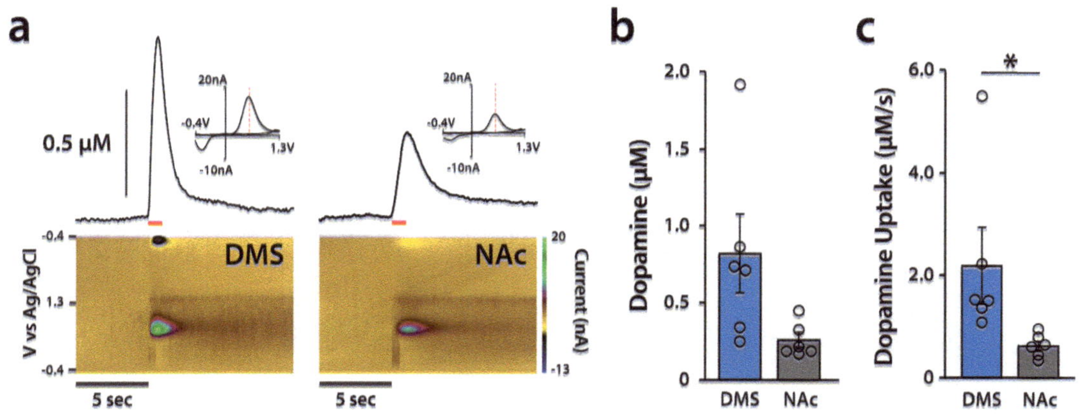

Fig. 8 DA release and uptake differ between the dorsomedial striatum and nucleus accumbens: FSCV in freely moving rats. (**a**) Colorplots and current vs time plots showing examples of electrically stimulated (30-pulse, 2 ms biphasic, ~400 μA) DA release in the dorsomedial (DMS) and NAc. Red bars indicate timing and length of electrical stimulation. Insets show cyclic voltammograms. Mean ± SEM: (**b**) DA release and (**c**) DA uptake. $N = 6$ males per group

Subheading 3.2.2. Following cycling, 30-pulse (60 Hz, 2 ms biphasic, 250 μA) stimulation trains were delivered every 5 min, until stimulation elicited robust DA release. After a minimum of three collections, the working electrode was lowered further into the NAc (+1.3 A, ±1.3 L, −6.5 V) where DA release was again maximized and at least three collections were obtained. DA concentrations and uptake rate in the DMS and NAc were analyzed as described in Subheading 4.2.1 and statistically compared using independent samples *t*-tests (SPSS). As shown in Fig. 8, the DMS displayed a strong trend for greater DA release ($t_{10} = 2.163$, $p = 0.056$) and significantly faster uptake ($t_{10} = 2.247$, $p < 0.05$) relative to the NAc—a result that has also been observed in brain slices [98, 104–106].

To examine spontaneous and cue-evoked DA release in the NAc of freely moving animals, rats were anesthetized with 2.5% isoflurane and first implanted with an i.v. catheter as previously described [1, 56, 64, 107]. Rats were then placed in a stereotaxic apparatus and chronic working electrodes were implanted into the NAc (+1.3 A, ±1.3 M/L, −7.5 D/V), and a reference electrode was implanted in the posterior cortex (−2.5 A, −2.5 L, −2.0 V). Electrodes were secured into place using dental acrylic cement and five screws driven into the skull. Rats received postsurgical antibiotic and analgesic and were allowed to recover for 21–28 days before FSCV experiments began [46].

For data shown in Fig. 9, rats were connected to FSCV equipment, electrodes were cycled, and spontaneous DA transient release was recorded continuously as 60 s files under baseline, nondrug conditions. After 30 min of recording, rats received a noncontingent, 0.75 mg/kg, i.v. cocaine injection and resulting changes in

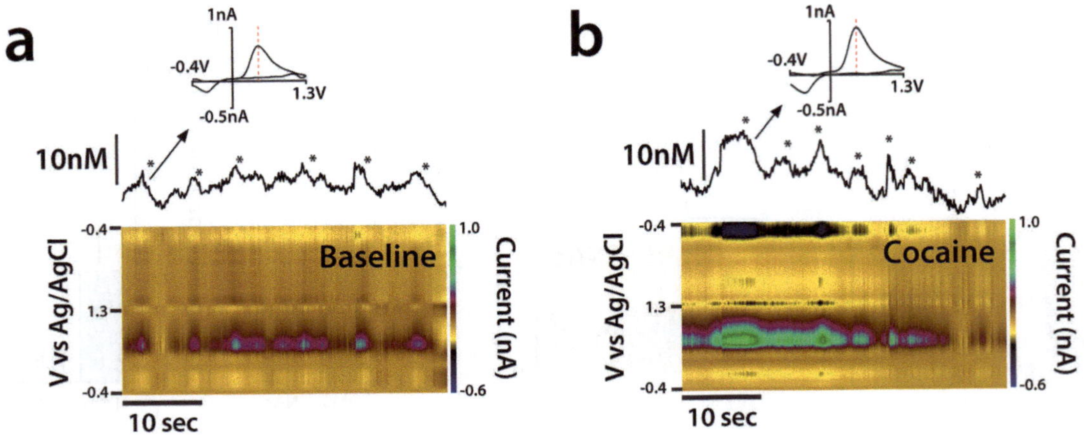

Fig. 9 Cocaine increases the amplitude of DA transient release: FSCV in freely moving rats. Colorplots and current vs time plots depicting spontaneous DA transient release in the NAc at (**a**) baseline and following (**b**) experimenter-delivered cocaine (0.75 mg/kg, i.v.). Insets show cyclic voltammograms and asterisks indicate individual transients

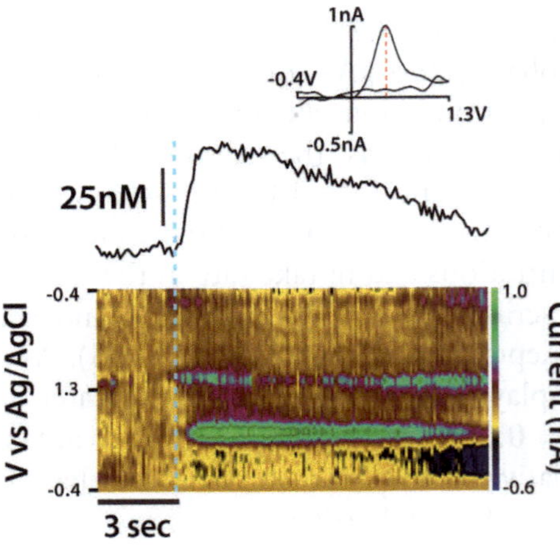

Fig. 10 Cue-evoked DA release during cocaine self-administration: FSCV in freely moving rats. Colorplot and current vs time plot depicting cue-evoked DA transient release in the NAc averaged across 13 self-administration trials from one session. Dashed blue line indicates time of lever response which resulted in immediate illumination of a cue light and initiation of 0.375 mg/kg cocaine delivery. Inset shows cyclic voltammogram

spontaneous DA transient release were recorded for another 30 min. Spontaneous DA transient release events were identified using Demon Voltammetry and Analysis software [68] and based on previously described approaches [47, 64, 80, 108]. Figure 9 shows example data from one rat displaying an increase in the amplitude of DA transient release following cocaine delivery.

For data shown in Fig. 10, rats were connected to FSCV equipment and electrodes were cycled as described above. After cycling, rats were allowed to self-administer cocaine on a progressive ratio schedule of reinforcement. During progressive ratio sessions, rats were given 6-h access to a response lever beginning at 10:00 h, and single 0.375 mg/kg cocaine injections were contingent upon increasing number of responses: 1, 2, 4, 6, 9, 12, 15, 20, 25, 32, 40, 50, 62, 77, 95, 118, 145, 178, 219, 268, 328, 402, 492, and 603 [109]. When the required lever response ratio was reached, the lever retracted, a cue light above the lever was illuminated, and cocaine was delivered over 3–5 s. DA transient release was recorded continuously as 60 s files throughout the self-administration session, and behaviorally relevant events (e.g., lever responses, cue presentation, and cocaine delivery) were time-stamped onto FSCV collection files. FSCV recordings were analyzed for cue-evoked DA release by creating 15 s, peri-event FSCV data files aligned to lever responses that resulted in a completed ratio, such that 3 s preceded, and 12 s followed the timepoint during which lever responses occurred. These peri-event files were then averaged across trials for each session. Figure 10 shows example data from one rat displaying DA transient release following lever presses that resulted in cue presentation and cocaine delivery.

5 Effects of Psychostimulants on DA Neurotransmission

5.1 Acute Actions of Dopamine Transporter Inhibitors

Psychostimulants such as cocaine and amphetamine interact with the DAT to increase extracellular DA levels across brain structures including the striatum. Although electrophysiology and microdialysis have been important approaches to show that the effects of psychostimulants span both somatodendritic and terminal compartments of the DA system [25, 110–115], FSCV has been integral to defining the temporal dynamics of release and uptake which underlie the effects of these drugs.

DA uptake inhibitors such as cocaine interact with the DAT to reduce translocation of DA from the extracellular space back into the terminal [116]. The slowing of uptake increases extracellular DA and produces greater activation of postsynaptic receptors. While microdialysis had defined the effects of cocaine on elevated DA levels, it was not until FSCV was employed in anesthetized rats that it was determined that varying doses of intravenous cocaine elicit significant DA uptake inhibition within a matter of seconds, and reach maximal inhibition of uptake within 30–60 s [117, 118]. Rapid inhibition of DA uptake was also observed with other DAT transporter inhibitors, such as PTT and WF23, which were posited to have slower onset and offset kinetics. Despite eliciting significant inhibition of DA uptake within seconds, it is important to note that maximal inhibition following PTT and WF23 took significantly longer than what is observed with cocaine [119].

FSCV in anesthetized preparations has also been used to assess the relative magnitude of DAT inhibition following single i.v. injections of cocaine, PTT, WF23, as well as following repeated cocaine treatments. Those experiments revealed that commonly self-administered doses of cocaine produce only modest inhibition of DA uptake following single i.v. delivery, and that repeated cocaine injections can significantly increase the magnitude of inhibition [120].

The effects of PTT and WF23 have also been examined using FSCV in brain slices, which revealed that these high affinity DAT inhibitors produce robust inhibition of DA uptake that far exceeds the effects produced by cocaine [92, 120, 121]. Further, FSCV experiments in slices helped to reveal that cocaine also influences DA release, which appears to be independent of direct DAT actions. For example, low concentrations of cocaine increased reserve pool release of DA in a synapsin-dependent manner [122], but high concentrations reduce DA release well below baseline levels [7, 92, 120, 123]. Importantly, PTT and WF23 do not appear to reduce stimulated release [120], suggesting that the effects of cocaine at reducing release may involve actions that extend beyond inhibition of the DAT.

With the advent of FSCV recordings in awake, freely moving animals [100], researchers began applying this technique to investigate the effects of cocaine on DA transmission. One early study found that 1.0 mg/kg i.v. cocaine increased the amplitude and decay-time of DA signals in freely moving rats for ~20 min, an effect mirrored by an increase in overall head movement for the same duration of time [124]. In a seminal publication, Phillips and colleagues show that DA transients in rats are evident immediately prior to lever pressing to receive a cocaine infusion [52]. Within each trial, a second transient was observed during and immediately after the cocaine injection itself. By electrically stimulating DA release, the researchers were able to elicit cocaine seeking, but only in rats that had previously been trained on the task. These data provided the first evidence of rapid DA transmission coinciding with drug seeking and drug delivery, and that rapid DA release can evoke drug seeking.

FSCV in freely moving animals has also been used to empirically test different theories of what DA signals represent, which is central to our understanding of DA's involvement in substance abuse. For example, the magnitude of DA responses to reward-predictive cues has been shown to scale with anticipated reward value, but may be unaffected by the cost associated with the rewards, indicating the DA signals only partially track the economic utility of a particular action [3]. Another study provides support for the theory that DA transients match positive and negative reward prediction errors [125], consistent with earlier electrophysiological studies [23, 24]. However, is has recently been demonstrated that in addition to carrying a reward prediction error, fast (phasic) DA signals may also drive motivation to work, supporting the notion that DA signals contribute to both learning and motivation [126]. Each of these studies used FSCV to probe real-time DA transmission and examine theories regarding the function of DA, and each of these results carry implications for targeting the DA system in substance abuse research and treatment.

5.2 Lasting Actions of Dopamine Transporter Inhibitors

Amassing evidence indicates that repeated exposure to psychostimulants exerts lasting alterations in the mesolimbic DA system, which may stem from alterations in DAT number or function [127–129]. The direction, magnitude, and longevity of DA plasticity changes are complex and appear to depend on cocaine exposure conditions. For example, using FSCV in anesthetized rats, it was shown that repeated daily i.p. injections of cocaine for 7 days increased inhibition of DA uptake in response to an acute cocaine challenge such that rats with prior exposure to cocaine showed a potentiated DA response in both the NAc core and shell [130].

Self-administration of cocaine has also been shown to confer changes in DA dynamics. FSCV in brain slices taken from rats after completing 5 days of long access cocaine self-administration show reduced DA release and uptake rate as well as decreased effects of cocaine on inhibition of DA uptake [128, 131]. Similar to the effects observed following long access, 24 h access to cocaine using a discrete trials schedule in which rats are allowed a single cocaine injection every 15 min displayed significant decreases in DA release and increases in DA uptake at baseline and decreased cocaine-mediated inhibition of DA uptake [129]. However, another study found that the effect of cocaine self-administration on DA dynamics differed depending on which self-administration protocol was used [127]. Notably, intermittent access to cocaine, a schedule thought to be representative of cocaine use patterns in humans, increased DA uptake rate and enhanced cocaine-mediated inhibition of the DAT in NAc slices [127]. Furthermore, including a period of abstinence following some cocaine self-administration schedules appears to reverse some of the aforementioned dopaminergic adaptations [123], while abstinence following other self-administration schedules exaggerates changes in DA uptake and sensitivity to cocaine [127]. Although many questions remain regarding the mechanisms underlying changes in DAT function following exposure to cocaine, these studies underscore the versatility of FSCV in brain slices to probe changes in DA transmission and highlight the complicated systems that contribute to dopaminergic adaptations following exposure to psychostimulants.

We recently used a predator odor stress model to examine if susceptibility to a traumatic stress predisposes rats to excessive cocaine intake and exaggerated DA responses to cocaine [7]. Susceptibility to stress has been shown to increase motivation for cocaine and is linked to changes in DA neuron physiology [132]. However, for years it was unclear whether such behavioral and physiological adaptations are tied to differences in DA transmission. Compared to control and resilient rats, we showed that susceptible rats display increased anxiety-like behavior, higher corticosterone levels, and greater motivation to self-administer cocaine compared to resilient and control rats. Further, FSCV experiments in NAc slices demonstrated that susceptible rats had reduced DA release, slower DA uptake at baseline, and increased inhibition of DA uptake following cocaine. Together, these data indicate that individual responses to trauma may confer susceptibility to substance use disorders via adaptations to the dopaminergic system.

FSCV experiments in behaving rats have also demonstrated significant changes in DA transmission following exposure to cocaine. For example, escalation of cocaine intake on a long-access schedule of reinforcement has been tied to reduced phasic DA signaling in the striatum of freely moving animals [133]. Furthermore, initial drug seeking is posited to be controlled by the NAc, but habitual drug seeking is governed principally by the dorsolateral striatum [134]. Further, while the DAT-inhibiting effects of cocaine are not easily modeled from data obtained from freely moving animals, relatively high doses of cocaine have been shown to increase the frequency and amplitude of naturally occurring DA signals [81].

As discussed above, we demonstrated that susceptibility to a predator odor stress promotes cocaine use and elicits exaggerated DA responses to cocaine [7]. We again used the predator odor stress model to examine DA dynamics in awake, freely moving rats trained to learn cocaine-associated cues [47]. In these studies, we showed greater frequency of spontaneous DA transients in stress-susceptible rats compared to stress-resilient rats, and that cue-evoked DA transients develop faster and are more persistent in susceptible rats. These data convey that behaviorally distinct profiles can be associated with differential DA dynamics, and further provide region-specific insight about DA transmission during behavior.

6 Leveraging FSCV with Other Techniques

6.1 Electrophysiology

FSCV provides functional information about DA transmission that can be leveraged with other techniques to provide a holistic picture of DA system function. For example, FSCV has been used with patch electrophysiology in the striatum to show that brief optogenetic stimulation of cholinergic interneurons elicits action potentials and concurrent release of DA [16]. Pharmacological blockade of nicotinic acetylcholine receptors reduced optogenetically stimulated DA release, suggesting that local acetylcholine release drives rapid DA efflux in the striatum [16].

Other studies have combined these techniques in vivo. In freely moving rats, simultaneous FSCV and electrophysiology recordings were made in the NAc during an intracranial self-stimulation experiment [135]. These data, collected from a single carbon fiber electrode, showed that cue-induced DA transients were consistently associated with either an increase or a decrease in neuronal firing, depending on the recording site. In cases where DA release was not elicited by reward-predictive cues, there was no change in neuronal firing, suggesting a relationship between phasic DA signaling and local network activity.

More recent work combining electrophysiology and FSCV suggests a disconnect between the activity of DA neurons and DA release at terminal sites [15]. In these studies, investigators demonstrated that DA neuron firing and DA release both increase in response to cues that predict rewards, but only DA release tracked reward expectations during behavioral trials. To date, the mechanisms that regulate DA release independent of DA neuron firing have not been completely elucidated.

6.2 Biochemistry

Biochemical techniques, such as western blotting, are critical to the study of biological systems. Using these techniques in combination with functional measures such as FSCV can greatly increase the scientific impact of a set of studies without major increases in time or resource demands. Our lab has used FSCV and western blotting from separate animals to show that conditions resulting in higher phosphorylated levels at the Threonine-53 site on the DAT also produce greater DA terminal responses to cocaine and often faster baseline uptake rates [92, 136], which is consistent with prior in vitro work, suggesting that phosphorylation of the Threonine-53 site promotes a cocaine-sensitive DAT conformation [137–139]. In ongoing studies, we are now using tissue from the same animal to examine both DAT function and protein expression using FSCV and western blotting. Such an approach not only offers an efficient use of animals and resources but it also allows DA neurotransmission to be correlated with protein expression [95, 96].

6.3 Optogenetics and Chemogenetics

With recent advances in selective targeting of neuronal populations, investigators have begun to incorporate new methodologies to activate DA neurons during FSCV recordings. The first implementations of optical stimulation of DA neurons to elicit DA release was conducted using transgenic mice or combinatorial viral approaches to selectively express channel rhodopsin (ChR2) in DA neurons of the VTA [140–142]. Using these approaches, investigators were able to demonstrate that optical stimulation results in robust DA release in the striatum using FSCV in anesthetized rats. Importantly, selective excitation of DA neurons in the VTA reduces recruitment of other cell types in this region [143], which can also release neurotransmitter in target structures.

The importance of selectively targeting stimulation to DA neurons was recently emphasized by a set of FSCV experiments in which optogenetic stimulation of DA terminals in the NAc core induced DA release with distinct characteristics compared to what is achieved with electrical stimulation. Some of the differences appear to be due to exclusion of local GABA and cholinergic influences when using optogenetic stimulation [144]. In another study, investigators demonstrated that phasic optogenetic stimulation of DA neurons in vivo induces robust DA transients in the striatum and elicits conditioned place preference [142]. These studies were notable as they were the first to confirm the causal role of phasic DA signals in behavioral conditioning.

In ongoing studies, we have begun to employ optogenetic stimulation to elicit DA release using FSCV in brain slices. In these experiments, male Long Evans-Tg3.1Deis rats [145] (Rat Resource and Research Center, Columbia, MO) were anesthetized with 2.5% isoflurane and injected bilaterally in the VTA with 1 μL of a Cre-dependent ChR2 adeno-associated virus (Addgene, Watertown, MA; # pAAV-EF1a-double floxed-hChR2(H134R)-EYFP-WPRE-HGHpA). After 4–5 weeks of viral incubation, rats were sacrificed and brain slices containing the NAc were prepared as described in Subheading 4.2.1. Once in the recording chamber, a glass working electrode was placed into the NAc and a fiber optic patch cable (Thor Labs, Newton, NJ; # FT200EMT) was placed <1 mm above the surface of the tissue. Using 473 nm DPSS Laser System (LaserGlow, North York, ON, Canada; # LRS-0473-PFM-00100-05), DA release was evoked every 3 min using a single light pulse (5 mW power, 4 ms) or 5 pulse trains (5 mW power, 4 ms, 20 Hz). Figure 11 shows example data using optogenetic activation of DA release obtained from FSCV studies in a slice preparation.

Chemogenetic approaches provide yet another route for delivering targeted stimulation to DA neurons. Although the temporal control of neuronal activity granted by chemogenetics is limited compared to optogenetics, chemogenetics has nevertheless been successfully leveraged in numerous studies to examine the behavioral and neurobiological function of DA circuitry. One recent publication used inhibitory Designer Receptors Exclusively Activated by Designer Drugs (DREADDs) to reduce activity of VTA DA neurons in rats performing a gambling task. These studies show that inhibiting DA neurons reduced risky decision-making in males, but increased risky decision-making in females, while increasing cocaine consumption in both sexes [146]. Another set of studies showed that female rats in estrus have increased cocaine-mediated inhibition of DA uptake compared to males and diestrus females. However, activating DA neurons with excitatory DREADDs increased cocaine-mediated inhibition of uptake in diestrus females to levels seen in the estrus females, suggesting that DA neuron activity may confer specific adaptations of the DAT that impact cocaine potency [8].

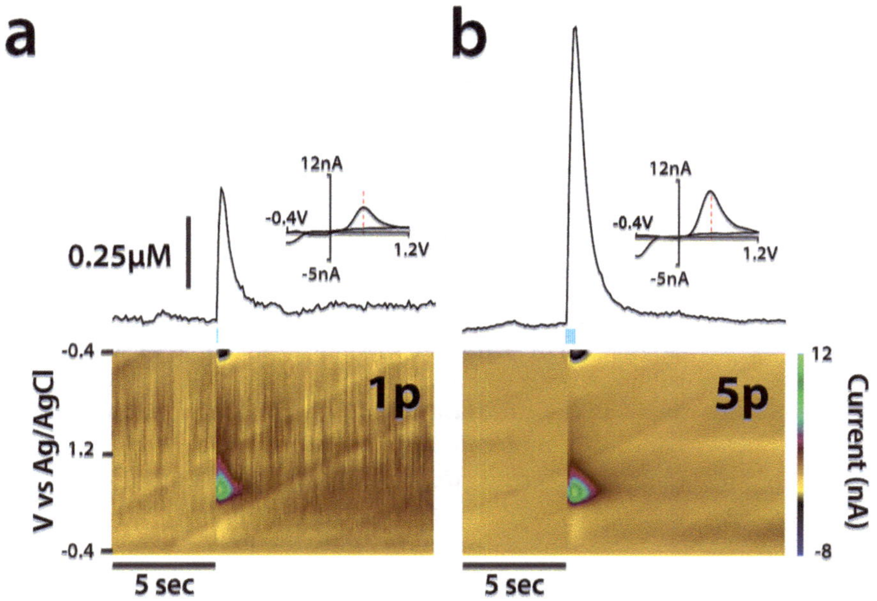

Fig. 11 Optogenetic stimulation of nucleus accumbens DA terminals: FSCV in brain slices. Cre-dependent ChR2 AAV was bilaterally infused into the VTA of TH-Cre$^+$ rats (1 μL, 100 nL/min) and given 4 weeks to reach full expression. (**a**) One pulse of 473 nm light (4 ms, 5 mW) elicited moderate DA release in the NAc whereas (**b**) a train of pulses (20 Hz, 5 pulses) elicited robust DA release. Blue vertical lines represent stimulation pulses

We recently used chemogenetics to show that activating VTA DA neurons with excitatory G_q DREADDs increased cocaine's behaviorally reinforcing effects, whereas inhibitory G_i DREADDs produced the opposite effect [147]. In a separate publication, we also demonstrated that activating VTA DA neurons with G_q DREADDs increased DAT phosphorylation at the Threonine-53 site and increased cocaine-mediated inhibition of DA uptake as measured by FSCV. The opposite effects were observed with G_i DREADDs, suggesting that biochemical and functional adaptations of the DAT may underlie the DREADDs-mediated alterations in the behavioral components of cocaine consumption, and suggest that the activity of DA neurons may govern DAT function [92]. Although investigators will undoubtedly continue to use electrical stimulation of DA neurons and terminals when performing FSCV experiments, there are advantages to using optogenetic and chemogenetic stimulation that often overwhelm the drawbacks of the added expense and time associated with such methods [148].

7 Conclusion

FSCV is a versatile technique with broad applications in neuroscience. Its high sampling rate and sensitivity for analytes such as DA have allowed scientists to examine release and uptake kinetics with a temporal resolution greater than what is possible with other common techniques. Different FSCV preparations allow researchers to answer distinct sets of questions and can be used in conjunction with other techniques to provide a holistic picture of DA system function. FSCV in anesthetized animals affords experimenters the opportunity to evaluate release and uptake kinetics in intact circuits, whereas FSCV in brain slices isolates local circuitry and enhances the ability to examine mechanistic changes in DA transmission. FSCV in awake, freely moving animals allows for monitoring of endogenous DA release, which can be correlated with behavioral events and presentation of environmental cues. Despite the technical limitations of FSCV, such as measuring only relative changes in analyte concentration and being limited to the measurement of a handful of molecules within the brain, this technique has come to define numerous areas of study and continues to be the gold-standard for measuring fast DA transmission.

Acknowledgments

We thank Bethan M. O'Connor for expert technical assistance. This work was supported by NIH grants DA031900 and DA043787 to R.A.E.

References

1. España RA, Oleson EB, Locke JL, Brookshire BR, Roberts DC, Jones SR (2010) The - hypocretin-orexin system regulates cocaine self-administration via actions on the mesolimbic dopamine system. Eur J Neurosci 31(2):336–348. PMID20039943; PMC2881680
2. Fordahl SC, Locke JL, Jones SR (2016) High fat diet augments amphetamine sensitization in mice: role of feeding pattern, obesity, and dopamine terminal changes. Neuropharmacology 109:170–182. PMID27267686; PMC4970886
3. Gan JO, Walton ME, Phillips PE (2010) Dissociable cost and benefit encoding of future rewards by mesolimbic dopamine. Nat Neurosci 13(1):25–27. PMID19904261; PMC2800310
4. Garris PA, Kilpatrick M, Bunin MA, Michael D, Walker QD, Wightman RM (1999) Dissociation of dopamine release in the nucleus accumbens from intracranial self-stimulation. Nature 398(6722):67–69. PMID10078530
5. Wightman RM, Kuhr WG, Ewing AG (1986) Voltammetric detection of dopamine release in the rat corpus striatum. Ann N Y Acad Sci 473:92–105. PMID3541742
6. Aggarwal S, Liu X, Rice C, Menell P, Clark PJ, Paparoidamis N, Xiao YC, Salvino JM, Fontana ACK, España RA, Kortagere S, Mortensen OV (2019) Identification of a novel allosteric modulator of the human dopamine transporter. ACS Chem Neurosci 10(8):3718–3730. PMID31184115; PMC6703927
7. Brodnik ZD, Black EM, Clark MJ, Kornsey KN, Snyder NW, España RA (2017) Susceptibility to traumatic stress sensitizes the dopaminergic response to cocaine and increases motivation for cocaine. Neuropharmacology 125:295–307. PMID28778834; PMC5585061
8. Calipari ES, Juarez B, Morel C, Walker DM, Cahill ME, Ribeiro E, Roman-Ortiz C, Ramakrishnan C, Deisseroth K, Han MH, Nestler EJ (2017) Dopaminergic dynamics underlying sex-specific cocaine reward. Nat Commun 8:13877. PMID28072417; PMC5234081
9. Ferris MJ, España RA, Locke JL, Konstantopoulos JK, Rose JH, Chen R, Jones SR

(2014) Dopamine transporters govern diurnal variation in extracellular dopamine tone. Proc Natl Acad Sci U S A 111(26): E2751–E2759. PMID24979798; PMC4084435
10. Schultz W (1998) Predictive reward signal of dopamine neurons. J Neurophysiol 80(1): 1–27. PMID9658025
11. Gainetdinov RR, Jones SR, Fumagalli F, Wightman RM, Caron MG (1998) Re-evaluation of the role of the dopamine transporter in dopamine system homeostasis. Brain Res Brain Res Rev 26(2–3):148–153. PMID9651511
12. Hamid AA, Frank MJ, Moore CI (2021) Wave-like dopamine dynamics as a mechanism for spatiotemporal credit assignment. Cell 184(10):2733–49.e16. PMID33861952; PMC8122079
13. Jones SR, Gainetdinov RR, Hu XT, Cooper DC, Wightman RM, White FJ, Caron MG (1999) Loss of autoreceptor functions in mice lacking the dopamine transporter. Nat Neurosci 2(7):649–655. PMID10404198
14. Mash DC, Pablo J, Ouyang Q, Hearn WL, Izenwasser S (2002) Dopamine transport function is elevated in cocaine users. J Neurochem 81(2):292–300. PMID12064476
15. Mohebi A, Pettibone JR, Hamid AA, Wong JT, Vinson LT, Patriarchi T, Tian L, Kennedy RT, Berke JD (2019) Dissociable dopamine dynamics for learning and motivation. Nature 570(7759):65–70. PMID31118513; PMC6555489
16. Threlfell S, Lalic T, Platt NJ, Jennings KA, Deisseroth K, Cragg SJ (2012) Striatal dopamine release is triggered by synchronized activity in cholinergic interneurons. Neuron 75(1):58–64. PMID22794260
17. Zahniser NR, Sorkin A (2004) Rapid regulation of the dopamine transporter: role in stimulant addiction? Neuropharmacology 47 (Suppl 1):80–91. PMID15464127
18. Grace AA, Bunney BS (1984) The control of firing pattern in nigral dopamine neurons: burst firing. J Neurosci 4(11):2877–2890. PMID6150071; PMC6564720
19. Grace AA, Bunney BS (1984) The control of firing pattern in nigral dopamine neurons: single spike firing. J Neurosci 4(11): 2866–2876. PMID6150070; PMC6564731
20. Korotkova TM, Sergeeva OA, Eriksson KS, Haas HL, Brown RE (2003) Excitation of ventral tegmental area dopaminergic and nondopaminergic neurons by orexins/hypocretins. J Neurosci 23(1):7–11. PMID12514194; PMC6742159
21. Margolis EB, Mitchell JM, Ishikawa J, Hjelmstad GO, Fields HL (2008) Midbrain dopamine neurons: projection target determines action potential duration and dopamine D (2) receptor inhibition. J Neurosci 28(36): 8908–8913. PMID18768684; PMC6670880
22. Margolis EB, Lock H, Hjelmstad GO, Fields HL (2006) The ventral tegmental area revisited: is there an electrophysiological marker for dopaminergic neurons? J Physiol 577 (Pt 3):907–924. PMID16959856; PMC1890372
23. Morris G, Nevet A, Arkadir D, Vaadia E, Bergman H (2006) Midbrain dopamine neurons encode decisions for future action. Nat Neurosci 9(8):1057–1063. PMID16862149
24. Schultz W, Dayan P, Montague PR (1997) A neural substrate of prediction and reward. Science 275(5306):1593–1599
25. Bocklisch C, Pascoli V, Wong JC, House DR, Yvon C, de Roo M, Tan KR, Luscher C (2013) Cocaine disinhibits dopamine neurons by potentiation of GABA transmission in the ventral tegmental area. Science 341(6153): 1521–1525. PMID24072923
26. Gessa GL, Muntoni F, Collu M, Vargiu L, Mereu G (1985) Low doses of ethanol activate dopaminergic neurons in the ventral tegmental area. Brain Res 348(1):201–203. PMID2998561
27. Mereu G, Yoon KW, Boi V, Gessa GL, Naes L, Westfall TC (1987) Preferential stimulation of ventral tegmental area dopaminergic neurons by nicotine. Eur J Pharmacol 141(3): 395–399. PMID3666033
28. Del Castillo J, Katz B (1954) Statistical factors involved in neuromuscular facilitation and depression. J Physiol 124(3):574–585. PMID13175200; PMC1366293
29. Katz B, Miledi R (1968) The role of calcium in neuromuscular facilitation. J Physiol 195(2):481–492. PMID4296699; PMC1351674
30. Cummings DD, Wilcox KS, Dichter MA (1996) Calcium-dependent paired-pulse facilitation of miniature EPSC frequency accompanies depression of EPSCs at hippocampal synapses in culture. J Neurosci 16(17): 5312–5323. PMID8757244; PMC6578891
31. Debanne D, Guerineau NC, Gahwiler BH, Thompson SM (1996) Paired-pulse facilitation and depression at unitary synapses in rat hippocampus: quantal fluctuation affects subsequent release. J Physiol 491(Pt 1): 163–176. PMID9011608; PMC1158767

32. Nicola SM, Surmeier J, Malenka RC (2000) Dopaminergic modulation of neuronal excitability in the striatum and nucleus accumbens. Annu Rev Neurosci 23:185–215. PMID10845063
33. Surmeier DJ, Ding J, Day M, Wang Z, Shen W (2007) D1 and D2 dopamine-receptor modulation of striatal glutamatergic signaling in striatal medium spiny neurons. Trends Neurosci 30(5):228–235. PMID17408758
34. Marcott PF, Mamaligas AA, Ford CP (2014) Phasic dopamine release drives rapid activation of striatal D2-receptors. Neuron 84(1): 164–176. PMID25242218; PMC4325987
35. Schultz KN, Kennedy RT (2008) Time-resolved microdialysis for in vivo neurochemical measurements and other applications. Annu Rev Anal Chem (Palo Alto Calif) 1: 627–661. PMID20636092
36. Wise RA, Newton P, Leeb K, Burnette B, Pocock D, Justice JB Jr (1995) Fluctuations in nucleus accumbens dopamine concentration during intravenous cocaine self-administration in rats. Psychopharmacology (Berl) 120(1):10–20. PMID7480530
37. Gilinsky MA, Faibushevish AA, Lunte CE (2001) Determination of myocardial norepinephrine in freely moving rats using in vivo microdialysis sampling and liquid chromatography with dual-electrode amperometric detection. J Pharm Biomed Anal 24(5–6): 929–935. PMID11248486; PMC2519812
38. Shackman HM, Shou M, Cellar NA, Watson CJ, Kennedy RT (2007) Microdialysis coupled on-line to capillary liquid chromatography with tandem mass spectrometry for monitoring acetylcholine in vivo. J Neurosci Methods 159(1):86–92. PMID16876256
39. Gunaydin LA, Grosenick L, Finkelstein JC, Kauvar IV, Fenno LE, Adhikari A, Lammel S, Mirzabekov JJ, Airan RD, Zalocusky KA, Tye KM, Anikeeva P, Malenka RC, Deisseroth K (2014) Natural neural projection dynamics underlying social behavior. Cell 157(7):1535–1551. PMID24949967; PMC4123133
40. Patriarchi T, Cho JR, Merten K, Howe MW, Marley A, Xiong WH, Folk RW, Broussard GJ, Liang R, Jang MJ, Zhong H, Dombeck D, von Zastrow M, Nimmerjahn A, Gradinaru V, Williams JT, Tian L (2018) Ultrafast neuronal imaging of dopamine dynamics with designed genetically encoded sensors. Science 360(6396): eaat4422. PMID29853555; PMC6287765
41. Sun F, Zeng J, Jing M, Zhou J, Feng J, Owen SF, Luo Y, Li F, Wang H, Yamaguchi T, Yong Z, Gao Y, Peng W, Wang L, Zhang S, Du J, Lin D, Xu M, Kreitzer AC, Cui G, Li Y (2018) A genetically encoded fluorescent sensor enables rapid and specific detection of dopamine in flies, fish, and mice. Cell 174(2):481–96.e19. PMID30007419; PMC6092020
42. Hawley MD, Tatawawadi SV, Piekarski S, Adams RN (1967) Electrochemical studies of the oxidation pathways of catecholamines. J Am Chem Soc 89(2):447–450. PMID6031636
43. Kissinger PT, Hart JB, Adams RN (1973) Voltammetry in brain tissue--a new neurophysiological measurement. Brain Res 55(1): 209–213. PMID4145914
44. Millar J, Stamford JA, Kruk ZL, Wightman RM (1985) Electrochemical, pharmacological and electrophysiological evidence of rapid dopamine release and removal in the rat caudate nucleus following electrical stimulation of the median forebrain bundle. Eur J Pharmacol 109(3):341–348. PMID3872803
45. McCreery RL (2008) Advanced carbon electrode materials for molecular electrochemistry. Chem Rev 108(7):2646–2687. PMID18557655
46. Clark JJ, Sandberg SG, Wanat MJ, Gan JO, Horne EA, Hart AS, Akers CA, Parker JG, Willuhn I, Martinez V, Evans SB, Stella N, Phillips PE (2010) Chronic microsensors for longitudinal, subsecond dopamine detection in behaving animals. Nat Methods 7(2): 126–129. PMID20037591; PMC2849934
47. Brodnik ZD, Black EM, España RA (2020) Accelerated development of cocaine-associated dopamine transients and cocaine use vulnerability following traumatic stress. Neuropsychopharmacology 45(3):472–481. PMID31539899; PMC6969179
48. Shaw JK, Ferris MJ, Locke JL, Brodnik ZD, Jones SR, España RA (2017) Hypocretin/orexin knock-out mice display disrupted behavioral and dopamine responses to cocaine. Addict Biol 22(6):1695–1705. PMID27480648; PMC5468487
49. Albulbul A (2016) Evaluating major electrode types for idle biological signal measurements for modern medical technology. Bioengineering (Basel) 3(3):20. PMID28952582; PMC5597189
50. Hashemi P, Walsh PL, Guillot TS, Gras-Najjar J, Takmakov P, Crews FT, Wightman RM (2011) Chronically implanted, Nafion-coated Ag/AgCl reference electrodes for neurochemical applications. ACS Chem Neurosci 2(11):658–666. PMID22125666; PMC3224088

51. Moussy F, Harrison DJ, Rajotte RV (1994) A miniaturized Nafion-based glucose sensor: in vitro and in vivo evaluation in dogs. Int J Artif Organs 17(2):88–94. PMID8039946
52. Phillips PE, Stuber GD, Heien ML, Wightman RM, Carelli RM (2003) Subsecond dopamine release promotes cocaine seeking. Nature 422(6932):614–618. PMID12687000
53. Baur JE, Kristensen EW, May LJ, Wiedemann DJ, Wightman RM (1988) Fast-scan voltammetry of biogenic amines. Anal Chem 60(13): 1268–1272. PMID3213946
54. Ewing AG, Dayton MA, Wightman RM (1981) Pulse voltammetry with microvoltammetric electrodes. Anal Chem 53:1842–1847
55. Kuhr WG, Ewing AG, Caudill WL, Wightman RM (1984) Monitoring the stimulated release of dopamine with in vivo voltammetry. I: Characterization of the response observed in the caudate nucleus of the rat. J Neurochem 43(2):560–569. PMID6736965
56. Brodnik ZD, España RA (2015) Dopamine uptake dynamics are preserved under isoflurane anesthesia. Neurosci Lett 606:129–134
57. Cid-Jofre V, Garate-Perez M, Clark PJ, Valero-Jara V, España RA, Sotomayor-Zarate-R, Cruz G, Renard GM (2021) Chronic modafinil administration to preadolescent rats impairs social play behavior and dopaminergic system. Neuropharmacology 183:108404. PMID33197467
58. Roberts JG, Sombers LA (2018) Fast-scan cyclic voltammetry: chemical sensing in the brain and beyond. Anal Chem 90(1): 490–504. PMID29182309; PMC5750125
59. Venton BJ, Cao Q (2020) Fundamentals of fast-scan cyclic voltammetry for dopamine detection. Analyst 145(4):1158–1168. PMID31922176; PMC7028514
60. Qi L, Thomas E, White SH, Smith SK, Lee CA, Wilson LR, Sombers LA (2016) Unmasking the effects of L-DOPA on rapid dopamine signaling with an improved approach for Nafion coating carbon-fiber microelectrodes. Anal Chem 88(16): 8129–8136. PMID27441547; PMC5226322
61. Takmakov P, Zachek MK, Keithley RB, Walsh PL, Donley C, McCarty GS, Wightman RM (2010) Carbon microelectrodes with a renewable surface. Anal Chem 82(5):2020–2028. PMID20146453; PMC2838506
62. Chang AY, Dutta G, Siddiqui S, Arumugam PU (2019) Surface fouling of ultrananocrystalline diamond microelectrodes during dopamine detection: improving lifetime via electrochemical cycling. ACS Chem Neurosci 10(1):313–322. PMID30285418
63. Heien ML, Phillips PE, Stuber GD, Seipel AT, Wightman RM (2003) Overoxidation of carbon-fiber microelectrodes enhances dopamine adsorption and increases sensitivity. Analyst 128(12):1413–1419. PMID14737224
64. Levy KA, Brodnik ZD, Shaw JK, Perrey DA, Zhang Y, España RA (2017) Hypocretin receptor 1 blockade produces bimodal modulation of cocaine-associated mesolimbic dopamine signaling. Psychopharmacology (Berl) 234(18):2761–2776. PMID28667509; PMC5709206
65. Meunier CJ, Roberts JG, McCarty GS, Sombers LA (2017) Background signal as an in situ predictor of dopamine oxidation potential: improving interpretation of fast-scan cyclic voltammetry data. ACS Chem Neurosci 8(2):411–419. PMID28044445; PMC5684890
66. Roberts JG, Toups JV, Eyualem E, McCarty GS, Sombers LA (2013) In situ electrode calibration strategy for voltammetric measurements in vivo. Anal Chem 85(23): 11568–11575. PMID24224460; PMC3935327
67. Alonso IP, Pino JA, Kortagere S, Torres GE, España RA (2021) Dopamine transporter function fluctuates across sleep/wake state: potential impact for addiction. Neuropsychopharmacology 46:699–708. PMID33032296
68. Yorgason JT, España RA, Jones SR (2011) Demon voltammetry and analysis software: analysis of cocaine-induced alterations in dopamine signaling using multiple kinetic measures. J Neurosci Methods 202(2): 158–164. PMID21392532; PMC3149733
69. Covey DP, Bunner KD, Schuweiler DR, Cheer JF, Garris PA (2016) Amphetamine elevates nucleus accumbens dopamine via an action potential-dependent mechanism that is modulated by endocannabinoids. Eur J Neurosci 43(12):1661–1673. PMID27038339; PMC5819353
70. Goodwin JS, Larson GA, Swant J, Sen N, Javitch JA, Zahniser NR, De Felice LJ, Khoshbouei H (2009) Amphetamine and methamphetamine differentially affect dopamine transporters in vitro and in vivo. J Biol Chem 284(5):2978–2989. PMID19047053; PMC2631950
71. Kutlu MG, Brady LJ, Peck EG, Hofford RS, Yorgason JT, Siciliano CA, Kiraly DD, Calipari ES (2018) Granulocyte colony stimulating factor enhances reward learning through potentiation of mesolimbic dopamine system

function. J Neurosci 38(41):8845–8859. PMID30150359; PMC6181308

72. Wightman RM, Heien ML, Wassum KM, Sombers LA, Aragona BJ, Khan AS, Ariansen JL, Cheer JF, Phillips PE, Carelli RM (2007) Dopamine release is heterogeneous within microenvironments of the rat nucleus accumbens. Eur J Neurosci 26(7):2046–2054. PMID17868375
73. Keithley RB, Heien ML, Wightman RM (2009) Multivariate concentration determination using principal component regression with residual analysis. Trends Analyt Chem 28(9):1127–1136. PMID20160977; PMC2760950
74. Rodeberg NT, Johnson JA, Cameron CM, Saddoris MP, Carelli RM, Wightman RM (2015) Construction of training sets for valid calibration of in vivo cyclic voltammetric data by principal component analysis. Anal Chem 87(22):11484–11491. PMID26477708; PMC5131642
75. Kuhr WG, Wightman RM (1986) Real-time measurement of dopamine release in rat brain. Brain Res 381(1):168–171. PMID3489505
76. Wightman RM, May LJ, Michael AC (1988) Detection of dopamine dynamics in the brain. Anal Chem 60(13):769A–779A. PMID3063135
77. Wightman RM, Zimmerman JB (1990) Control of dopamine extracellular concentration in rat striatum by impulse flow and uptake. Brain Res Brain Res Rev 15(2):135–144. PMID2282449
78. Budygin EA, Phillips PE, Wightman RM, Jones SR (2001) Terminal effects of ethanol on dopamine dynamics in rat nucleus accumbens: an in vitro voltammetric study. Synapse 42(2):77–79. PMID11574942
79. Park J, Takmakov P, Wightman RM (2011) In vivo comparison of norepinephrine and dopamine release in rat brain by simultaneous measurements with fast-scan cyclic voltammetry. J Neurochem 119(5):932–944. PMID21933188; PMC3217157
80. Robinson DL, Venton BJ, Heien ML, Wightman RM (2003) Detecting subsecond dopamine release with fast-scan cyclic voltammetry in vivo. Clin Chem 49(10):1763–1773. PMID14500617
81. Stuber GD, Roitman MF, Phillips PE, Carelli RM, Wightman RM (2005) Rapid dopamine signaling in the nucleus accumbens during contingent and noncontingent cocaine administration. Neuropsychopharmacology 30(5):853–863. PMID15549053
82. Garris PA, Ciolkowski EL, Pastore P, Wightman RM (1994) Efflux of dopamine from the synaptic cleft in the nucleus accumbens of the rat brain. J Neurosci 14(10):6084–6093. PMID7931564; PMC6577011
83. Garris PA, Wightman RM (1995) Distinct pharmacological regulation of evoked dopamine efflux in the amygdala and striatum of the rat in vivo. Synapse 20(3):269–279. PMID7570359
84. May LJ, Wightman RM (1989) Heterogeneity of stimulated dopamine overflow within rat striatum as observed with in vivo voltammetry. Brain Res 487(2):311–320. PMID2786444
85. Walker QD, Rooney MB, Wightman RM, Kuhn CM (2000) Dopamine release and uptake are greater in female than male rat striatum as measured by fast cyclic voltammetry. Neuroscience 95(4):1061–1070. PMID10682713
86. Shaw JK, Pamela Alonso I, Lewandowski SI, Scott MO, O'Connor BM, Aggarwal S, De Biasi M, Mortensen OV, España RA (2021) Individual differences in dopamine uptake in the dorsomedial striatum prior to cocaine exposure predict motivation for cocaine in male rats. Neuropsychopharmacology 46(10):1757–1767. PMID33953341
87. Kelly RS, Wightman RM (1987) Detection of dopamine overflow and diffusion with voltammetry in slices of rat brain. Brain Res 423(1–2):79–87. PMID3676822
88. Brodnik ZD, Batra A, Oleson EB, España RA (2019) Local GABAA receptor-mediated suppression of dopamine release within the nucleus accumbens. ACS Chem Neurosci 10(4):1978–1985. PMID30253088
89. Jones SR, Gainetdinov RR, Jaber M, Giros B, Wightman RM, Caron MG (1998) Profound neuronal plasticity in response to inactivation of the dopamine transporter. Proc Natl Acad Sci U S A 95(7):4029–4034. PMID9520487; PMC19957
90. Jones SR, Gainetdinov RR, Wightman RM, Caron MG (1998) Mechanisms of amphetamine action revealed in mice lacking the dopamine transporter. J Neurosci 18(6): 1979–1986. PMID9482784; PMC6792915
91. Schmitz Y, Schmauss C, Sulzer D (2002) Altered dopamine release and uptake kinetics in mice lacking D2 receptors. J Neurosci 22(18):8002–8009. PMID12223553; PMC6758092
92. Brodnik ZD, Xu W, Batra A, Lewandowski SI, Ruiz CM, Mortensen OV, Kortagere S, Mahler SV, España RA (2020) Chemogenetic

manipulation of dopamine neurons dictates cocaine potency at distal dopamine transporters. J Neurosci 40(45):8767–8779. PMID33046544; PMC7643298

93. Jones SR, Bowman BP, Kuhn CM, Wightman RM (1996) Development of dopamine neurotransmission and uptake inhibition in the caudate nucleus as measured by fast-cyclic voltammetry. Synapse 24(3):305–307. PMID8923671
94. Jones SR, Garris PA, Kilts CD, Wightman RM (1995) Comparison of dopamine uptake in the basolateral amygdaloid nucleus, caudate-putamen, and nucleus accumbens of the rat. J Neurochem 64(6):2581–2589. PMID7760038
95. Anzalone A, Lizardi-Ortiz JE, Ramos M, De Mei C, Hopf FW, Iaccarino C, Halbout B, Jacobsen J, Kinoshita C, Welter M, Caron MG, Bonci A, Sulzer D, Borrelli E (2012) Dual control of dopamine synthesis and release by presynaptic and postsynaptic dopamine D2 receptors. J Neurosci 32(26): 9023–9034. PMID22745501; PMC3752062
96. Bello EP, Mateo Y, Gelman DM, Noain D, Shin JH, Low MJ, Alvarez VA, Lovinger DM, Rubinstein M (2011) Cocaine supersensitivity and enhanced motivation for reward in mice lacking dopamine D2 autoreceptors. Nat Neurosci 14(8):1033–1038. PMID21743470; PMC3175737
97. Kennedy RT, Jones SR, Wightman RM (1992) Dynamic observation of dopamine autoreceptor effects in rat striatal slices. J Neurochem 59(2):449–455. PMID1352798
98. Calipari ES, Huggins KN, Mathews TA, Jones SR (2012) Conserved dorsal-ventral gradient of dopamine release and uptake rate in mice, rats and rhesus macaques. Neurochem Int 61(7):986–991. PMID22819794; PMC3593229
99. Berke JD (2018) What does dopamine mean? Nat Neurosci 21(6):787–793. PMID29760524; PMC6358212
100. Garris PA, Christensen JR, Rebec GV, Wightman RM (1997) Real-time measurement of electrically evoked extracellular dopamine in the striatum of freely moving rats. J Neurochem 68(1):152–161. PMID8978721
101. Robinson DL, Phillips PE, Budygin EA, Trafton BJ, Garris PA, Wightman RM (2001) Sub-second changes in accumbal dopamine during sexual behavior in male rats. Neuroreport 12(11):2549–2552. PMID11496146
102. Zhang L, Doyon WM, Clark JJ, Phillips PE, Dani JA (2009) Controls of tonic and phasic dopamine transmission in the dorsal and ventral striatum. Mol Pharmacol 76(2):396–404. PMID19460877; PMC2713129
103. Roberts JG, Lugo-Morales LZ, Loziuk PL, Sombers LA (2013) Real-time chemical measurements of dopamine release in the brain. Methods Mol Biol 964:275–294. PMID23296789; PMC5259961
104. Cragg SJ, Hille CJ, Greenfield SA (2000) Dopamine release and uptake dynamics within nonhuman primate striatum in vitro. J Neurosci 20(21):8209–8217. PMID11050144; PMC6772736
105. Siciliano CA, Calipari ES, Jones SR (2014) Amphetamine potency varies with dopamine uptake rate across striatal subregions. J Neurochem 131(3):348–355. PMID24988947; PMC4205180
106. Yorgason JT, Calipari ES, Ferris MJ, Karkhanis AN, Fordahl SC, Weiner JL, Jones SR (2016) Social isolation rearing increases dopamine uptake and psychostimulant potency in the striatum. Neuropharmacology 101:471–479. PMID26525189; PMC4681685
107. Brodnik ZD, Bernstein DL, Prince CD, España RA (2015) Hypocretin receptor 1 blockade preferentially reduces high effort responding for cocaine without promoting sleep. Behav Brain Res 291:377–384. PMID26049058; PMC4734652
108. Shnitko TA, Robinson DL (2015) Regional variation in phasic dopamine release during alcohol and sucrose self-administration in rats. ACS Chem Neurosci 6(1):147–154. PMID25493956; PMC4304482
109. Richardson NR, Roberts DCS (1996) Progressive ratio schedules in drug self-administration studies in rats: a method to evaluate reinforcing efficacy. J Neurosci Methods 66(1):1–11
110. Borgland SL, Malenka RC, Bonci A (2004) Acute and chronic cocaine-induced potentiation of synaptic strength in the ventral tegmental area: electrophysiological and behavioral correlates in individual rats. J Neurosci 24(34):7482–7490. PMID15329395; PMC6729639
111. Chen BT, Bowers MS, Martin M, Hopf FW, Guillory AM, Carelli RM, Chou JK, Bonci A (2008) Cocaine but not natural reward self-administration nor passive cocaine infusion produces persistent LTP in the VTA. Neuron 59(2):288–297. PMID18667156; PMC2593405
112. Hurd YL, Ungerstedt U (1989) In vivo neurochemical profile of dopamine uptake

inhibitors and releasers in rat caudate-putamen. Eur J Pharmacol 166(2):251–260. PMID2477259

113. Kuczenski R, Segal DS, Aizenstein ML (1991) Amphetamine, cocaine, and fencamfamine: relationship between locomotor and stereotypy response profiles and caudate and accumbens dopamine dynamics. J Neurosci 11(9):2703–2712. PMID1715389; PMC6575248

114. Moghaddam B, Bunney BS (1989) Differential effect of cocaine on extracellular dopamine levels in rat medial prefrontal cortex and nucleus accumbens: comparison to amphetamine. Synapse 4(2):156–161. PMID2781466

115. Weiss F, Paulus MP, Lorang MT, Koob GF (1992) Increases in extracellular dopamine in the nucleus accumbens by cocaine are inversely related to basal levels: effects of acute and repeated administration. J Neurosci 12(11):4372–4380. PMID1432099; PMC6576001

116. Beuming T, Kniazeff J, Bergmann ML, Shi L, Gracia L, Raniszewska K, Newman AH, Javitch JA, Weinstein H, Gether U, Loland CJ (2008) The binding sites for cocaine and dopamine in the dopamine transporter overlap. Nat Neurosci 11(7):780–789. PMID18568020; PMC2692229

117. España RA, Roberts DC, Jones SR (2008) Short-acting cocaine and long-acting GBR-12909 both elicit rapid dopamine uptake inhibition following intravenous delivery. Neuroscience 155(1):250–257. PMID18597947; PMC2538802

118. Mateo Y, Budygin EA, Morgan D, Roberts DC, Jones SR (2004) Fast onset of dopamine uptake inhibition by intravenous cocaine. Eur J Neurosci 20(10):2838–2842. PMID15548229

119. Yorgason JT, Jones SR, España RA (2011) Low and high affinity dopamine transporter inhibitors block dopamine uptake within 5 sec of intravenous injection. Neuroscience 182: 125–132. PMID21402130; PMC3085557

120. Brodnik ZD, Ferris MJ, Jones SR, España RA (2017) Reinforcing doses of intravenous cocaine produce only modest dopamine uptake inhibition. ACS Chem Neurosci 8(2): 281–289. PMID27936579; PMC5553220

121. John CE, Jones SR (2007) Voltammetric characterization of the effect of monoamine uptake inhibitors and releasers on dopamine and serotonin uptake in mouse caudate-putamen and substantia nigra slices. Neuropharmacology 52(8):1596–1605. PMID17459426; PMC2041899

122. Venton BJ, Seipel AT, Phillips PE, Wetsel WC, Gitler D, Greengard P, Augustine GJ, Wightman RM (2006) Cocaine increases dopamine release by mobilization of a synapsin-dependent reserve pool. J Neurosci 26(12):3206–3209. PMID16554471; PMC6674099

123. Siciliano CA, Fordahl SC, Jones SR (2016) Cocaine self-administration produces long-lasting alterations in dopamine transporter responses to cocaine. J Neurosci 36(30): 7807–7816. PMID27466327; PMC4961771

124. Kiyatkin EA, Kiyatkin DE, Rebec GV (2000) Phasic inhibition of dopamine uptake in nucleus accumbens induced by intravenous cocaine in freely behaving rats. Neuroscience 98(4):729–741. PMID10891616

125. Hart AS, Rutledge RB, Glimcher PW, Phillips PE (2014) Phasic dopamine release in the rat nucleus accumbens symmetrically encodes a reward prediction error term. J Neurosci 34(3):698–704. PMID24431428; PMC3891951

126. Hamid AA, Pettibone JR, Mabrouk OS, Hetrick VL, Schmidt R, Vander Weele CM, Kennedy RT, Aragona BJ, Berke JD (2016) Mesolimbic dopamine signals the value of work. Nat Neurosci 19(1):117–126. PMID26595651; PMC4696912

127. Calipari ES, Ferris MJ, Zimmer BA, Roberts DC, Jones SR (2013) Temporal pattern of cocaine intake determines tolerance vs sensitization of cocaine effects at the dopamine transporter. Neuropsychopharmacology 38(12):2385–2392. PMID23719505; PMC3799057

128. Ferris MJ, Calipari ES, Mateo Y, Melchior JR, Roberts DC, Jones SR (2012) Cocaine self-administration produces pharmacodynamic tolerance: differential effects on the potency of dopamine transporter blockers, releasers, and methylphenidate. Neuropsychopharmacology 37(7):1708–1716. PMID22395730; PMC3358740

129. Mateo Y, Lack CM, Morgan D, Roberts DC, Jones SR (2005) Reduced dopamine terminal function and insensitivity to cocaine following cocaine binge self-administration and deprivation. Neuropsychopharmacology 30(8): 1455–1463. PMID15702135

130. Addy NA, Daberkow DP, Ford JN, Garris PA, Wightman RM (2010) Sensitization of rapid dopamine signaling in the nucleus accumbens core and shell after repeated cocaine in rats. J Neurophysiol 104(2):922–931

131. Calipari ES, Ferris MJ, Jones SR (2014) Extended access of cocaine self-

administration results in tolerance to the dopamine-elevating and locomotor-stimulating effects of cocaine. J Neurochem 128(2):224–232. PMID24102293; PMC3947316

132. Saal D, Dong Y, Bonci A, Malenka RC (2003) Drugs of abuse and stress trigger a common synaptic adaptation in dopamine neurons. Neuron 37(4):577–582. PMID12597856
133. Willuhn I, Burgeno LM, Groblewski PA, Phillips PE (2014) Excessive cocaine use results from decreased phasic dopamine signaling in the striatum. Nat Neurosci 17(5): 704–709. PMID24705184; PMC4714770
134. Willuhn I, Burgeno LM, Everitt BJ, Phillips PE (2012) Hierarchical recruitment of phasic dopamine signaling in the striatum during the progression of cocaine use. Proc Natl Acad Sci U S A 109(50):20703–20708. PMID23184975; PMC3528544
135. Cheer JF, Aragona BJ, Heien ML, Seipel AT, Carelli RM, Wightman RM (2007) Coordinated accumbal dopamine release and neural activity drive goal-directed behavior. Neuron 54(2):237–244. PMID17442245
136. Brodnik ZD, Alonso IP, Xu W, Zhang Y, Kortagere S, España RA (2020) Hypocretin receptor 1 involvement in cocaine-associated behavior: therapeutic potential and novel mechanistic insights. Brain Res 1731: 145894. PMID30071195
137. Challasivakanaka S, Zhen J, Smith ME, Reith MEA, Foster JD, Vaughan RA (2017) Dopamine transporter phosphorylation site threonine 53 is stimulated by amphetamines and regulates dopamine transport, efflux, and cocaine analog binding. J Biol Chem 292(46):19066–19075. PMID28939767; PMC5704487
138. Foster JD, Yang JW, Moritz AE, Challasivakanaka S, Smith MA, Holy M, Wilebski K, Sitte HH, Vaughan RA (2012) Dopamine transporter phosphorylation site threonine 53 regulates substrate reuptake and amphetamine-stimulated efflux. J Biol Chem 287(35):29702–29712. PMID22722938; PMC3436161
139. Schmitt KC, Reith ME (2010) Regulation of the dopamine transporter: aspects relevant to psychostimulant drugs of abuse. Ann N Y Acad Sci 1187:316–340. PMID20201860
140. Bass CE, Grinevich VP, Vance ZB, Sullivan RP, Bonin KD, Budygin EA (2010) Optogenetic control of striatal dopamine release in rats. J Neurochem 114(5):1344–1352. PMID20534006; PMC2923233
141. Gompf HS, Budygin EA, Fuller PM, Bass CE (2015) Targeted genetic manipulations of neuronal subtypes using promoter-specific combinatorial AAVs in wild-type animals. Front Behav Neurosci 9:152. PMID26190981; PMC4488755
142. Tsai HC, Zhang F, Adamantidis A, Stuber GD, Bonci A, de Lecea L, Deisseroth K (2009) Phasic firing in dopaminergic neurons is sufficient for behavioral conditioning. Science 324(5930):1080–1084. PMID19389999; PMC5262197
143. Van Bockstaele EJ, Pickel VM (1995) GABA-containing neurons in the ventral tegmental area project to the nucleus accumbens in rat brain. Brain Res 682(1–2):215–221. PMID7552315
144. Melchior JR, Ferris MJ, Stuber GD, Riddle DR, Jones SR (2015) Optogenetic versus electrical stimulation of dopamine terminals in the nucleus accumbens reveals local modulation of presynaptic release. J Neurochem 134(5):833–844. PMID26011081; PMC4537642
145. Witten IB, Steinberg EE, Lee SY, Davidson TJ, Zalocusky KA, Brodsky M, Yizhar O, Cho SL, Gong S, Ramakrishnan C, Stuber GD, Tye KM, Janak PH, Deisseroth K (2011) Recombinase-driver rat lines: tools, techniques, and optogenetic application to dopamine-mediated reinforcement. Neuron 72(5):721–733. PMID22153370; PMC3282061
146. Hynes TJ, Hrelja KM, Hathaway BA, Hounjet CD, Chernoff CS, Ebsary SA, Betts GD, Russell B, Ma L, Kaur S, Winstanley CA (2021) Dopamine neurons gate the intersection of cocaine use, decision making, and impulsivity. Addict Biol 26:e13022. PMID33559379
147. Mahler SV, Brodnik ZD, Cox BM, Buchta WC, Bentzley BS, Quintanilla J, Cope ZA, Lin EC, Riedy MD, Scofield MD, Messinger J, Ruiz CM, Riegel AC, España RA, Aston-Jones G (2019) Chemogenetic manipulations of ventral tegmental area dopamine neurons reveal multifaceted roles in cocaine abuse. J Neurosci 39(3):503–518. PMID30446532; PMC6335749
148. McElligott Z (2015) Optogenetic and chemogenetic approaches to advance monitoring molecules. ACS Chem Neurosci 6(7): 944–947. PMID25791746

Chapter 12

Multi-circuit Recording in Animal Models of Parkinson's Disease

Ximena Paz Garcia-Nuñez, Carolina Astudillo-Valenzuela, and Romulo Fuentes-Flores

Abstract

One of the main features of Parkinson's disease is the neurodegeneration of the dopaminergic neurons of the substantia nigra pars compacta which causes a fall of the levels of dopamine in the striatum. This leads to alterations of the neural activity in all the structures of the loop circuit formed by the cortex, the basal ganglia, and thalamus. These alterations are believed to underlie the bradykinesia and hypokinesia observed in Parkinson's disease. To study the correlation or the causality between circuitry neural activity and specific parkinsonian motor symptoms, the circuit can be interrogated by simultaneous extracellular recordings in diverse brain structures in behaving freely moving animals modeling Parkinson's disease. Hereby, we describe the procedure to obtain the classical 6-hydroxydopamine Parkinson's model combined with multi-area extracellular recordings in small behaving animals to access the activity of the parkinsonian cortico-basal ganglia-thalamo-cortical loop. Given that both the dopaminergic lesion caused by 6-hydroxydopamine, and the quality of the extracellular recordings have temporal dynamics of their own, the relative timing of both procedures is essential to obtain successful recording with high-quality signals.

Key words Extracellular recordings, Striatum, Cortex, Thalamus, Parkinson's disease, 6-hydroxydopamine, Electrode array chronic implant

1 Introduction

Parkinson's disease is a neurodegenerative disorder involving the progressive loss of the dopaminergic neurons of the substantia nigra pars compacta [1] which supply dopamine to the basal ganglia, mainly to the striatum. In the striatum, dopamine differentially modulates the excitability of two main striatal neuronal populations: the spiny projection neurons of the direct pathway – expressing D1 dopamine receptors, and the spiny projection neurons of the indirect pathway – expressing D2 dopamine receptors [2, 3]. Normally regarded as having opposing actions, the direct pathway is labeled as pro-kinetic and the indirect pathways as anti-kinetic [4]; current view considers that the coordinated activity of

José Antonio Fuentealba-Evans and Pablo Henny (eds.), *Dopaminergic System Function and Dysfunction: Experimental Approaches*, Neuromethods, vol. 193, https://doi.org/10.1007/978-1-0716-2799-0_12,

these two pathways is necessary for proper motor performance, including selection, initiation, and termination of motor programs [5]. The direct and indirect pathways of the striatum/basal ganglia are within a major neural circuit which starts in the cortex, then projects to the basal ganglia, then to thalamus, and finally back to cortex, the so-called cortico-basal ganglia-thalamo-cortical loop (CBTC).

When the progressive dopaminergic neuron loss in Parkinson's disease reaches a certain level, the excitability of the direct and indirect pathways is altered in such magnitude that the compensations made by the circuit are no longer capable of maintaining proper motor control, and the inability to select and initiate body motor programs becomes apparent in the form of bradykinesia, hypokinesia, and rigidity. While important progress has been made in understanding the many molecular, cellular, and local circuit level events mediating the decrease of dopamine and the motor symptoms, the impact on the complete CBTC loop is still obscure, especially regarding what appear to be a robust but rather poorly understood finding: the increase of synchronized neural activity in both parkinsonian models and Parkinson's disease (for a review *see* [6]).

A way to evaluate the impact of dopaminergic lesions on the system or circuit level is the simultaneous *in vivo* extracellular electrophysiological recording of several areas of the CBTC loop. This can be achieved by the chronic implant of electrodes in selected CBTC areas. The signal obtained by the electrodes can be amplified, digitized at >30 kHz, and saved for off-line analysis. The saved signal can be split to a low-frequency (<300 Hz) component known as local field potential (LFP) and a high frequency component (>5 kHz) from which the precise firing of single neurons can be extracted. Oscillations of the LFP at different frequencies are regarded as the synchronic activity of neuron populations. Figure 1 shows and example of an extracellular recording and its processing to obtain single neuron and population activity. For a detailed discussion on the origin of the signals from extracellular recordings, *see* [7].

Here we will detail the combination of two procedures: the bilateral dopaminergic chronic striatal lesion with 6-hydroxydopamine (6-OHDA) and the chronic implant of electrodes in the CBTC loop for in vivo extracellular recording. While other variations of the dopaminergic lesion with 6-OHDA, for example the unilateral medial forebrain bundle 6-OHDA injection, have been extensively characterized and used for preclinical research on Parkinson's disease (for a review *see* [8]), the bilateral lesion model is not often used. Yet, it is considered that the striatal lesions are more like the actual disease than medial forebrain bundle or substantia

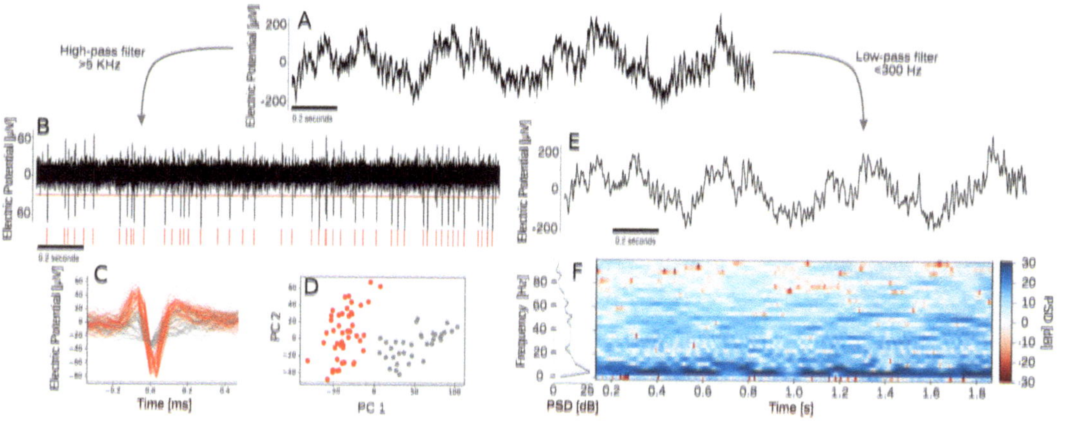

Fig. 1 Example of extracellular electrophysiological recording from a single channel. (**a**) Wide band signal digitized at 30 kHz. (**b**) The signal resulting from high pass filtering the signal in A contains the actual action potentials of the neurons close to the electrode. These can be detected and extracted with a threshold (red horizontal line). (**c**) Every time the signal crosses the threshold, that portion (typically a 1-ms window centered at the threshold event) is saved as a putative action potential named waveform. Waveforms can be assigned as belonging to single neurons (red traces) and separated from noise (gray traces) in a process known as "spike sorting." (**d**) This process typically involves the clustering of waveforms, in the example, Principal Component Analysis was performed over the waveforms and the 1st and 2nd principal components (PC) plotted, which allow the separation into two clusters. The red cluster corresponds to actual action potentials likely from a single neuron. The gray cluster corresponds to noise (threshold was set low on the purpose of capturing some noise). Once the waveforms are sorted, the precise firing in time of that neuron can be reconstructed, as seen in the sequence of red vertical lines at the bottom of B. (**e**) The low frequency component of the wide band signal is known as local field potential (LFP). (**f**) The most basic analysis of LFP consists of detecting the intensity of the oscillations, for which several methods and representations are available. The colored image is a spectrogram, which represents the power spectral density (PSD) (color coded as indicated in the bar to the right) of the LFP at each frequency (*y*-axis) in time (*x*-axis). The black trace to the left of the spectrogram is the PSD averaged across time. The spectral analysis reveals that the most powerful oscillations are below 20 Hz, moderate intensity between ~20 and ~60 Hz, and low intensity above ~70 Hz

nigra pars compacta lesions, while bilateral lesions mimics better the actual motor impairment found in patients (*see* review in [9]). On the other hand, *in vivo* recording of different brain areas is a powerful tool to access the dynamics of brain activity and the interaction between areas in the behaving animal; nonetheless, compared to other techniques, relatively few studies have been combining in vivo electrophysiological recordings on Parkinson's models. These recordings include the dorsolateral striatum [10–14], dorsomedial striatum [14], substantia nigra pars reticulata [14, 15], subthalamic nucleus [14, 16], medial prefrontal cortex [16], primary motor cortex [12–14, 16], rodent supplementary motor area (rostral forelimb area) [14], globus pallidus [14], and ventrolateral/ventroanterior nuclei of the thalamus [14].

2 Materials

This protocol is useful for small mammals, including mice and rats [11, 17]. Here we describe it for adult male rats 6–8 weeks of age, weighing between 250 and 350 g.

2.1 Striatal Injection of 6-Hydroxydopamine

1. 6-Hydroxydopamine hydrochloride (Sigma- Aldrich Cat: H4381), powder, and solution are sensitive to light; store at −20 °C in the dark.
2. L-Ascorbic acid (Sigma-Aldrich Cat: A44544).
3. Sterile sodium chloride ampoule 0.9%.
4. Veterinary anesthesia machine (Dre veterinary, compact 150, ID10150) .
5. Induction chamber (Dre veterinary, Cat# 12568).
6. Isoflurane 100% (Baxter Cat# 218082).
7. Medicinal compressed oxygen (Linde, Cat# Nu1072).,
8. Desipramine hydrochloride (Sigma-Aldrich Cat: D3900), powder, and solution are sensitive to light; store at −20 ° C in the dark.
9. Human Ophthalmic Gel or Chloramphenicol ointment.
10. Heat blanket (Beurer, HK Comfort).
11. Pediatric pulse oximeter (Heal Force® prince 100i).
12. Iodine soap (Povisept®, Difem Pharma Cat# 2400061).
13. Lidocaine (Laboratorio Sanderson Cat# 75IG1218).
14. Hydrogen peroxide (Difem Pharma Cat# 2400440).
15. Surgical microscope (Stemi 305 with articulated arm U, Zeizz Cat# 37081).
16. Rat Stereotaxic frame (RWI, Cat# 68005).
17. Hair clipper (Pichara® model RF6608).
18. Depilation cream (Veet®).
19. Microdrill (Rwi, Cat# 78001).
20. Low-Cost Cauterizer Kit (FST, Cat# 18010-00)
21. Retractor, Speculum Eye Plain Wire, 5 cm (World Precision Inst Cat#14130).
22. Syringe Holder (RWI, Cat# 68218).
23. Microinjection syringe, 5 μl (Hamilton Cat# 87900).
24. Polyglycolic acid suture 4–0 (ATRAMAT, Cat# R2694-75)
25. Polyglycolic acid suture 3–0 (ATRAMAT, Cat# R2693-75)
26. Enrofloxacin 5% injectable solution (Rostrum®, Drag Pharma, Cat#1130).

27. Tramadol hydrochloride 5% injectable solution (Triamcol®, Drag Pharma, Cat#1091).
28. Ketoprofen Injectable Solution (NAXPET®, Drag Pharma Cat#1004).
29. Ketoprofen Oral suspension (NAXPET®, Drag Pharma Cat#1006).
30. Infant Ibuprofen 100 mg/5 ml Human Use.

2.2 Electrode Implants

1. Stereotaxic arm or Holder (RWI, Cat# 68222).
2. Mini Screws stainless steel (Antrin Miniature Specialties, Inc Cat# NAS723CE140-090).
3. Silver Colloidal paint (PELCO®, Cat# 16031).
4. Dental acrylic resin, acrylic powder methyl methacrylate monomer and solvent (Marche®).
5. Antibiotic topic with Neomycin Sulfate, Clotrimazole, Benzocaine, and Prednisolone Acetate (MIXANTIP® PLUS, Drag Pharma).
6. Tungsten-coated fine wire, ~50 mm (California Fine Wire Co. Cat# CFW2030836).
7. Silver wire (Bak Electronics, Inc. Cat# IMP-2A).

3 Methods

All procedures including animals should be previously approved by your institutional bioethics committee.

3.1 Striatal Injection of 6-Hydroxydopamine

1. Weight the animal on a scale. Save the weight value as it will serve to calculate the drug doses to be delivered during and after the procedure.
2. Prepare 1 ml of a solution of 6-hydroxydopamine (6-OHDA) 3.3 mg/ml, 0.1% ascorbic acid in saline as follows. Equipped with a lab coat, gloves, and a mask, place a 1 ml Eppendorf tube on an analytical balance and tare. Put 6-hydroxydopamine hydrochloride into the tube up to 4 mg. Carefully remove the tube from the balance and fill it with 1 ml of sterile ascorbic acid 0.1% saline. Wrap the tube completely with aluminum foil to protect the solution from oxidizing from light and shake gently. Place the tube in a bucket with ice to preserve for up to 3 h.
3. Anesthetize the animal by putting it within an acrylic chamber with 93% O_2 (g) and 5% isoflurane. Always observe the animal. In the next minutes, when the animal loses muscular tone, remove it from the chamber and place on a preparation table

or surface with an anesthesia mask delivering continuous O_2 (g) with 1–2% isoflurane.

4. Deliver to the animal a dose of desipramine (25 mg/kg) by intraperitoneal (i.p.) injection. Prepare the desipramine solution at a concentration such that the injected volume is minimal (0.1–0.3 ml). Dissolving the desipramine might be difficult, mild heating (37 °C) might help.
5. Fully cover the animal's eyes with eye ointment to protect them and prevent dryness. Shave the fur of the head with a small electrical shaver or with depilation cream. Be extremely careful not to touch the eyes of the animal.
6. Maintaining the anesthesia mask, place the animal on the stereotaxic frame with its body over a heat blanket with temperature set on 37 °C. Fix the animal head to the stereotaxic frame.
7. Attach a pediatric pulse oximeter to the forelimb of the animal to monitor oxygen saturation levels and heart rate during the surgery. Also use a rectal temperature probe to monitor the body temperature.
8. Scrub the shaved and surrounding area with iodine soap or chlorhexidine soap. Remove the soap with ethanol 70%. Be extremely careful not to put soap or ethanol in the eyes.
9. Wearing sterile coat, mask, headdress, put sterile gloves in your hands. From now on, all materials used should be sterile and kept during the surgery in a sterile field.
10. Using a 1 ml syringe, inject saline with lidocaine 1% v/v subcutaneously in the head, to achieve a local anesthetic effect.
11. Make an incision in the skin of the head and expose the skull with a surgical blunt separator. Gently scrub the exposed portion of the skull with a cotton swab soaked with hydrogen peroxide 3%, rinse with saline, and dry with a gauze. In case of excessive bleeding from the skull, use a cauterizer in the points the bleeding occurs.
12. Mount a fine tip marker in the stereotaxic arm. Using the surgical microscope, locate bregma and lambda on the animal' skull. Bregma is the point where the sagittal suture and the best fit of the coronal suture intersect; lambda is the midpoint of best fit curve along the lambdoid suture [18]. To confirm the skull is not tilted in the anteroposterior axis, use the tip of the marker to check that bregma and lambda are at the same level in the dorsoventral axis. If not, loose the incisor bar and carefully regulate the height of it until bregma and lambda are leveled. Gently secure the incisor bar.
13. Using bregma as the origin of coordinates, gently draw marks on the skull that indicate the position (anteroposterior and mediolateral coordinates) for the injections.

14. With a dental drill, start making small holes through the skull at each point marked in the previous step (*see* **Note 1**). If excessive bleeding occurs, gently apply a gauze or sterile compressed sponge for a few minutes until bleeding stops. While drilling, and to prevent damage to the brain, be careful not to put the bead in contact with the dura. After the hole is made, gently clean the area with a squirt of saline from a syringe. Do not direct the squirt directly to the hole. Remove excess saline with a gauze.
15. Mount a clean Hamilton precision syringe (5 μl) on the stereotaxic arm. The arm should be at the other side of the stereotaxic frame, to ensure easy access and view of the skull.
16. The handling of the 6-OHDA solution should be done with the minimum light possible. Thus, you should work ideally in a room without natural light or with windows properly covered. Room lights should be turned off during the injection. Use a small lantern or your cellphone to provide enough light for you to operate. Using a micropipette, place a 4–5 μl 6-OHDA solution previously prepared in a ~2 cm × ~2 cm piece of parafilm. Rinse the Hamilton syringe with the 6-OHDA solution by filling it directly from the drop in the parafilm and discarding the solution. Fill the syringe again and empty it gently until it contains sharp 2 μl solution to deliver. Make sure there are no air bubbles in the syringe or the needle.
17. Insert the Hamilton needle slowly and carefully in the burred hole, with pauses of up to 5 min between each descending movement. When the desired dorsoventral coordinate is reached, stop descending and pause for a minute or two to allow the tissue to accommodate after the insertion. Slowly and gently inject the solution at 0.5 μl/min until completing 2 μl. If you are not using a programmed pump, just inject 0.5 μl and pause for a minute between each time. Once the injection is finished, to avoid the solution to flow back as the needle is withdrawn, wait between 3 and 5 min before start taking out the needle.
18. Repeat **steps 15** and **16** for each of injection point (*see* **Note 2**).
19. Suture the skin of the head with 4–0 PGA thread and clean the area around the incision with iodized soap or chlorhexidine soap. Rinse with saline and optionally put antibiotic topic ointment around the incision.
20. Remove the temperature probe and the pulse oximeter. Remove the animal from the stereotaxic frame.
21. Inject the animal with appropriate doses of analgesic drugs (e.g., ketoprofen 2.5–5 mg/kg, SC) and antibiotics (enroflaxine 10 mg/kg, IM) and place it in a clean box next to a source

of moderate heat. Be careful to put the heat source in such a way that the animal has the option to stay close or far from the source. The heat source can be a lamp that should focus only half of the cage. Observe the animal every 15 min until it comes out of anesthesia. Food pellets moistened with water are placed within reach when the animal is fully awake.

22. During the days following the 6-OHDA injection, complete the animal's analgesic and/or prophylactic antibiotic treatment (ketoprofen 2.5 mg/kg for 4–5 days and enroflaxine 5 mg/kg for 3–5 days). Let the animal to recover at least 2 weeks with food and water *ad libitum* before the next procedure. According to the guidelines of your bioethical committee, make available to the animal high-caloric palatable food to minimize weight loss. Everyday weight the animal and record the values.

3.2 Electrode Implants

1. Before the surgery, plan carefully the coordinates of each electrode array. Each array should be inserted into the proper position in the brain, for which a craniotomy is needed for each (although if two arrays are rather close, 1 mm or less, a single craniotomy for both can be made). If the array is composed of 1–4 electrodes, a craniotomy is not needed, but a single hole will accommodate the electrodes. A set of coordinates for four areas of the CBTC is included as example (*see* Table 1). The position of each electrode of these coordinates is represented in Fig. 2a with other important elements of the implant procedure.
2. Repeat **steps 1** and **3** to **12** of the previous section (*see* Subheading 3.1).
3. Using bregma as the origin of coordinates (Fig. 2a), gently draw marks on the skull in the coordinates that define each craniotomy window where the electrodes will enter the brain (*see* **Note 3** and Fig. 2b.i). These coordinates are those of the electrodes in the corners of the craniotomy you want to perform (*see* Fig. 2b.i). For maximum precision of the brain area targeted, the electrode array to be implanted should fit tight the craniotomy. Do not plan smaller or wider craniotomies.
4. Once you have marked the place for craniotomies, do no proceed to drilling yet, as you want to minimize the time the brain is exposed. Therefore, first proceed to fix three anchoring mini screws to the skull, two at the most posterior and one in the most anterior end of the exposed skull (Fig. 2a). The position of the screws should be as far away as possible from the windows for the craniotomy, so they do not obstacle the drilling and the insertion and final positioning of the electrode array. The two screws in the back should be on each side at the most lateral position available. The anterior screw should be

Table 1
Stereotactic coordinates (in millimeters) for the electrodes in Fig. 2c

Electrode	Area	Anteroposterior (AP)	Mediolateral (ML)	Dorsoventral (DV)
Frontal	Caudal forelimb area (CFA)	1.8	±2.8	1.5
		1.5	±2.5	1.4
		1.5	±2.8	1.4
		1.2	±2.5	1.5
		1.2	±2.8	1.35
		0.9	±2.2	1.3
		0.9	±2.5	1.4
		0.9	±2.8	1.4
		0.6	±2.5	1.3
	Dorsolateral striatum (STR)	0.8	±3.5	2.65
		0.5	±3.2	2.65
		0.5	±3.5	2.55
		0.2	±3.2	2.7
		0.2	±3.5	2.6
		−0.1	±3.2	2.8
		−0.1	±3.5	2.7
Caudal	Ventral posterolateral nucleus of the thalamus (VPL)	−2.28	±3.1	5.25
		−2.28	±3.4	5.3
		−2.58	±3.2	5.25
		−2.58	±3.5	5.2
	Parafascicular thalamic nucleus (Pf)	−4.05	±1.1	5.5
		−4.05	±1.4	5.5
		−4.05	±1.7	5.5
		−4.35	±0.8	5.5
		−4.35	±1.1	5.5
		−4.35	±1.4	5.5
		−4.35	±1.7	5.5

The bundles of electrodes targeting the same area are referred as an array. ML coordinates should be used with negative and positive values to target both hemispheres

positioned closer to the medial line. Avoid putting the screws on skull sutures or above areas where big vein vases are known to be located. Depending on the windows position, the screw position might vary. Remember that the aim of the screw is to provide an electrical ground and importantly, a firm anchor to attach the electrode arrays; so when deciding their position, do so thinking of each screw as the point where the legs of tripod touch ground to give maximum support. For each screw, drill a hole in the skull with the dentist's drill with a bit of diameter slightly smaller than the diameter of the screw, keeping the handpiece on the plane of the horizontal surface of the skull (*see* **Note 1**). To fix each screw, hold it with fine tip forceps just under the head and position it with the tip at the entrance of the hole and turn it with a mini screwdriver. Do not screw all the way down and make sure that the tip of the screw does not

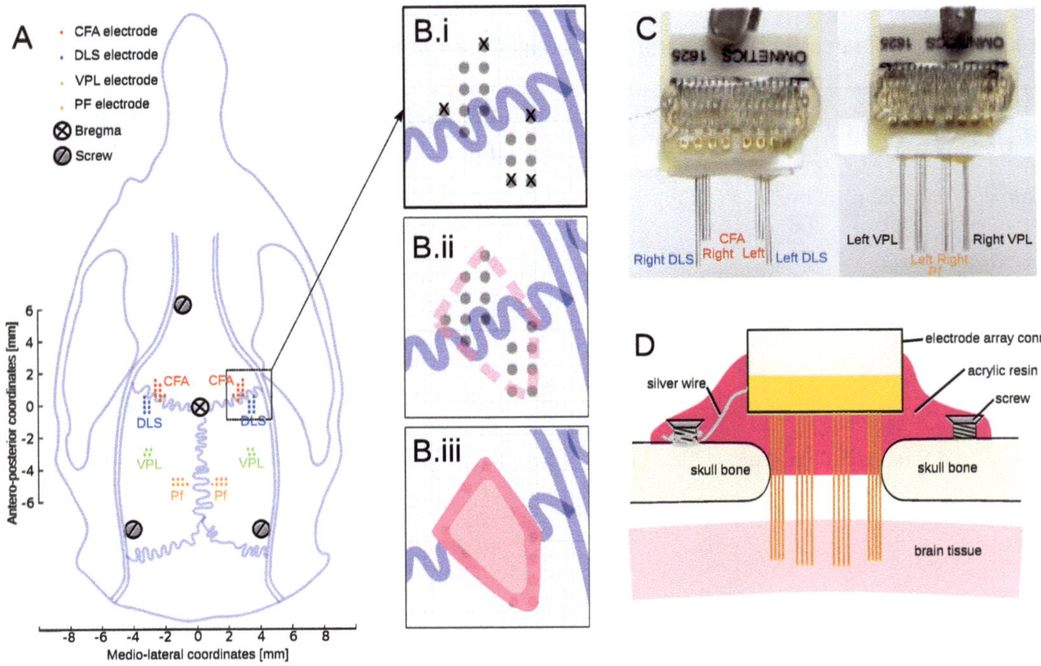

Fig. 2 Schematic for the implant of electrodes arrays in caudal forelimb area (CFA), dorsolateral striatum (DLS), ventral posterolateral thalamic nucleus (VPL), and parafascicular thalamic nucleus (Pf). (**a**) Dorsal view of a rat skull showing bregma, the final position of the electrodes in each area (red dots: CFA, blue: DLS, green: VPL, orange: Pf) and the position of the anchoring/ground screws. The stereotactic coordinates of each electrode are detailed in Table 1. (**b**) Detail of the area of the craniotomy for the arrays targeting CFA and DLS. (**b**.i) The positions chosen to define the drilling are actual electrode positions (marked with a cross). (**b**.ii) The dotted pink trace represents the movement of the drill to finally remove a small bone piece that will allow the entrance of the electrodes to the brain. (**b**.iii) Craniotomy finished. (**c**) Two electrodes arrays are used for this implant, a frontal one (left) targeting the right and left CFA and DLS with a total of 32 microwires, and a caudal one (right), targeting the right and left VPL and Pf with a total of 22 microwires. (**d**) View of the caudal electrode in its final position, attached to the skull with dental acrylic that in turn is attached to the screws. Note how the skull craniotomy is sealed with the resin. Silver wire for ground is represented as connected to only one screw but should be to the three of them

exceed the plane given by the underside of the bone, so as not to put pressure on the brain. Once screwed in, check its tightness by exerting a slight force from one side to the other on the screw with the forceps. If it is properly fixed, it should not move at all.

5. To perform each craniotomy, gently apply the dental drill with back and forward movements between any two contiguous marks draw previously in **step 3** (*see* **Note 1** and Fig. 2b.ii), repeat for all contiguous pairs of markers. This should create the contour of the craniotomy. Gently deepen the contour evenly until the bone portion defined by it comes out with little or no effort (Fig. 2b.iii). Gently remove the bone and rinse the area with saline without directly targeting the exposed

brain. Make sure the exposed brain is always humid with saline. Repeat the procedure for each craniotomy.

6. Mount the electrode array on the stereotaxic arm. Unless the electrode was designed to entry the brain with an angle other than 90°, make sure the electrodes are not tilted or rotated in the dorsoventral axis. This is crucial to reach the brain targets as planned according to the electrode array design. Bring the electrode array over the craniotomy exactly in the position where they are to descend and check there are no obstacles such as the anchor screws or another array for the electrodes to enter the craniotomy and be lowered to its final dorsoventral position. If the electrode has multiples arrays, make sure each array is properly oriented to its corresponding craniotomy. Lower the electrode array with the longest electrode wire around 1 cm from the brain surface, to allow manipulation in the exposed brain.
7. Attach the ground wire of the electrode array to the three anchor screws (*see* **Note 4**) by winding one turn of the wire just below the head of each screw. Make sure the wire is away from any of the craniotomies and does not obstruct the working space on the skull surface and the access of the electrodes to their target. Put a drop of silver paint on each screw to ensure a proper electrical connection between the wire and the screw (*see* **Note 5**).
8. Bend the tip of the bevel of a fine syringe needle and put the needle on a cotton swab to manipulate it. Remove the dura by slightly pressing it with the blunt part of the bend needle trying to hook the dura with the sharp tip pointing upward. Do not expect to do it on the first trial. Once the dura is torn, expose the brain with gentle movements with the blunt part of the hook from the center to the periphery (*see* **Note 6**).
9. Check that the electrode array is in proper orientation in the dorsal–ventral axis, as they could have moved during the connection of the ground wire. Position the array over the craniotomy and start lowering it manually or with an automatic stereotaxic frame until the longest electrode touches the surface of the brain. In case the brain popped out after the removal of the dura, try to bring the electrode to your best guess of the original level of the brain. Take note of the dorsoventral level. Start lowering the array. In case you are using an automatic arm, make it at speed of 0.25–0.5 mm/min. In case you are doing it manually, after each downward movement, wait 1–2 min for the tissue to reaccommodate. When other, shorter, electrode arrays touch the brain surface, take note on the dorsoventral level. Stop descending when desired dorsal–ventral level is reached within the brain for the longest array. Check

that the other arrays are also in the correct dorsoventral coordinate. If this is not the case, and depending on the magnitude of the difference, you can either prioritize the correct position of the array of your most interest or a position where the discrepancy in dorsoventral coordinates for all arrays is minimized (*see* **Note 7**).

10. Attach the electrode array to the skull with dental acrylic resin. Soak a fine brush in solvent for dental acrylic, remove excess solvent, and put the tip of the brush on the acrylic powder. Put the powder on the screws, covering the silver paint, if needed add a little additional solvent with the brush. Remember that acrylic does not glue to anything but acrylic, so it should fill every space between the screw and the bone to be effective as an anchor. Clean the brush and using the same procedure, start "bridging" with acrylic each screw with the electrode (*see* Fig. 2d). Cover the ground wire too. Do not allow empty spaces within the acrylic, as they are prone to bacterial growth; therefore, the space between the electrodes over the brain is critical to fill with acrylic (*see* Fig. 2d). Fully cover with acrylic the electrode array and the area of the skull bellow, including the screws and the ground wire with acrylic. Be careful not to put acrylic on the electrode array connector or the parts of the connector that will act as attachment during the recording. It is also good idea to attach in the acrylic a small stick (could be a piece of stick of a cotton swab), bellow the connector, horizontally from side to side, to later allow securing the recording connector with rubber bands during the recording sessions. To perform **step 10** in an alternative way, mix the monomer and solvent of the dental acrylic on a rubber container and put it on the electrodes and screw with a wooden or plastic stick.
11. Repeat **steps 6–10** with each electrode to be implanted (in this example there are two electrodes to be implanted, *see* Fig. 2c).
12. Remove the separators, make sure the acrylic is not over the skin, and make sure the acrylic has no cavities where dirt can accumulate. Bring the skin of the head to the borders of the acrylic. If necessary, suture the skin in the front or in the back, so it is tight against the implant. Clean the area around the incision with iodized soap or chlorhexidine soap. Rinse with saline and optionally put antibiotic topic ointment around the incision.
13. Repeat **steps 20** and **21** of the previous section (*see* Subheading 3.1).
14. Allow the animal to rest for a week before starting recording sessions (*see* **Note 8**).

4 Notes

1. For drilling the skull bone, ideally use both hands to manipulate the dental drill. In the case you are drilling a hole for injection or screw, the handpiece should be as perpendicular as possible to the plane of the drilling point. While drilling, make sure your elbows and wrists, if possible, are supported by a firm surface.
2. For rats, three injections per hemisphere are typically used. Suggested coordinates for striatum are [1.0, ±3.0, −5.0]; [−0.1, ±3.7, −5.0]; [−1.2, ±4.5, −5.0] (format [AP, ML, DV] in mm). An alternative to the 6-OHDA lesion for modeling Parkinson's disease is the overexpression of the alpha-synuclein protein using recombinant adeno-associated viral vectors injected to the substantia nigra [19]. The injections, even if done in the same place that later will receive the electrode implant, do not interfere with it.
3. Electrodes arrays for this type of procedure can be custom made or acquired. They consist in bundles of microwires arranged in geometrical shapes, normally rectangular (e.g., 4 × 4 electrodes with separation of 300 micrometers between each), but if custom made, any arrange can be designed to proper target desired brain areas. A common material for the electrodes is tungsten-coated fine wire, ~50 μm diameter for rats and ~30 μm diameter for mice. Taking into account the coordinates of your target areas and the shape and size of the electrode array, carefully plan the size and position of the craniotomy(ies) in the skull. Craniotomy windows should be the precise size for the array to enter. As example, electrodes and coordinates targeting caudal forelimb area (CFA), dorsolateral striatum (DLS), ventral posterolateral thalamic nucleus (VPL), and parafascicular thalamic nucleus (Pf) are included (Fig. 2 and Table 1).
4. A common reason for a poor electrophysiological signal is a loose or poorly connected ground. For a proper ground you need the screws tightly attached to the bone and the wire tightly attached to the screw. Screws might loosen with time and electrical connections might also fail. One way to ensure good signal is to have more than one ground connection, so one may fail but the others work. In this protocol it is recommended to connect the ground to the three anchor screws. If connection to all three is not possible, two is good too, but a single connection is not recommended.
5. Due to its high conductivity and mechanical properties, silver wire provides an excellent ground connection, and it is easy to work with during the implant.

6. In mice there is no need to remove the dura for the electrodes to enter the brain.
7. There is no straightforward solution for this case. While some brain areas are big enough so the electrode can be hundreds of micrometers away from target and still be within the area, some areas are too small or too important for your research aims to tolerate any error in position.
8. Timing between the 6-OHDA injection and the recording is the essence for the success of this protocol. The dopaminergic lesion achieves its peak effect around 21 days after the 6-OHDA injection. Depending on the brain areas and the species, the quality of the signals from the chronically implanted electrodes will peak 2 or 3 weeks after the implant (although it can be of good quality for months). If you performed the 6-OHDA lesion at day 0 and the electrode array implant at day 14, the best period to perform recordings are the days 21–28.

References

1. Hornykiewicz O (1966) Dopamine (3-hydroxytyramine) and brain function. Pharmacol Rev 18:925–964
2. Yung KKL, Bolam JP, Smith AD et al (1995) Immunocytochemical localization of D1 and D2 dopamine receptors in the basal ganglia of the rat: light and electron microscopy. Neuroscience 65:709–730
3. Surmeier DJ, Ding J, Day M et al (2007) D1 and D2 dopamine-receptor modulation of striatal glutamatergic signaling in striatal medium spiny neurons. Trends Neurosci 30: 228–235
4. Albin RL, Young AB, Penney JB (1989) The functional anatomy of basal ganglia disorders. Trends Neurosci 12:366–375
5. Cui G, Jun SB, Jin X et al (2013) Concurrent activation of striatal direct and indirect pathways during action initiation. Nature 494: 238–242
6. Halje P, Brys I, Mariman JJ et al (2019) Oscillations in cortico-basal ganglia circuits: implications for Parkinson's disease and other neurologic and psychiatric conditions. J Neurophysiol 122:203–231
7. Buzsáki G, Anastassiou CA, Koch C (2012) The origin of extracellular fields and currents--EEG, ECoG, LFP and spikes. Nat Rev Neurosci 13:407–420
8. Schwarting RK, Huston JP (1996) The unilateral 6-hydroxydopamine lesion model in behavioral brain research. Analysis of functional deficits, recovery and treatments. Prog Neurobiol 50:275–331
9. Deumens R, Blokland A, Prickaerts J (2002) Modeling Parkinson's disease in rats: an evaluation of 6-OHDA lesions of the nigrostriatal pathway. Exp Neurol 175:303–317
10. Costa RM, Lin S, Sotnikova TD et al (2006) Rapid alterations in corticostriatal ensemble coordination during acute dopamine-dependent motor dysfunction. Neuron 52: 359–369
11. Fuentes R, Petterson P, Siesser WB et al (2009) Spinal cord stimulation restores locomotion in animal models of Parkinson's disease. Science (80-) 323:1578–1582
12. Moënne-Loccoz C, Astudillo-Valenzuela C, Skovgård K et al (2020) Cortico-striatal oscillations are correlated to motor activity levels in both physiological and Parkinsonian conditions. Front Syst Neurosci 14:56
13. Brys I, Nunes J, Fuentes R (2017) Motor deficits and beta oscillations are dissociable in an alpha-synuclein model of Parkinson's disease. Eur J Neurosci 46:1906–1917
14. Brys I, Halje P, Scheffer-Teixeira R et al (2018) Neurophysiological effects in cortico-basal ganglia-thalamic circuits of antidyskinetic treatment with 5-HT1A receptor biased agonists. Exp Neurol 302:155–168
15. Avila I, Parr-Brownlie LC, Brazhnik E et al (2010) Beta frequency synchronization in basal ganglia output during rest and walk in a

hemiparkinsonian rat. Exp Neurol 221: 307–319

16. Delaville C, McCoy AJ, Gerber CM et al (2015) Subthalamic nucleus activity in the awake Hemiparkinsonian rat: relationships with motor and cognitive networks. J Neurosci 35:6918–6930
17. Yadav AP, Fuentes R, Zhang H et al (2014) Chronic spinal cord electrical stimulation protects against 6-hydroxydopamine lesions. Sci Rep 4:3839
18. Paxinos G, Watson C (2007) The rat brain in stereotaxic coordinates. Elsevier, San Diego
19. Brys I, Bobela W, Schneider BL et al (2016) Spinal cord stimulation improves forelimb use in an alpha-synuclein animal model of Parkinson's disease. Int J Neurosci 7454:1–9

Chapter 13

Neuroimaging Human Dopamine-Related Neurophysiology Across Development

Ashley Parr, Bart Larsen, Finnegan Calabro, Brenden Tervo-Clemmens, and Beatriz Luna

Abstract

Measures of dopamine (DA) in vivo in humans have been obtained using positron emission tomography (PET), however, its invasive nature and exposure to radioactivity has limited its application to populations such as pediatric cohorts. Brain tissue iron measured noninvasively with magnetic resonance imaging (MRI) has been found to be linked to DA neurophysiology in the basal ganglia, providing a novel approach for in vivo human characterization of DA function. This chapter describes the links between brain tissue iron and DA neurophysiology, the MRI approaches available to measure it, and application to understanding development and disease. The literature characterizing brain tissue iron in DA neurophysiology and new advances in MRI approaches to measure it support this as an innovative and promising approach to understand the DAergic mechanisms underlying lifespan development and disease.

Key words Dopamine, Tissue iron, T2*, R2*, R2′, QSM, MRI, Basal ganglia, Development, Cognition

1 Introduction

The development of the striatal dopamine (DA) system is thought to underlie the normative maturation of a variety of cognitive- and reward- functions throughout childhood and adolescence. For example, animal models provide evidence of increasing DA availability from childhood to adolescence, which may drive increases in reward-driven behaviors during the same period [1, 2]. Furthermore, striatal DA has been implicated in multiple psychiatric disorders that first emerge during adolescence, including schizophrenia [3, 4], depression [5, 6], and substance use disorder [7–10], suggesting a prominent role of striatal DA in normative and clinical development. In order to investigate both the normative and clinical development of the human DA system in youth,

José Antonio Fuentealba-Evans and Pablo Henny (eds.), *Dopaminergic System Function and Dysfunction: Experimental Approaches*, Neuromethods, vol. 193, https://doi.org/10.1007/978-1-0716-2799-0_13,

there is a need for noninvasive, in vivo markers of DAergic neurobiology that can be applied to pediatric populations.

The current best available technique to assess striatal DA neurobiology in vivo in humans is positron emission tomography (PET). PET studies of DA utilize positron-emitting radiotracers tailored to specific aspects of DA transmission, e.g., DAergic receptor subtypes D2/D3 [11, 12] and DA and monoamine transporters [13], to measure their concentrations in the brain. Although this provides highly precise estimates of specific DA targets, PET imaging of DA is an invasive technique that requires intravenous injection of a radiotracer, involving exposure to a small dose of ionizing radiation. For these reasons, PET imaging may not be suitable for all populations, including children and adolescents, limiting our ability to capture developmental changes in DA processes during these critical periods. As such there is a need for alternative biomarkers that can be applied to these populations.

Prior noninvasive in vivo markers of the human mesocortical and mesolimbic DA system have been proposed. A significant focus has been placed on measuring blood oxygen level dependent (BOLD) activation in the striatum and midbrain using fMRI as a proxy for DA function [14], which is interpreted as a general index of adolescent reward sensitivity [15–23]. However, in general, the BOLD signal reflects changes in deoxyhemoglobin, resulting from fluctuations in blood flow and neurovascular coupling that occur in response to neuronal activity [24, 25]. As such, the cellular and molecular mechanisms linking BOLD to DA and neuronal activity more broadly are unclear [25], and multimodal studies linking striatal BOLD and in vivo measures of DA function (e.g., measured using PET) are sparse and inconsistent [26–28]. Furthermore, several lines of evidence suggest that BOLD activation may reflect primarily non-DAergic signaling [29], including being driven more dominantly by glutamatergic firing [30, 31], complicating the use of BOLD as an indicator of DAergic pathways.

Behavioral markers indexing underlying DAergic function have also been proposed, for example, eye blink rate (EBR) [32]. However, these behavioral markers have not been validated as viable measures reflecting DAergic function and are further limited by the inability to directly assess whether DA drives the particular behavioral correlate being used.

In this chapter, we present evidence suggesting that MRI measures of brain tissue iron may be an alternative, safe, and scalable indirect measure of DA function [33], and we will describe various existing methodologies developed to map tissue iron in the brain relevant for normative and clinical neurodevelopmental research.

2 Overview of Iron in the Brain

Tissue iron is heterogeneously distributed throughout the brain with the greatest iron concentration occurring within dopaminergic (DA) brain regions including the midbrain and the basal ganglia [34, 35]. The trafficking, use, and storage of iron is tightly regulated. Iron is transported across the blood–brain barrier and into neurons, oligodendrocytes, and astrocytes via transferrin and is either used or stored as ferritin [34, 36, 37]. Within these cells, iron supports several critical processes, including cellular respiration, myelination (maintenance and synthesis), and monoamine neurotransmitter synthesis [34–36]. Importantly, iron is a cofactor for tyrosine hydroxylase [38] and monoamine oxidase [39–41], two necessary enzymes for DA synthesis and catabolism, respectively [38, 42]. Further, iron co-localizes with DA vesicles [43] and cytosol within DA neurons. Indeed, MRI measures of tissue iron show specific concentration in the basal ganglia [44–46]. The predominance of iron in the basal ganglia, its importance for DA production, and its co-localization with DA provide compelling evidence that tissue iron may serve as an indirect marker of DA-related striatal neurophysiology.

Further evidence for a role of tissue iron in DA neuroanatomy and neurophysiology comes from studies in neurological disorders that affect the mesolimbic and nigrostriatal DA systems, for example, Parkinson's disease, Huntington's disease, restless leg syndrome (RLS), and attention deficit hyperactivity disorder (ADHD), which are consistently associated with atypical iron concentration [35, 42, 47–51]. Reduced iron concentrations in these disorders have been linked to DAergic alterations, including reduced dopamine transporter (DAT) function and decreased D2 receptor density [52–55]. Cerebrospinal fluid (CSF) iron levels, which correlate with CSF DA levels, are reduced in patients with Parkinson's disease with comorbid RLS [51], and patients with ADHD have reductions in iron levels that can be rescued following treatment with DA reuptake inhibitors [47]. Similarly, increased iron levels are found in the basal ganglia in individuals with cocaine addiction [56], which together with the findings in ADHD, suggest that pharmacological upregulation of DA may result in increased iron concentrations.

Animal studies have corroborated the relationship between iron and DA, consistent with what has been observed in human neurological disorders. Iron-deficient diets in rodents lead to reduced striatal tissue iron concentrations as well as decreased DAT function, D1 and D2 receptor density, and intracellular DA concentrations [57–64]. Iron deficiency further leads to reductions in the expression of genes controlling midbrain DA excitability [61]. Critically, iron concentrations can be restored in iron-deficient animals via iron infusion into either the ventral midbrain or striatum [55], and striatal D2 receptor density can be restored by introducing a high iron diet [63].

Research in humans has indicated that brain tissue iron concentration accumulates throughout the lifespan, increasing rapidly throughout childhood and adolescence before beginning to stabilize in adulthood [33, 45, 65–69]. These findings indicate that tissue iron is sensitive to developmental changes in the first two decades of life, and coupled with evidence suggesting links with DA concentrations, suggests that tissue iron neuroimaging reflects a viable marker for the development of DAergic processes. In this chapter, we review work validating and utilizing striatal tissue iron as an indirect measure of striatal DA and provide step-by-step guidance and methodological considerations for its use to assess striatal neurophysiology in human neurodevelopmental studies.

3 Materials and Methods

3.1 MRI

Due to its paramagnetic properties, tissue iron can be estimated noninvasively using magnetic resonance imaging (MRI). Specifically, the paramagnetism of iron strongly influences the effective transverse relaxation rate and magnetic susceptibility of brain tissue [70–73]. Several MR acquisitions have been developed to quantitatively assess brain iron content in vivo, including T2* relaxometry and quantitative susceptibility mapping (QSM) [74–77], which can be applied in human neurodevelopmental studies to capture, in part, changes in dopamine (DA) neurophysiology.

3.2 Relaxometry-Based Approaches

3.2.1 R2

The effective transverse relaxation time (T2*) is affected by magnetic field inhomogeneities induced by paramagnetic iron concentrations [74, 78]. Due to its magnetic susceptibility, iron results in a shortening of T2*, increasing the rate of transverse relaxation, R2* (R2* = 1/T2*) (Fig. 1) [74, 78]. The T2* signal decay can be measured using multi-echo gradient echo (mGRE) sequences to sample the decay curve directly at different echo times (TE) [79, 80]. Acquisition of additional TEs (i.e., three or more) allows direct estimation of the relaxation curve via voxelwise fitting to an exponential decay curve [78, 81, 82], which provides a robust estimation of R2*. The resulting T2* images are then spatially normalized to a study-specific template, and R2* is calculated as 1/T2*.

R2* is linearly associated with postmortem tissue iron in the basal ganglia [72, 78], and has been used to characterize the development of tissue iron in the striatum and gray matter structures across the lifespan [65, 75, 78] including during adolescence [67]. R2* has been used to assess pathological iron accumulation in nigrostriatal regions in Parkinson's disease (PD) [35, 83–89],

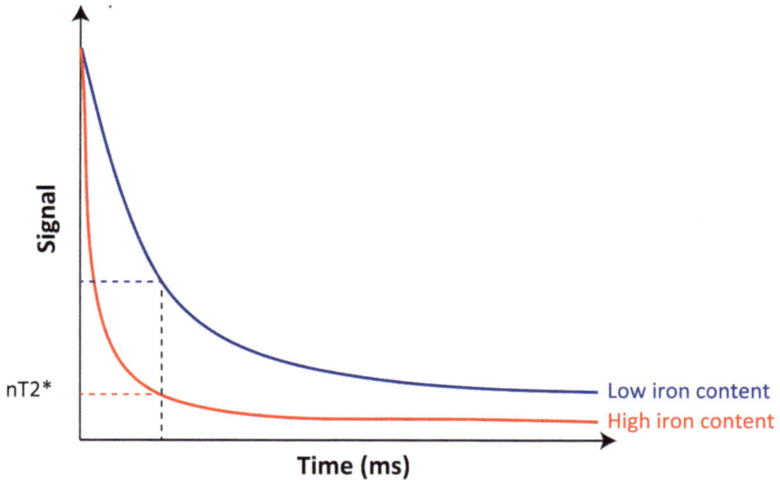

Fig. 1 Signal basis for methods for iron quantification. The signal intensity on T2*-weighted images decreases as a function of increased iron content. Relaxometry-based methods (i.e., R2* and R2′) sample multiple points along the decay curve using multi-echo acquisitions in order to model the exponential rate of decay. nT2* relies on a single echo, which samples the signal at a specific point on the curve (black dotted line), allowing for the analysis of relative signal decay at a fixed time. However, the specific echo time will affect the magnitude of the measured signal which is why nT2* values are interpreted relative to the rest of the brain

lesions in multiple sclerosis (MS) [90], and changes in iron concentration in ageing populations [91]. More recently, Brammerloh et al., 2021 combined 3D iron histology and biophysical modeling with R2* to assess the basis of substantia nigra iron content in postmortem tissue, and showed that cellular iron distributions in the neuromelanin of DA neurons was a primary contributor to the R2* signal, demonstrating not only its utility for monitoring cell loss as an early marker of PD pathology but also the sensitivity of R2* to DAergic neurons [92].

3.2.2 R2′

R2′ can be obtained by estimating R2 (1/T2) and R2* (1/T2*), which can be done separately using multi-echo turbo spin echo (mTSE) and mGRE sequences, respectively, or simultaneously using gradient echo sampling of the free induction decay echo (GESFIDE), gradient echo sampling of a spin echo (GESSE) [80, 93–95], or partially refocused interleaved multiple echo (PRIME) sequences [96, 97]. R2′ reflects the reversible transverse relaxation rate (1/T2′) calculated as the difference between the effective (R2* = 1/T2*) and irreversible (R2 = 1/T2) relaxation rates [74, 78, 95, 96]. R2* is acquired as described in the preceding section, and R2 images are registered to standard space, allowing the R2′ image to be calculated as R2′ = R2* − R2.

Like R2*, R2′ has been shown to be linearly related to iron concentrations [74, 78, 93] and has been used to probe the integrity of basal ganglia physiology in patients with dysfunction of dopaminergic systems, including Parkinson's disease [96] and in ageing [78]. Relative to R2*, R2′ has been shown to be more sensitive to iron in deep gray matter structures [78], potentially owing to the fact that R2′ attempts to remove R2 component from R2*. R2 is affected by sources induced magnetic field inhomogeneity, such as lipids, macromolecules, and water concentration [96, 98, 99], and the removal of this component isolates the sources of the R2* signal that affect magnetic susceptibility, such as local iron concentration. Importantly, as discussed above, Brammerloh et al., 2021 showed that the R2* signal in DA neurons is primarily driven by the magnetic susceptibility of the iron-rich DA neurons relative to neighboring cells (affecting R2′) rather than nanoscale spin–spin interactions of iron with water (affecting R2), thus removing R2 further isolates the source of the signal that is sensitive to iron in DA neurons [92], making R2′ potentially more sensitive to DAergic neurophysiology.

Methodological Considerations for R2* and R2′

R2* and R2′ images are susceptible to several artifacts, including motion artifacts, macroscale susceptibility artifacts, extreme value outliers, and T2* hyperintensities that impact the signal in the basal ganglia. In particular, relaxation rates within regions that are proximal to air–tissue interfaces such as the sinuses and nasal cavities can be influenced by global magnetic field inhomogeneities [93]. Given these susceptibilities, quality assessments are needed to minimize their impact on data quality (e.g., by visually inspecting each image to assure that susceptibility artifacts do not affect regions of interest, such as the basal ganglia (Fig. 2), manually placing region of interest (ROI) masks [100], and outlier exclusion [101]). The nucleus accumbens, which is of particular interest for developmental neuroscientists due to its role in reward processing, can be particularly susceptible to this artifact, owing to its small size and proximity to the sinuses. Quality assurance for R2* and R2′ can be time consuming, often requiring multiple raters to ensure unbiased and reliable assessments, and typically results in the exclusion of a significant amount of data [33, 102]. Further, relaxation-based measures of tissue iron are sensitive to magnetic field inhomogeneities that are induced by diamagnetic sources, such as water content and myelin [78, 99]. Additionally, partial volume effects may bias relaxation rates in small brain structures, such as the substantia nigra, and in areas in close proximity to regions with high or low relaxation rates, such as cortical gray matter [78, 98, 103]. R2* (1/T2*) is most strongly related to transverse (spin–spin) relaxation time, magnetic susceptibility of tissue, and magnetic field inhomogeneity. Thus, tissue-iron (nonheme) concentration and

Fig. 2 Outcome of quality assurance procedures for R2. The left panel shows the voxelwise mean of all R2* scans that passed quality assurance, and the right panel shows the voxelwise mean of all R2* scans that failed quality assurance. (Reprinted with permission from Larsen et al. [67] (permissions to be obtained))

myelin concentration are the tissue types that contribute most strongly to the R2* signal [65, 104–107]. Both tissue-iron and myelin have long transverse relaxation times, causing a hypo-intense T2* signal [79, 108, 109]; therefore, a major challenge in interpreting relaxation-based estimates is teasing out the relative contributions of tissue iron and myelin. However, myelin is dia-magnetic and tissue-iron is paramagnetic, and tissue-iron therefore has a greater contribution to T2* (greater hypo-intensity) as a consequence of its magnetic susceptibility and effect on magnetic field inhomogeneity [72, 107]. Therefore, although tissue-iron and myelin both contribute to T2* in deep gray matter regions of the brain (like the basal ganglia and substantia nigra), iron concentration is very high and the signal should be most strongly influenced by tissue-iron properties [72, 75].

However, in other areas of the cortex where myelin is more abundant, these tissues may have more relative influence on the signal [78, 110], and gaining reliable relaxation-based quantifications of tissue iron is challenging and these values should be interpreted with caution [78]. It has been shown that iron estimates of R2′ and R2* within cortical gray matter are overestimated, likely due to diffusion of magnetic field inhomogeneities caused by gray matter–white matter boundaries, in addition to draining veins between cortical surfaces inside the sulci [78, 103]. Furthermore, R2* and R2′ are sensitive to fiber orientation relative to the magnetic field [78, 98, 110–115], and therefore, interpreting values in regions heavy in white matter may require modeling macromolecular contributions to the signal as a means of correcting for myelin content [78, 116, 117]. We therefore recommend that analyses are restricted to areas that are known to be high in tissue iron (i.e., the basal ganglia and midbrain) for ease of interpretability of the neurobiological mechanisms contributing to R2*.

3.3 A Single-Echo Approximation of T2 Relaxation

3.3.1 Normalized T2*-Weighted Imaging

Iron's effect on T2* is evident in T2*-weighted images [65, 68, 105–107]. The T2* decay rate across the brain can be measured at a given echo time (TE). Given that iron contributes to faster relaxation rates, areas with greater iron concentration (i.e., the basal ganglia) will undergo greater T2* relaxation relative to other areas in the brain with less iron content. This will be evident in the hypointensities in the T2* signal in iron-rich areas [68]. Because this is a relative measure, the absolute values will depend on the acquisition parameters used (e.g., the TE), and may vary across sessions and participants. Therefore, it is critical to normalize the measure to allow for comparisons.

T2*-weighted imaging has been used as an indicator of age-related changes in basal ganglia iron content through adolescence [68, 69]. Estimates of T2* can be obtained by normalizing and time-averaging T2*-weighted images (nT2*w) in the following way: First, T2* images are processed using minimal functional neuroimaging preprocessing steps, including 4D slice-timing and head motion correction, skull stripping, co-registration to the structural image, and nonlinear warping to standard space (e.g., Montreal Neurological Institute (MNI)). Next, each volume is normalized to the whole brain mean or to a reference region, and this normalized signal is then averaged across all volumes which results in one normalized T2*-weighted image for each participant. The normalization step results in the T2* decay in the basal ganglia relative to the rest of the brain (or to a reference region such as the corpus callosum; [69]), allowing for comparisons across participants. Averaging across multiple timepoints increases the signal-to-noise ratio (SNR) relative to obtaining one value from one or few timepoints. A similar approach using relative T2*-weighted signal contrast has been used to assess iron content in the heart and liver [118, 119].

One major advantage of using nT2* to characterize tissue iron is that it can be derived from existing T2*-weighted echo planar imaging (EPI) scans that are commonly acquired as part of standard functional neuroimaging protocols (e.g., fMRI and resting-state), thus making this technique widely available and cost-effective for researchers and clinicians who may already have existing datasets from which these estimates can be derived. An advantage of these datasets is that they typically include multiple acquired volumes used for BOLD analyses (on the level of hundreds, albeit at lower resolution than typical quantitative relaxometry-based acquisitions), which can be used to increase the SNR by means of averaging [79]. Furthermore, nT2*w methods allow for the identification of high-motion volumes as a result of image co-registration. These high-motion volumes can then be omitted from the volume-wise average. This is beneficial in developmental studies and certain clinical populations to overcome and minimize the impact of motion artifacts. Importantly, legacy EPI data can be used to obtain estimates of brain tissue iron.

3.3.2 Methodological Considerations for nT2*-Weighted Imaging

In contrast to relaxometry-based methods (R2* and R2′), nT2*w imaging is not a quantitative measure of T2*, as it does not sample the exponential signal decay rate (as in multi-echo GRE sequences discussed above). Furthermore, nT2*w measures T2* decay relative to the rest of the brain [68] or to a reference region [69]. Because nT2*w values depend on the scan parameters, particularly the echo time (TE; *see* [98] for expanded discussion), in addition to the strength of the main magnetic field [98, 120], this method may be limited in its ability to compare values across longitudinal visits, scanners, and studies using different imaging parameters. Finally, nT2*w estimates of tissue iron have not yet been validated (e.g., with postmortem tissue iron concentrations). Therefore, researchers may also wish to apply quantitative methods that have been validated as biomarkers of iron content, such as R2′, R2*, or Quantitative Susceptibility Mapping (QSM) [78, 121].

3.4 Susceptibility-Based Approaches

Susceptibility weighted imaging (SWI) enhances image contrast by revealing differences in magnetic susceptibility of various compounds and tissues, including deoxygenated blood, iron, and calcium [98, 122, 123]. SWI is obtained by comparing both magnitude and phase in a single image (T2*-weighted gradient-echo), and has been used to identify brain iron [122]. However, SWI does not provide quantitative measures of magnetic susceptibility, as signal phase is orientation-dependent (relative to the static main magnetic field (B0)), and is nonlocal in that the phase reflects magnetic properties in the surrounding tissue (i.e., nearby air–tissue interfaces), in addition to the local tissue magnetic properties [77, 122, 124], limiting its reproducibility and sensitivity.

3.4.1 Quantitative Susceptibility Mapping (QSM)

QSM is a relatively modern post-processing technique that uses the same phase signal as is obtained by conventional dual- or multi-echo GRE sequences [120, 124]. QSM overcomes the issue of estimating the source of local tissue magnetic susceptibility by deconvolving the field, allowing for quantitative assessments of magnetic susceptibility distributions across brain regions [77, 124]. The inversion from the local field (phase) to magnetic susceptibility is “ill-posed” [77, 120, 124–126] as there are multiple susceptibility distributions that contribute to the same measured field [124]. To overcome this ill-posed inverse problem, QSM utilizes regularization algorithms that integrate prior information about the effect of tissue anatomical structures (i.e., air/tissue interfaces) on measured phase changes, allowing for a better understanding of the potential distortions and signal loss arising from these effects and providing more direct information about the source of magnetic susceptibility in tissues [77, 120, 124–126]. Briefly, processing proceeds in the following manner: Multi-echo phase images undergo linear regression, phase unwrapping, background field removal, and local field images, then undergo dipole inversion (detailed methods are found in refs. 120, 124).

QSM estimates have been shown to have a linear relationship with iron content [98, 106, 127–129], and provide robust estimates for tissue iron concentration in midbrain, basal ganglia, and dentate nucleus (DN) structures [81, 106, 128, 130–133]. QSM has been used to characterize changes in tissue iron concentration across the lifespan [69, 134, 135], has been used extensively in clinical settings to characterize changes in iron in patients with PD [130, 136–143], and to enhance image contrast between lesions and normal tissue in multiple sclerosis (MS) [144–148].

3.5 QSM Versus Relaxometry-Based Methods

One advantage of QSM and traditional relaxometry-based methods (R2* and R2′) is that they are quantitative in nature, as compared to nT2*w imaging, which may be sensitive to changes in imaging parameters such as the main field strength and echo time (TE) [98, 120]. This facilitates greater reproducibility and comparison across longitudinal visits (particularly important for developmental studies), scanners, sites, and studies with different imaging parameters. Furthermore, the susceptibility of iron-rich structures (like the basal ganglia and midbrain structures) is not dependent on the orientation relative to the magnetic field [98, 149, 150], in contrast to white matter susceptibility [151–153].

Importantly, like R2*, R2′, and nT2*w estimates, QSM-derived estimates of tissue iron are still sensitive to the contribution of myelin and diamagnetic tissue properties [98, 154–157], and therefore it is still recommended to limit analyses to regions rich in iron content. However, signed QSM can resolve the polarity of the QSM effects [91, 98, 158], providing information as to whether a given structure is paramagnetic (i.e., iron) or diamagnetic (i.e., myelin), reflecting an advantage over traditional SWI techniques and relaxometry-based techniques [91, 98, 158].

Deistung et al., 2013 compared multiple contrasts (T2-weighted images, longitudinal relaxation rate (R1) maps, effective transverse relaxation rate (R2*) maps, and susceptibility maps) at 7T and assessed the ability of the various methods to delineate brainstem structures by comparing to histological staining and anatomical atlases. They showed that in comparison to T2-weighted images and longitudinal relaxation rate maps (R1), R2* and QSM were able to anatomically delineate brainstem nuclei with greater precision [81]. Studies comparing R2* and QSM have shown that QSM may be a more accurate (provide better anatomical contrast) and a more sensitive marker of brain iron within midbrain structures [91, 120, 154, 159], and also shows improved source localization, boundary definition, and is sensitive to lower levels of iron [75, 77, 120, 159]. Additionally, studies comparing QSM and R2* estimates of tissue iron in PD patients have found improved sensitivity and improved diagnostic accuracy using QSM (i.e., illuminating greater differences between control participants and PD patients) [130, 137, 142, 160]. This apparent superiority of QSM in this regard may be due, in part, to the fact that the R2* signal is sensitive to differences in water content in addition to iron content, while QSM-based estimates are not confounded by water content as it does not affect the susceptibility of iron deposition [98].

Nevertheless, strategies combining both relaxometry (R2*) and QSM have been shown to be superior to either technique alone [81, 98, 161], particularly when quantifying cerebral iron content [72, 98, 162–165] and for separating the effects of iron and myelin [161, 166]. R2* and QSM can be obtained simultaneously with the same acquisition by using an mGRE [91, 130, 167], the advantages of which have been demonstrated in clinical studies of MS lesions [90, 168, 169].

In conclusion, there are several MRI approaches that provide characterization of brain tissue iron. New advances support QSM as a more precise approach, but all approaches provide important measures of tissue iron. nT2*w has the particular advantage that it can be derived from existing EPI data (task- and rest-fMRI), allowing for ready investigations and comparison with other acquisitions (e.g., task BOLD, functional connectivity, brain structure) for further probing of brain mechanisms underlying development and disease.

3.6 Field Strength Considerations

Field strength affects the ability to visualize brain tissue, including tissue iron contrast, with higher strengths affording improved spatial resolution, greater sensitivity to susceptibility sources, and increased SNR [81, 98, 170]. Imaging at higher field strengths (i.e., Bo $\geq$ 7T) therefore reflects a promising new direction for mapping iron concentration across small deep gray matter nuclei, including the basal ganglia, with unprecedented precision.

4 Applications

4.1 Linking Tissue Iron to Striatal Dopamine

Recent evidence demonstrated correspondence between Positron Emission Tomography (PET) indices of presynaptic DA availability, specifically Vesicular Monoamine Transporter 2 (VMAT2) estimated using [11C]dihydrotetrabenazine (DTBZ), and tissue iron concentrations in the nucleus accumbens, measured using R2′

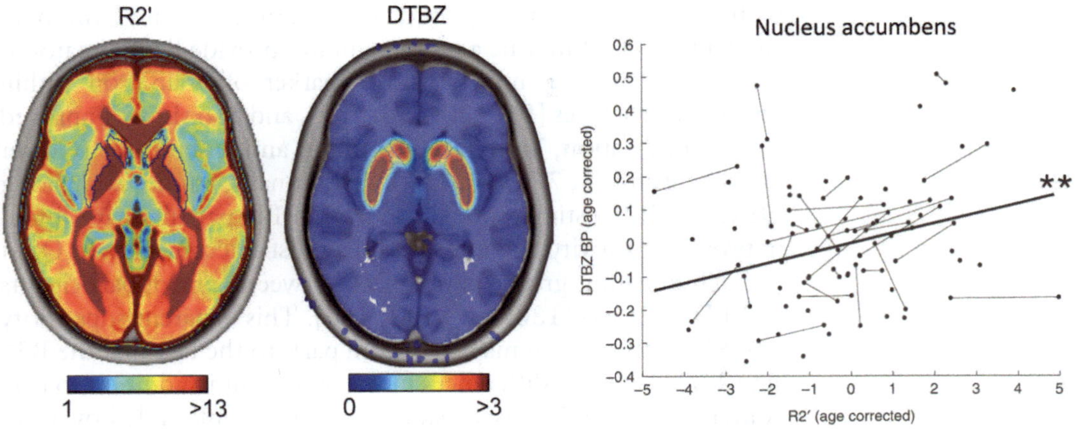

Fig. 3 Correspondence between R2′ and [11C]dihydrotetrabenazine binding potential (DTBZ BP). (**a**) Images for R2′ and DTBZ averaged across participants. (**b**) The association between R2′ and DTBZ in the nucleus accumbens (corrected for age). (Reprinted with permission from Larsen et al. [33] (permissions to be obtained))

(Fig. 3; [33]). This relationship was evident at both the between-subject level (cross-sectional) as well as the within-subject level (longitudinal) and remained after regressing out the effects of age from each measure to control for age-related accumulation of tissue iron.

VMAT2 is involved in packaging and transporting vesicular DA, and reflects presynaptic DA availability [13]. R2′ was not related to [11C]raclopride (RAC) across striatal regions, which reflects postsynaptic D2/D3 receptor availability [171], demonstrating relative specificity with presynaptic aspects of DA concentrations. These results provide support for the hypothesis that tissue iron (R2′) may reflect, in part, important aspects of vesicular DA concentration [33, 43], and variation in tissue iron may reflect individual differences in VMAT2 with respect to age. This is in agreement with tissue iron's role in supporting DA production. Further, this study may suggest that tissue iron may be a sensitive measure to capture developmental changes relevant to presynaptic vesicular DA concentrations across adolescence.

4.2 Linking Tissue Iron to Cognition Across the Lifespan

Tissue iron concentrations increase throughout childhood and adolescence, reaching stabilization into adulthood [33, 45, 65, 66, 68, 69, 78, 91, 105, 135], although some studies show continued accumulation into the third and fourth decades of life [65, 67, 172]. Iron deficiency in early development results in abnormally low tissue iron concentrations [57, 59, 60], which has been shown to correspond to atypical basal ganglia DAergic function [60–62, 173]. Iron deficiency in infancy and childhood has been associated with poorer outcomes in terms of cognitive function both cross-sectionally and later in adolescence (*see* Table 1 for a

Table 1
Findings relating iron to cognition in healthy participants

Findings relating iron to cognition					
Study reference	**Age range**	***N***	**Cognitive assessment**	**Relationship to cognition**	**Imaging method**
Tissue iron studies					
Darki et al. (2016) [201]	6–7 and 21–40 YO	46	Visuospatial working memory	+	QSM
Hect et al. (2018) [66]	7–16 YO	57	General intelligence, processing speed	+	R2*
Carpenter et al. (2016) [202]	7–11 YO	39	Spatial IQ	+	QSM
Larsen et al. (2020) [67]	8–26 YO	922	General and complex cognition	+	R2*
Steiger et al. (2016) [203]	18–32 and 56–78 YO	62	Verbal learning memory	–	R2*
Daugherty et al. (2015) [204]	19–77 YO	218[a]	Verbal working memory	–	R2*
Daugherty et al. (2019) [100]	20–94 YO	183	Cognitive switching	–	R2*
Kalpouzos et al. (2017) [197]	26–42 and 65–74 YO	37	Memory	–	R2*
Persson et al. (2020) [198]	26–42 and 64–77 YO	36	Implicit sequence learning	+	QSM
Ghadery et al. (2015) [189]	38–86 YO	381	Memory, executive function, psychomotor speed	–	R2*
Biel et al. (2021) [205]	50–80 YO	79	Verbal memory, executive function	–	R2*
Bartzokis et al. (2011) [206]	55–76 YO	63	Verbal working memory	–	FDRI
Sullivan et al. (2009) [207]	65–79 YO	10	Attention, dementia rating scale	–	FDRI
Valdés Hernández et al. (2015) [208]	71–74 YO	622	Fluid intelligence, processing speed, memory	–	T2*w and FLAIR
Iron-deficiency studies					
Burden et al. (2007) [209]	9–12 months	15	Attention	+	N/A

(continued)

Table 1
(continued)

Findings relating iron to cognition					
Study reference	**Age range**	***N***	**Cognitive assessment**	**Relationship to cognition**	**Imaging method**
Carter et al. (2010) [210]	9 and 12 months	113[a]	Object permanence, memory, socioemotional	+	N/A
Lozoff et al. (1987) [211]	12–23 months	191	Mental development index	+	N/A
Lozoff et al. (1998) [212]	12–23 months	191	Affect and activity levels	+	N/A
Deinard et al. (1986) [213]	18–60 months	25	Mental development and responsiveness	+	N/A
Lozoff et al. (2000) [182]	5 and 11–14 YO	191[a]	Mental and motor function, arithmetic, spatial memory, recall	+	N/A
Pollitt et al. (1983) [214]	3–6 YO	30	Learning and attention	+	N/A
Halterman et al. (2001) [215]	6–16 YO	5398	Math ability	+	N/A
Pollitt et al. (1989) [216]	9–11 YO	1358	IQ	+	N/A
Scott et al. (2017) [217]	18–35 YO	127	Attention, inhibitory control, planning	+	N/A
Lukowski et al. (2010) [218]	19 YO	114[a]	Inhibitory control, set shifting, planning, memory	+	N/A
Lozoff et al. (2006) [219]	19 YO	185[a]	General cognition	+	N/A
Wenger et al. (2019) [220]	College age	210	Visual short-term memory	+	N/A

FDRI field-dependent relaxation increase, *FLAIR* fluid-attenuated inversion recovery, *MRI* magnetic resonance imaging, *IQ* intelligence quotient, *GPI* globus pallidus, *SN* substantia nigra, *QSM* quantitative susceptibility mapping
[a]Denotes longitudinal studies, in which case *N* reflects number of participants at baseline. "+" indicates that increased iron had a positive relationship with cognition, while "−" indicates that increased iron had a negative relationship with cognition

summary, and [174, 175] for review), and increased risk for psychiatric disorders [176–186]. Likewise, normatively lower levels of basal ganglia tissue iron in childhood through adolescence are associated with poorer cognitive outcomes (*see* Table 1 and [187] for a review); however, conversely, in ageing populations, abnormally elevated tissue iron has been associated with neurodegeneration and cognitive decline (*see* Table 1 for a summary in healthy ageing and [35]). These seemingly opposing effects of iron concentration on cognition across the lifespan may be driven by differential metabolic requirements and mechanisms across stages of life [42]. Given the role of brain iron in critical neurobiological processes that undergo substantive changes throughout adolescence, including myelination [34, 99] and neurotransmitter synthesis [39, 41], particularly DA, abnormally low levels may impair neurodevelopmental processes that result in impairments in cognition into adulthood [67]. On the other hand, abnormally high iron content in aging may reflect disruptions in critical homeostatic processes involved in iron storage and metabolism, which may accelerate pathological neurodegenerative processes leading to cognitive decline (i.e., via increased iron-related oxidative stress) [35, 187–193]. These findings are consistent with the literature showing that the relationship between DA and cognition is best depicted as an inverted U, where too little or too much DA is suboptimal [194, 195].

4.3 Linking Tissue Iron to Functional Changes in the Brain

Although the mechanisms through which iron affects cognition and its developmental time course remain unclear, it is possible that tissue iron influences the organization and specialization of functional networks in the brain. Striatal tissue iron has been linked to the modulation of frontostriatal BOLD activation during working memory [196], memory recall [197], and implicit learning [198]. Furthermore, striatal tissue iron has been shown to modulate connectivity patterns in the brain [101, 102], reflecting a potential mediator for iron–behavior relationships. Developmentally, recent evidence shows a role for striatal tissue iron (R2′), specifically within the nucleus accumbens, in the development of frontostriatal connectivity throughout adolescence [102]. Age-related decreases in connectivity between the ventromedial prefrontal cortex and the nucleus accumbens, a critical DAergic pathway involved in reward processing and risk taking [199, 200], were associated with developmental increases in tissue iron [102]. These results held after controlling for changes in myelination assessed with magnetic transfer ratio (MTR). Furthermore, these decreases in connectivity were associated with improvements in reward learning and decreased risk-taking behavior [102]. These initial results provide evidence that DA-related striatal properties, reflected in tissue iron concentration, contribute to developmental specialization across reward networks in the brain.

5 Summary and Future Directions

Studies in animal models and clinical populations have demonstrated promising links between tissue iron and DAergic processes at the mechanistic and correlational level, respectively. Importantly, tissue iron has been found to be associated with DA availability using PET in humans [33]. Developmental studies are beginning to shed light on the role of tissue iron in the maturation of cognitive and reward processes, setting up future work to identify factors that lead to individual differences in developmental trajectories across the lifespan. The ability to measure brain iron is a promising new avenue for investigating DA-related neurophysiology in a noninvasive manner in populations in which PET cannot be applied. Future directions include further confirmation that these MRI approaches are measuring critical aspects of DA-related function, that these measures are independent from other factors that are supported by tissue iron, and validation across in vivo neuroimaging methods (R2*, R2′, QSM, nT2*w) in order to extend our understanding of the nature of tissue iron and its sensitivity to DAergic sources. In conclusion, MRI approaches that noninvasively measure tissue iron in the basal ganglia can provide a new line of inquiry elucidating DA-related processes underlying typical and atypical development, aging, neurological, and psychiatric disorders linked to impairment in DA function.

References

1. Luciana M, Wahlstrom D, Porter JN, Collins PF (2012) Dopaminergic modulation of incentive motivation in adolescence: age-related changes in signaling, individual differences, and implications for the development of self-regulation. Dev Psychol 48(3): 844–861. https://doi.org/10.1037/a0027432
2. Wahlstrom D, White T, Luciana M (2010) Neurobehavioral evidence for changes in dopamine system activity during adolescence. Neurosci Biobehav Rev 34(5):631–648. https://doi.org/10.1016/j.neubiorev.2009.12.007
3. Howes OD, Kambeitz J, Kim E, Stahl D, Slifstein M, Abi-Dargham A, Kapur S (2012) The nature of dopamine dysfunction in schizophrenia and what this means for treatment: meta-analysis of imaging studies. Arch Gen Psychiatry 69(8):776–786. https://doi.org/10.1001/archgenpsychiatry.2012.169
4. Paus T, Keshavan M, Giedd JN (2008) Why do many psychiatric disorders emerge during adolescence? Nat Rev Neurosci 9(12): 947–957. https://doi.org/10.1038/nrn2513
5. Davey CG, Yücel M, Allen NB (2008) The emergence of depression in adolescence: development of the prefrontal cortex and the representation of reward. Neurosci Biobehav Rev 32(1):1–19. https://doi.org/10.1016/j.neubiorev.2007.04.016
6. Diehl DJ, Gershon S (1992) The role of dopamine in mood disorders. Compr Psychiatry 33(2):115–120. https://doi.org/10.1016/0010-440X(92)90007-D
7. Dionelis K, Meng C, Craig K, Shabbir S, Fineberg N, Sahakian B, Suckling J, Bullmore E, Robbins T, Ersche K (2019) Dopaminergic modulation of frontostriatal networks in disorders of addiction and compulsion. Eur Neuropsychopharmacol 29: S490–S491. https://doi.org/10.17863/CAM.48474
8. Ernst M, Luciana M (2015) Neuroimaging of the dopamine/reward system in adolescent drug use. CNS Spectr 20(4):427–441. https://doi.org/10.1017/S1092852915000395
9. Luciana M, Bjork JM, Nagel BJ, Barch DM, Gonzalez R, Nixon SJ, Banich MT (2018) Adolescent neurocognitive development and impacts of substance use: overview of the adolescent brain cognitive development (ABCD) baseline neurocognition battery. Dev Cogn Neurosci 32:67–79. https://doi.org/10.1016/j.dcn.2018.02.006
10. Tervo-Clemmens B, Quach A, Calabro FJ, Foran W, Luna B (2020) Meta-analysis and review of functional neuroimaging differences underlying adolescent vulnerability to substance use. NeuroImage 209:116476. https://doi.org/10.1016/j.neuroimage.2019.116476

11. Alakurtti K, Johansson JJ, Joutsa J, Laine M, Bäckman L, Nyberg L, Rinne JO (2015) Long-term test–retest reliability of striatal and extrastriatal dopamine D2/3 receptor binding: study with [11C]raclopride and high-resolution PET. J Cereb Blood Flow Metab 35(7):1199–1205. https://doi.org/10.1038/jcbfm.2015.53
12. Farde L, Hall H, Pauli S, Halldin C (1995) Variability in D2-dopamine receptor density and affinity: a PET study with [11C]raclopride in man. Synapse 20(3):200–208. https://doi.org/10.1002/syn.890200303
13. Kilbourn MR (2014) Radioligands for imaging vesicular monoamine transporters. In: Dierckx RAJO, Otte A, de Vries EFJ, van Waarde A, Luiten PGM (eds) PET and SPECT of neurobiological systems. Springer, pp 765–790. https://doi.org/10.1007/978-3-642-42014-6_27
14. Knutson B, Gibbs SEB (2007) Linking nucleus accumbens dopamine and blood oxygenation. Psychopharmacology 191(3):813–822. https://doi.org/10.1007/s00213-006-0686-7
15. Bjork JM, Knutson B, Fong GW, Caggiano DM, Bennett SM, Hommer DW (2004) Incentive-elicited brain activation in adolescents: similarities and differences from young adults. J Neurosci 24(8):1793–1802. https://doi.org/10.1523/JNEUROSCI.4862-03.2004
16. Bjork JM, Smith AR, Chen G, Hommer DW (2010) Adolescents, adults and rewards: comparing motivational neurocircuitry recruitment using fMRI. PLoS One 5(7):e11440. https://doi.org/10.1371/journal.pone.0011440
17. Braams BR, van Duijvenvoorde ACK, Peper JS, Crone EA (2015) Longitudinal changes in adolescent risk-taking: a comprehensive study of neural responses to rewards, pubertal development, and risk-taking behavior. J Neurosci 35(18):7226–7238. https://doi.org/10.1523/JNEUROSCI.4764-14.2015
18. Ernst M, Nelson EE, Jazbec S, McClure EB, Monk CS, Leibenluft E, Blair J, Pine DS (2005) Amygdala and nucleus accumbens in responses to receipt and omission of gains in adults and adolescents. NeuroImage 25(4):1279–1291. https://doi.org/10.1016/j.neuroimage.2004.12.038
19. Galvan A, Hare TA, Parra CE, Penn J, Voss H, Glover G, Casey BJ (2006) Earlier development of the accumbens relative to orbitofrontal cortex might underlie risk-taking behavior in adolescents. J Neurosci 26(25):6885–6892. https://doi.org/10.1523/JNEUROSCI.1062-06.2006
20. Geier CF, Terwilliger R, Teslovich T, Velanova K, Luna B (2010) Immaturities in reward processing and its influence on inhibitory control in adolescence. Cereb Cortex 20(7):1613–1629. https://doi.org/10.1093/cercor/bhp225
21. Luna B, Paulsen DJ, Padmanabhan A, Geier C (2013) Cognitive control and motivation. Curr Dir Psychol Sci 22(2):94–100. https://doi.org/10.1177/0963721413478416
22. Padmanabhan A (2011) Developmental changes in brain function underlying the influence of reward processing on inhibitory control. Dev Cogn Neurosci 1:517–529
23. Paulsen DJ, Hallquist MN, Geier CF, Luna B (2015) Effects of incentives, age, and behavior on brain activation during inhibitory control: a longitudinal fMRI study. Dev Cogn Neurosci 11:105–115. https://doi.org/10.1016/j.dcn.2014.09.003
24. Hillman EMC (2014) Coupling mechanism and significance of the BOLD signal: a status report. Annu Rev Neurosci 37(1):161–181. https://doi.org/10.1146/annurev-neuro-071013-014111
25. Logothetis NK, Wandell BA (2004) Interpreting the BOLD signal. Annu Rev Physiol 66(1):735–769. https://doi.org/10.1146/annurev.physiol.66.082602.092845
26. Heinz A, Siessmeier T, Wrase J, Hermann D, Klein S, Grüsser-Sinopoli SM, Flor H, Braus DF, Buchholz HG, Gründer G, Schreckenberger M, Smolka MN, Rösch F, Mann K, Bartenstein P (2004) Correlation between dopamine D2 receptors in the ventral striatum and central processing of alcohol cues and craving. Am J Psychiatr 161(10):1783–1789. https://doi.org/10.1176/ajp.161.10.1783
27. Judenhofer MS, Wehrl HF, Newport DF, Catana C, Siegel SB, Becker M, Thielscher A, Kneilling M, Lichy MP, Eichner M, Klingel K, Reischl G, Widmaier S, Röcken M, Nutt RE, Machulla H-J, Uludag K, Cherry SR, Claussen CD, Pichler BJ (2008) Simultaneous PET-MRI: a new approach for functional and morphological imaging. Nat Med 14(4):459–465. https://doi.org/10.1038/nm1700
28. Schott BH, Minuzzi L, Krebs RM, Elmenhorst D, Lang M, Winz OH, Seidenbecher CI, Coenen HH, Heinze H-J, Zilles K, Düzel E, Bauer A (2008) Mesolimbic functional magnetic resonance imaging activations during reward anticipation correlate with reward-related ventral striatal dopamine release. J Neurosci 28(52):14311–14319. https://doi.org/10.1523/JNEUROSCI.2058-08.2008

29. Brocka M, Helbing C, Vincenz D, Scherf T, Montag D, Goldschmidt J, Angenstein F, Lippert M (2018) Contributions of dopaminergic and non-dopaminergic neurons to VTA-stimulation induced neurovascular responses in brain reward circuits. NeuroImage 177:88–97. https://doi.org/10.1016/j.neuroimage.2018.04.059
30. Attwell D, Iadecola C (2002) The neural basis of functional brain imaging signals. Trends Neurosci 25(12):621–625. https://doi.org/10.1016/S0166-2236(02)02264-6
31. Logothetis NK (2003) The underpinnings of the BOLD functional magnetic resonance imaging signal. J Neurosci 23(10):3963–3971. https://doi.org/10.1523/JNEUROSCI.23-10-03963.2003
32. Jongkees BJ, Colzato LS (2016) Spontaneous eye blink rate as predictor of dopamine-related cognitive function—a review. Neurosci Biobehav Rev 71:58–82. https://doi.org/10.1016/j.neubiorev.2016.08.020
33. Larsen B, Olafsson V, Calabro F, Laymon C, Tervo-Clemmens B, Campbell E, Minhas D, Montez D, Price J, Luna B (2020) Maturation of the human striatal dopamine system revealed by PET and quantitative MRI. Nat Commun 11(1):846. https://doi.org/10.1038/s41467-020-14693-3
34. Connor JR, Menzies SL (1996) Relationship of iron to oligodendrocytes and myelination. Glia 17(2):83–93. https://doi.org/10.1002/(SICI)1098-1136(199606)17:2<83::AID-GLIA1>3.0.CO;2-7
35. Ward RJ, Zucca FA, Duyn JH, Crichton RR, Zecca L (2014) The role of iron in brain ageing and neurodegenerative disorders. Lancet Neurol 13(10):1045–1060. https://doi.org/10.1016/S1474-4422(14)70117-6
36. Rouault TA (2013) Iron metabolism in the CNS: implications for neurodegenerative diseases. Nat Rev Neurosci 14(8):551–564. https://doi.org/10.1038/nrn3453
37. Todorich B, Pasquini JM, Garcia CI, Paez PM, Connor JR (2009) Oligodendrocytes and myelination: the role of iron. Glia 57(5):467–478. https://doi.org/10.1002/glia.20784
38. Ramsey AJ, Hillas PJ, Fitzpatrick PF (1996) Characterization of the active site iron in tyrosine hydroxylase. J Biol Chem 271(40):24395–24400. https://doi.org/10.1074/jbc.271.40.24395
39. Lu H, Chen J, Huang H, Zhou M, Zhu Q, Yao SQ, Chai Z, Hu Y (2017) Iron modulates the activity of monoamine oxidase B in SH-SY5Y cells. Biometals 30(4):599–607. https://doi.org/10.1007/s10534-017-0030-1
40. Youdim MBH (2018) Monoamine oxidase inhibitors, and iron chelators in depressive illness and neurodegenerative diseases. J Neural Transm 125(11):1719–1733. https://doi.org/10.1007/s00702-018-1942-9
41. Youdim MBH, Grahame-Smith DG, Woods HF (1976) Some properties of human platelet monoamine oxidase in iron-deficiency anaemia. Clin Sci Mol Med 50(6):479–485. https://doi.org/10.1042/cs0500479
42. Zucca FA, Segura-Aguilar J, Ferrari E, Muñoz P, Paris I, Sulzer D, Sarna T, Casella L, Zecca L (2017) Interactions of iron, dopamine and neuromelanin pathways in brain aging and Parkinson's disease. Prog Neurobiol 155:96–119. https://doi.org/10.1016/j.pneurobio.2015.09.012
43. Ortega R, Cloetens P, Devès G, Carmona A, Bohic S (2007) Iron storage within dopamine neurovesicles revealed by chemical nano-imaging. PLoS One 2(9):e925. https://doi.org/10.1371/journal.pone.0000925
44. Brass SD, Chen N, Mulkern RV, Bakshi R (2006) Magnetic resonance imaging of iron deposition in neurological disorders. Top Magn Reson Imaging 17(1):31–40. https://doi.org/10.1097/01.rmr.0000245459.82782.e4
45. Hallgren B, Sourander P (1958) The effect of age on the non-haemin iron in the human brain. J Neurochem 3(1):41–51. https://doi.org/10.1111/j.1471-4159.1958.tb12607.x
46. Thomas LO, Boyko OB, Anthony DC, Burger PC (1993) MR detection of brain iron. Am J Neuroradiol 14(5):1043–1048
47. Adisetiyo V, Jensen JH, Tabesh A, Deardorff RL, Fieremans E, Di Martino A, Gray KM, Castellanos FX, Helpern JA (2014) Multimodal MR imaging of brain iron in attention deficit hyperactivity disorder: a noninvasive biomarker that responds to psychostimulant treatment? Radiology 272(2):524–532. https://doi.org/10.1148/radiol.14140047
48. Allen RP, Earley CJ (2007) The role of iron in restless legs syndrome. Mov Disord 22(Suppl 18):S440–S448. https://doi.org/10.1002/mds.21607
49. Bartzokis G, Tishler TA, Shin I-S, Lu PH, Cummings JL (2004) Brain ferritin iron as a risk factor for age at onset in neurodegenerative diseases. Ann N Y Acad Sci 1012(1):224–236. https://doi.org/10.1196/annals.1306.019

50. Khan FH, Ahlberg CD, Chow CA, Shah DR, Koo BB (2017) Iron, dopamine, genetics, and hormones in the pathophysiology of restless legs syndrome. J Neurol 264(8):1634–1641. https://doi.org/10.1007/s00415-017-8431-1
51. Piao Y-S, Lian T-H, Hu Y, Zuo L-J, Guo P, Yu S-Y, Liu L, Jin Z, Zhao H, Li L-X, Yu Q-J, Wang R-D, Chen S-D, Chan P, Wang X-M, Zhang W (2017) Restless legs syndrome in Parkinson disease: clinical characteristics, abnormal iron metabolism and altered neurotransmitters. Sci Rep 7:10547. https://doi.org/10.1038/s41598-017-10593-7
52. Connor JR, Wang X-S, Allen RP, Beard JL, Wiesinger JA, Felt BT, Earley CJ (2009) Altered dopaminergic profile in the putamen and substantia nigra in restless leg syndrome. Brain 132(9):2403–2412. https://doi.org/10.1093/brain/awp125
53. Earley CJ, Kuwabara H, Wong DF, Gamaldo C, Salas R, Brasic J, Ravert HT, Dannals RF, Allen RP (2011) The dopamine transporter is decreased in the striatum of subjects with restless legs syndrome. Sleep 34(3):341–347. https://doi.org/10.1093/sleep/34.3.341
54. Earley CJ, Connor J, Garcia-Borreguero D, Jenner P, Winkelman J, Zee PC, Allen R (2014) Altered brain iron homeostasis and dopaminergic function in restless legs syndrome (Willis–Ekbom disease). Sleep Med 15(11):1288–1301. https://doi.org/10.1016/j.sleep.2014.05.009
55. Unger EL, Bianco LE, Jones BC, Allen RP, Earley CJ (2014) Low brain iron effects and reversibility on striatal dopamine dynamics. Exp Neurol 261:462–468. https://doi.org/10.1016/j.expneurol.2014.06.023
56. Ersche KD, Acosta-Cabronero J, Jones PS, Ziauddeen H, van Swelm RPL, Laarakkers CMM, Raha-Chowdhury R, Williams GB (2017) Disrupted iron regulation in the brain and periphery in cocaine addiction. Transl Psychiatry 7(2):e1040. https://doi.org/10.1038/tp.2016.271
57. Beard J (2003) Iron deficiency alters brain development and functioning. J Nutr 133(5):1468S–1472S. https://doi.org/10.1093/jn/133.5.1468S
58. Beard JL, Erikson KM, Jones BC (2002) Neurobehavioral analysis of developmental iron deficiency in rats. Behav Brain Res 134(1):517–524. https://doi.org/10.1016/S0166-4328(02)00092-X
59. Erikson KM, Jones BC, Beard JL (2000) Iron deficiency alters dopamine transporter functioning in rat striatum. J Nutr 130(11):2831–2837. https://doi.org/10.1093/jn/130.11.2831
60. Erikson KM, Jones BC, Hess EJ, Zhang Q, Beard JL (2001) Iron deficiency decreases dopamine D1 and D2 receptors in rat brain. Pharmacol Biochem Behav 69(3):409–418. https://doi.org/10.1016/S0091-3057(01)00563-9
61. Jellen LC, Lu L, Wang X, Unger EL, Earley CJ, Allen RP, Williams RW, Jones BC (2013) Iron deficiency alters expression of dopamine-related genes in the ventral midbrain in mice. Neuroscience 252:13–23. https://doi.org/10.1016/j.neuroscience.2013.07.058
62. Lozoff B (2011) Early iron deficiency has brain and behavior effects consistent with dopaminergic dysfunction. J Nutr 141(4):740S–746S. https://doi.org/10.3945/jn.110.131169
63. Unger EL, Wiesinger JA, Hao L, Beard JL (2008) Dopamine D2 receptor expression is altered by changes in cellular iron levels in PC12 cells and rat brain tissue. J Nutr 138(12):2487–2494. https://doi.org/10.3945/jn.108.095224
64. Wiesinger JA, Buwen JP, Cifelli CJ, Unger EL, Jones BC, Beard JL (2007) Down-regulation of dopamine transporter by iron chelation in vitro is mediated by altered trafficking, not synthesis. J Neurochem 100(1):167–179. https://doi.org/10.1111/j.1471-4159.2006.04175.x
65. Aquino D, Bizzi A, Grisoli M, Garavaglia B, Bruzzone MG, Nardocci N, Savoiardo M, Chiapparini L (2009) Age-related iron deposition in the basal ganglia: quantitative analysis in healthy subjects. Radiology 252(1):165–172. https://doi.org/10.1148/radiol.2522081399
66. Hect JL, Daugherty AM, Hermez KM, Thomason ME (2018) Developmental variation in regional brain iron and its relation to cognitive functions in childhood. Dev Cogn Neurosci 34:18–26. https://doi.org/10.1016/j.dcn.2018.05.004
67. Larsen B, Bourque J, Moore TM, Adebimpe A, Calkins ME, Elliott MA, Gur RC, Gur RE, Moberg PJ, Roalf DR, Ruparel K, Turetsky BI, Vandekar SN, Wolf DH, Shinohara RT, Satterthwaite TD (2020) Longitudinal development of brain iron is linked to cognition in youth. J Neurosci 40(9):1810–1818. https://doi.org/10.1523/JNEUROSCI.2434-19.2020
68. Larsen B, Luna B (2015) In vivo evidence of neurophysiological maturation of the human adolescent striatum. Dev Cogn Neurosci 12:

74–85. https://doi.org/10.1016/j.dcn.2014.12.003

69. Peterson ET, Kwon D, Luna B, Larsen B, Prouty D, Bellis MDD, Voyvodic J, Liu C, Li W, Pohl KM, Sullivan EV, Pfefferbaum A (2019) Distribution of brain iron accrual in adolescence: evidence from cross-sectional and longitudinal analysis. Hum Brain Mapp 40(5):1480–1495. https://doi.org/10.1002/hbm.24461
70. Drayer B, Burger P, Darwin R, Riederer S, Herfkens R, Johnson GA (1986) MRI of brain iron. AJR Am J Roentgenol 147(1): 103–110. https://doi.org/10.2214/ajr.147.1.103
71. Hametner S, Endmayr V, Deistung A, Palmrich P, Prihoda M, Haimburger E, Menard C, Feng X, Haider T, Leisser M, Köck U, Kaider A, Höftberger R, Robinson S, Reichenbach JR, Lassmann H, Traxler H, Trattnig S, Grabner G (2018) The influence of brain iron and myelin on magnetic susceptibility and effective transverse relaxation—a biochemical and histological validation study. NeuroImage 179:117–133. https://doi.org/10.1016/j.neuroimage.2018.06.007
72. Langkammer C, Krebs N, Goessler W, Scheurer E, Ebner F, Yen K, Fazekas F, Ropele S (2010) Quantitative MR imaging of brain iron: a postmortem validation study. Radiology 257(2):455–462. https://doi.org/10.1148/radiol.10100495
73. Stüber C, Morawski M, Schäfer A, Labadie C, Wähnert M, Leuze C, Streicher M, Barapatre N, Reimann K, Geyer S, Spemann D, Turner R (2014) Myelin and iron concentration in the human brain: a quantitative study of MRI contrast. NeuroImage 93:95–106. https://doi.org/10.1016/j.neuroimage.2014.02.026
74. Haacke EM, Cheng NYC, House MJ, Liu Q, Neelavalli J, Ogg RJ, Khan A, Ayaz M, Kirsch W, Obenaus A (2005) Imaging iron stores in the brain using magnetic resonance imaging. Magn Reson Imaging 23(1):1–25. https://doi.org/10.1016/j.mri.2004.10.001
75. Haacke EM, Miao Y, Liu M, Habib CA, Katkuri Y, Liu T, Yang Z, Lang Z, Hu J, Wu J (2010) Correlation of putative iron content as represented by changes in R2* and phase with age in deep gray matter of healthy adults. J Magn Reson Imaging 32(3):561–576. https://doi.org/10.1002/jmri.22293
76. Ma J, Wehrli FW (1996) Method for image-based measurement of the reversible and irreversible contribution to the transverse-relaxation rate. J Magn Reson Ser B 111(1): 61–69. https://doi.org/10.1006/jmrb.1996.0060
77. Wang Y, Liu T (2015) Quantitative susceptibility mapping (QSM): decoding MRI data for a tissue magnetic biomarker. Magn Reson Med 73(1):82–101. https://doi.org/10.1002/mrm.25358
78. Sedlacik J, Boelmans K, Löbel U, Holst B, Siemonsen S, Fiehler J (2014) Reversible, irreversible and effective transverse relaxation rates in normal aging brain at 3T. NeuroImage 84:1032–1041. https://doi.org/10.1016/j.neuroimage.2013.08.051
79. Chavhan GB, Babyn PS, Thomas B, Shroff MM, Haacke EM (2009) Principles, techniques, and applications of T2*-based MR imaging and its special applications. Radiographics 29(5):1433–1449. https://doi.org/10.1148/rg.295095034
80. Yablonskiy DA, Haacke EM (1994) Theory of NMR signal behavior in magnetically inhomogeneous tissues: the static dephasing regime. Magn Reson Med 32(6):749–763. https://doi.org/10.1002/mrm.1910320610
81. Deistung A, Schäfer A, Schweser F, Biedermann U, Güllmar D, Trampel R, Turner R, Reichenbach JR (2013) High-resolution MR imaging of the human brainstem in vivo at 7 Tesla. Front Hum Neurosci 7:710. https://doi.org/10.3389/fnhum.2013.00710
82. Péran P, Hagberg G, Luccichenti G, Cherubini A, Brainovich V, Celsis P, Caltagirone C, Sabatini U (2007) Voxel-based analysis of R2* maps in the healthy human brain. J Magn Reson Imaging 26(6): 1413–1420. https://doi.org/10.1002/jmri.21204
83. Esterhammer R, Seppi K, Reiter E, Pinter B, Mueller C, Kremser C, Zitzelsberger T, Nocker M, Scherfler C, Poewe W, Schocke M (2015) Potential of diffusion tensor imaging and relaxometry for the detection of specific pathological alterations in Parkinson's disease (PD). PLoS One 10(12):e0145493. https://doi.org/10.1371/journal.pone.0145493
84. Friedrich I, Reimann K, Jankuhn S, Kirilina E, Stieler J, Sonntag M, Meijer J, Weiskopf N, Reinert T, Arendt T, Morawski M (2021) Cell specific quantitative iron mapping on brain slices by immuno-µPIXE in healthy elderly and Parkinson's disease. Acta Neuropathol Commun 9(1):47. https://doi.org/10.1186/s40478-021-01145-2

85. Gorell JM, Ordidge RJ, Brown GG, Deniau J-C, Buderer NM, Helpern JA (1995) Increased iron-related MRI contrast in the substantia nigra in Parkinson's disease. Neurology 45(6):1138–1143. https://doi.org/10.1212/WNL.45.6.1138
86. Hopes L, Grolez G, Moreau C, Lopes R, Ryckewaert G, Carrière N, Auger F, Laloux C, Petrault M, Devedjian J-C, Bordet R, Defebvre L, Jissendi P, Delmaire C, Devos D (2016) Magnetic resonance imaging features of the nigrostriatal system: biomarkers of Parkinson's disease stages? PLoS One 11(4):e0147947. https://doi.org/10.1371/journal.pone.0147947
87. Rossi M, Ruottinen H, Soimakallio S, Elovaara I, Dastidar P (2013) Clinical MRI for iron detection in Parkinson's disease. Clin Imaging 37(4):631–636. https://doi.org/10.1016/j.clinimag.2013.02.001
88. Ulla M, Bonny JM, Ouchchane L, Rieu I, Claise B, Durif F (2013) Is R2* a new MRI biomarker for the progression of Parkinson's disease? A longitudinal follow-up. PLoS One 8(3):e57904. https://doi.org/10.1371/journal.pone.0057904
89. Wieler M, Gee M, Martin WRW (2015) Longitudinal midbrain changes in early Parkinson's disease: iron content estimated from R2*/MRI. Parkinsonism Relat Disord 21(3):179–183. https://doi.org/10.1016/j.parkreldis.2014.11.017
90. Zhang Y, Gauthier SA, Gupta A, Chen W, Comunale J, Chiang GC-Y, Zhou D, Askin G, Zhu W, Pitt D, Wang Y (2016) Quantitative susceptibility mapping and R2* measured changes during white matter lesion development in multiple sclerosis: myelin breakdown, myelin debris degradation and removal, and iron accumulation. Am J Neuroradiol 37(9):1629–1635. https://doi.org/10.3174/ajnr.A4825
91. Betts MJ, Acosta-Cabronero J, Cardenas-Blanco A, Nestor PJ, Düzel E (2016) High-resolution characterisation of the aging brain using simultaneous quantitative susceptibility mapping (QSM) and R2* measurements at 7T. NeuroImage 138:43–63. https://doi.org/10.1016/j.neuroimage.2016.05.024
92. Brammerloh M, Morawski M, Friedrich I, Reinert T, Lange C, Pelicon P, Vavpetič P, Jankuhn S, Jäger C, Alkemade A, Balesar R, Pine K, Gavriilidis F, Trampel R, Reimer E, Arendt T, Weiskopf N, Kirilina E (2021) Measuring the iron content of dopaminergic neurons in substantia nigra with MRI relaxometry. NeuroImage 239:118255. https://doi.org/10.1016/j.neuroimage.2021.118255
93. Gelman N, Gorell JM, Barker PB, Savage RM, Spickler EM, Windham JP, Knight RA (1999) MR imaging of human brain at 3.0 T: preliminary report on transverse relaxation rates and relation to estimated iron content. Radiology 210(3):759–767. https://doi.org/10.1148/radiology.210.3.r99fe41759
94. Wehrli FW, MacFall JR, Glover GH, Grigsby N, Haughton V, Johanson J (1984) The dependence of nuclear magnetic resonance (NMR) image contrast on intrinsic and pulse sequence timing parameters. Magn Reson Imaging 2(1):3–16. https://doi.org/10.1016/0730-725X(84)90119-X
95. Yablonskiy DA (1998) Quantitation of intrinsic magnetic susceptibility-related effects in a tissue matrix. Phantom study. Magn Reson Med 39(3):417–428. https://doi.org/10.1002/mrm.1910390312
96. Graham JM, Paley MNJ, Grünewald RA, Hoggard N, Griffiths PD (2000) Brain iron deposition in Parkinson's disease imaged using the PRIME magnetic resonance sequence. Brain 123(12):2423–2431. https://doi.org/10.1093/brain/123.12.2423
97. Miszkiel KA, Paley MNJ, Wilkinson ID, Hall-Craggs MA, Ordidge R, Kendall BE, Miller RF, Harrison MJG (1997) The measurement of R2, R2* and R2′ in HIV-infected patients using the prime sequence as a measure of brain iron deposition. Magn Reson Imaging 15(10):1113–1119. https://doi.org/10.1016/S0730-725X(97)00089-1
98. Ghassaban K, Liu S, Jiang C, Haacke EM (2019) Quantifying iron content in magnetic resonance imaging. NeuroImage 187:77–92. https://doi.org/10.1016/j.neuroimage.2018.04.047
99. Möller HE, Bossoni L, Connor JR, Crichton RR, Does MD, Ward RJ, Zecca L, Zucca FA, Ronen I (2019) Iron, myelin, and the brain: neuroimaging meets neurobiology. Trends Neurosci 42(6):384–401. https://doi.org/10.1016/j.tins.2019.03.009
100. Daugherty AM, Hoagey DA, Kennedy KM, Rodrigue KM (2019) Genetic predisposition for inflammation exacerbates effects of striatal iron content on cognitive switching ability in healthy aging. NeuroImage 185:471–478. https://doi.org/10.1016/j.neuroimage.2018.10.064
101. Salami A, Avelar-Pereira B, Garzón B, Sitnikov R, Kalpouzos G (2018) Functional coherence of striatal resting-state networks is modulated by striatal iron content. NeuroImage 183:495–503. https://doi.org/10.1016/j.neuroimage.2018.08.036

102. Parr AC, Calabro F, Larsen B, Tervo-Clemmens B, Elliot S, Foran W, Olafsson V, Luna B (2021) Dopamine-related striatal neurophysiology is associated with specialization of frontostriatal reward circuitry through adolescence. Prog Neurobiol 201:101997. https://doi.org/10.1016/j.pneurobio.2021.101997
103. Deistung A, Rauscher A, Sedlacik J, Stadler J, Witoszynskyj S, Reichenbach JR (2008) Susceptibility weighted imaging at ultra high magnetic field strengths: theoretical considerations and experimental results. Magn Reson Med 60(5):1155–1168. https://doi.org/10.1002/mrm.21754
104. Bender B, Klose U (2010) The in vivo influence of white matter fiber orientation towards B0 on T2* in the human brain. NMR Biomed 23(9):1071–1076. https://doi.org/10.1002/nbm.1534
105. Daugherty A, Raz N (2013) Age-related differences in iron content of subcortical nuclei observed in vivo: a meta-analysis. NeuroImage 70:113–121. https://doi.org/10.1016/j.neuroimage.2012.12.040
106. Langkammer C, Schweser F, Krebs N, Deistung A, Goessler W, Scheurer E, Sommer K, Reishofer G, Yen K, Fazekas F, Ropele S, Reichenbach JR (2012) Quantitative susceptibility mapping (QSM) as a means to measure brain iron? A post mortem validation study. NeuroImage 62(3):1593–1599. https://doi.org/10.1016/j.neuroimage.2012.05.049
107. Schenck JF (2003) Magnetic resonance imaging of brain iron. J Neurol Sci 207(1):99–102. https://doi.org/10.1016/S0022-510X(02)00431-8
108. Aoki S, Okada Y, Nishimura K, Barkovich AJ, Kjos BO, Brasch RC, Norman D (1989) Normal deposition of brain iron in childhood and adolescence: MR imaging at 1.5 T. Radiology 172(2):381–385. https://doi.org/10.1148/radiology.172.2.2748819
109. Gossuin Y, Muller RN, Gillis P (2004) Relaxation induced by ferritin: a better understanding for an improved MRI iron quantification. NMR Biomed 17(7):427–432. https://doi.org/10.1002/nbm.903
110. Denk C, Torres EH, MacKay A, Rauscher A (2011) The influence of white matter fibre orientation on MR signal phase and decay. NMR Biomed 24(3):246–252. https://doi.org/10.1002/nbm.1581
111. He X, Yablonskiy DA (2009) Biophysical mechanisms of phase contrast in gradient echo MRI. Proc Natl Acad Sci 106(32):13558–13563. https://doi.org/10.1073/pnas.0904899106
112. Lee J, van Gelderen P, Kuo L-W, Merkle H, Silva AC, Duyn JH (2011) T2*-based fiber orientation mapping. NeuroImage 57(1):225–234. https://doi.org/10.1016/j.neuroimage.2011.04.026
113. Li W, Wu B, Avram AV, Liu C (2012) Magnetic susceptibility anisotropy of human brain in vivo and its molecular underpinnings. NeuroImage 59(3):2088–2097. https://doi.org/10.1016/j.neuroimage.2011.10.038
114. Rudko DA, Solovey I, Gati JS, Kremenchutzky M, Menon RS (2014) Multiple sclerosis: improved identification of disease-relevant changes in gray and white matter by using susceptibility-based MR imaging. Radiology 272(3):851–864. https://doi.org/10.1148/radiol.14132475
115. Wharton S, Bowtell R (2013) Gradient echo based fiber orientation mapping using R2* and frequency difference measurements. NeuroImage 83:1011–1023. https://doi.org/10.1016/j.neuroimage.2013.07.054
116. Mitsumori F, Watanabe H, Takaya N (2009) Estimation of brain iron concentration in vivo using a linear relationship between regional iron and apparent transverse relaxation rate of the tissue water at 4.7T. Magn Reson Med 62(5):1326–1330. https://doi.org/10.1002/mrm.22097
117. Mitsumori F, Watanabe H, Takaya N, Garwood M, Auerbach EJ, Michaeli S, Mangia S (2012) Toward understanding transverse relaxation in human brain through its field dependence. Magn Reson Med 68(3):947–953. https://doi.org/10.1002/mrm.23301
118. Anderson LJ, Holden S, Davis B, Prescott E, Charrier CC, Bunce NH, Firmin DN, Wonke B, Porter J, Walker JM, Pennell DJ (2001) Cardiovascular T2-star (T2*) magnetic resonance for the early diagnosis of myocardial iron overload. Eur Heart J 22(23):2171–2179. https://doi.org/10.1053/euhj.2001.2822
119. Gandon Y, Olivié D, Guyader D, Aubé C, Oberti F, Sebille V, Deugnier Y (2004) Non-invasive assessment of hepatic iron stores by MRI. Lancet 363(9406):357–362. https://doi.org/10.1016/S0140-6736(04)15436-6
120. Haacke EM, Liu S, Buch S, Zheng W, Wu D, Ye Y (2015) Quantitative susceptibility mapping: current status and future directions. Magn Reson Imaging 33(1):1–25. https://doi.org/10.1016/j.mri.2014.09.004
121. Yao B, Li T-Q, van Gelderen P, Shmueli K, de Zwart JA, Duyn JH (2009) Susceptibility contrast in high field MRI of human brain as a function of tissue iron content. NeuroImage

44(4):1259–1266. https://doi.org/10.1016/j.neuroimage.2008.10.029
122. Haacke EM, Xu Y, Cheng Y-CN, Reichenbach JR (2004) Susceptibility weighted imaging (SWI). Magn Reson Med 52(3):612–618. https://doi.org/10.1002/mrm.20198
123. Halefoglu AM, Yousem DM (2018) Susceptibility weighted imaging: clinical applications and future directions. World J Radiol 10(4):30–45. https://doi.org/10.4329/wjr.v10.i4.30
124. Reichenbach JR, Schweser F, Serres B, Deistung A (2015) Quantitative susceptibility mapping: concepts and applications. Clin Neuroradiol 25(2):225–230. https://doi.org/10.1007/s00062-015-0432-9
125. Iyer SK, Moon BF, Josselyn N, Ruparel K, Roalf D, Song JW, Guiry S, Ware JB, Kurtz RM, Chawla S, Nabavizadeh SA, Witschey WR (2020) Data-driven quantitative susceptibility mapping using loss adaptive dipole inversion (LADI). J Magn Reson Imaging 52(3):823–835. https://doi.org/10.1002/jmri.27103
126. Li W, Wu B, Liu C (2011) Quantitative susceptibility mapping of human brain reflects spatial variation in tissue composition. NeuroImage 55(4):1645–1656. https://doi.org/10.1016/j.neuroimage.2010.11.088
127. Hopp K, Popescu BFG, McCrea RPE, Harder SL, Robinson CA, Haacke ME, Rajput AH, Rajput A, Nichol H (2010) Brain iron detected by SWI high pass filtered phase calibrated with synchrotron X-ray fluorescence. J Magn Reson Imaging 31(6):1346–1354. https://doi.org/10.1002/jmri.22201
128. Liu M, Liu S, Ghassaban K, Zheng W, Dicicco D, Miao Y, Habib C, Jazmati T, Haacke EM (2016) Assessing global and regional iron content in deep gray matter as a function of age using susceptibility mapping. J Magn Reson Imaging 44(1):59–71. https://doi.org/10.1002/jmri.25130
129. Zheng W, Nichol H, Liu S, Cheng Y-CN, Haacke EM (2013) Measuring iron in the brain using quantitative susceptibility mapping and X-ray fluorescence imaging. NeuroImage 78:68–74. https://doi.org/10.1016/j.neuroimage.2013.04.022
130. Barbosa JHO, Santos AC, Tumas V, Liu M, Zheng W, Haacke EM, Salmon CEG (2015) Quantifying brain iron deposition in patients with Parkinson's disease using quantitative susceptibility mapping, R2 and R2*. Magn Reson Imaging 33(5):559–565. https://doi.org/10.1016/j.mri.2015.02.021
131. Li W, Avram AV, Wu B, Xiao X, Liu C (2014) Integrated Laplacian-based phase unwrapping and background phase removal for quantitative susceptibility mapping. NMR Biomed 27(2):219–227. https://doi.org/10.1002/nbm.3056
132. Li X, Chen L, Kutten K, Ceritoglu C, Li Y, Kang N, Hsu JT, Qiao Y, Wei H, Liu C, Miller MI, Mori S, Yousem DM, van Zijl PCM, Faria AV (2019) Multi-atlas tool for automated segmentation of brain gray matter nuclei and quantification of their magnetic susceptibility. NeuroImage 191:337–349. https://doi.org/10.1016/j.neuroimage.2019.02.016
133. Schweser F, Deistung A, Lehr BW, Reichenbach JR (2011) Quantitative imaging of intrinsic magnetic tissue properties using MRI signal phase: an approach to in vivo brain iron metabolism? NeuroImage 54(4):2789–2807. https://doi.org/10.1016/j.neuroimage.2010.10.070
134. Acosta-Cabronero J, Betts MJ, Cardenas-Blanco A, Yang S, Nestor PJ (2016) In vivo MRI mapping of brain iron deposition across the adult lifespan. J Neurosci 36(2):364–374. https://doi.org/10.1523/JNEUROSCI.1907-15.2016
135. Bilgic B, Pfefferbaum A, Rohlfing T, Sullivan EV, Adalsteinsson E (2012) MRI estimates of brain iron concentration in normal aging using quantitative susceptibility mapping. NeuroImage 59(3):2625–2635. https://doi.org/10.1016/j.neuroimage.2011.08.077
136. Dashtipour K, Liu M, Kani C, Dalaie P, Obenaus A, Simmons D, Gatto NM, Zarifi M (2015) Iron accumulation is not homogenous among patients with Parkinson's disease. Parkinsons Dis 2015:e324843. https://doi.org/10.1155/2015/324843
137. Du G, Liu T, Lewis MM, Kong L, Wang Y, Connor J, Mailman RB, Huang X (2016) Quantitative susceptibility mapping of the midbrain in Parkinson's disease. Mov Disord 31(3):317–324. https://doi.org/10.1002/mds.26417
138. Guan X, Xuan M, Gu Q, Huang P, Liu C, Wang N, Xu X, Luo W, Zhang M (2017) Regionally progressive accumulation of iron in Parkinson's disease as measured by quantitative susceptibility mapping. NMR Biomed 30(4):e3489. https://doi.org/10.1002/nbm.3489
139. Guan X, Xu X, Zhang M (2017) Region-specific iron measured by MRI as a biomarker for Parkinson's disease. Neurosci Bull 33(5):561–567. https://doi.org/10.1007/s12264-017-0138-x
140. He N, Huang P, Ling H, Langley J, Liu C, Ding B, Huang J, Xu H, Zhang Y, Zhang Z, Hu X, Chen S, Yan F (2017) Dentate nucleus iron deposition is a potential biomarker for

tremor-dominant Parkinson's disease. NMR Biomed 30(4):e3554. https://doi.org/10.1002/nbm.3554

141. Lotfipour AK, Wharton S, Schwarz ST, Gontu V, Schäfer A, Peters AM, Bowtell RW, Auer DP, Gowland PA, Bajaj NPS (2012) High resolution magnetic susceptibility mapping of the substantia nigra in Parkinson's disease. J Magn Reson Imaging 35(1): 48–55. https://doi.org/10.1002/jmri.22752

142. Murakami Y, Kakeda S, Watanabe K, Ueda I, Ogasawara A, Moriya J, Ide S, Futatsuya K, Sato T, Okada K, Uozumi T, Tsuji S, Liu T, Wang Y, Korogi Y (2015) Usefulness of quantitative susceptibility mapping for the diagnosis of Parkinson disease. Am J Neuroradiol 36(6):1102–1108. https://doi.org/10.3174/ajnr.A4260

143. Wang Y, Butros SR, Shuai X, Dai Y, Chen C, Liu M, Haacke EM, Hu J, Xu H (2012) Different iron-deposition patterns of multiple system atrophy with predominant parkinsonism and idiopathetic Parkinson diseases demonstrated by phase-corrected susceptibility-weighted imaging. Am J Neuroradiol 33(2):266–273. https://doi.org/10.3174/ajnr.A2765

144. Bian W, Harter K, Hammond-Rosenbluth KE, Lupo JM, Xu D, Kelley DA, Vigneron DB, Nelson SJ, Pelletier D (2013) A serial in vivo 7T magnetic resonance phase imaging study of white matter lesions in multiple sclerosis. Mult Scler J 19(1):69–75. https://doi.org/10.1177/1352458512447870

145. Eissa A, Lebel RM, Korzan JR, Zavodni AE, Warren KG, Catz I, Emery DJ, Wilman AH (2009) Detecting lesions in multiple sclerosis at 4.7 Tesla using phase susceptibility-weighting and T2-weighting. J Magn Reson Imaging 30(4):737–742. https://doi.org/10.1002/jmri.21926

146. Haacke EM, Makki M, Ge Y, Maheshwari M, Sehgal V, Hu J, Selvan M, Wu Z, Latif Z, Xuan Y, Khan O, Garbern J, Grossman RI (2009) Characterizing iron deposition in multiple sclerosis lesions using susceptibility weighted imaging. J Magn Reson Imaging 29(3):537–544. https://doi.org/10.1002/jmri.21676

147. Hammond KE, Metcalf M, Carvajal L, Okuda DT, Srinivasan R, Vigneron D, Nelson SJ, Pelletier D (2008) Quantitative in vivo magnetic resonance imaging of multiple sclerosis at 7 Tesla with sensitivity to iron. Ann Neurol 64(6):707–713. https://doi.org/10.1002/ana.21582

148. Rumzan R, Wang J, Zeng C, Chen X, Li Y, Luo T, Lv F, Wang Z, Hou H, Huang F (2013) Iron deposition in the precentral grey matter in patients with multiple sclerosis: a quantitative study using susceptibility-weighted imaging. Eur J Radiol 82(2):e95–e99. https://doi.org/10.1016/j.ejrad.2012.09.006

149. Liu S, Mok K, Neelavalli J, Cheng Y-CN, Tang J, Ye Y, Haacke EM (2014) Improved MR venography using quantitative susceptibility-weighted imaging. J Magn Reson Imaging 40(3):698–708. https://doi.org/10.1002/jmri.24413

150. Liu C, Li W, Tong KA, Yeom KW, Kuzminski S (2015) Susceptibility-weighted imaging and quantitative susceptibility mapping in the brain. J Magn Reson Imaging 42(1):23–41. https://doi.org/10.1002/jmri.24768

151. Deistung A, Schweser F, Reichenbach JR (2017) Overview of quantitative susceptibility mapping. NMR Biomed 30(4):e3569. https://doi.org/10.1002/nbm.3569

152. Wharton S, Bowtell R (2015) Effects of white matter microstructure on phase and susceptibility maps. Magn Reson Med 73(3): 1258–1269. https://doi.org/10.1002/mrm.25189

153. Yablonskiy DA, Sukstanskii AL (2015) Generalized Lorentzian Tensor Approach (GLTA) as a biophysical background for quantitative susceptibility mapping. Magn Reson Med 73(2):757–764. https://doi.org/10.1002/mrm.25538

154. Li J, Chang S, Liu T, Wang Q, Cui D, Chen X, Jin M, Wang B, Pei M, Wisnieff C, Spincemaille P, Zhang M, Wang Y (2012) Reducing the object orientation dependence of susceptibility effects in gradient echo MRI through quantitative susceptibility mapping. Magn Reson Med 68(5):1563–1569. https://doi.org/10.1002/mrm.24135

155. Luo S, Yang L, Wang L (2015) Comparison of susceptibility-weighted and perfusion-weighted magnetic resonance imaging in the detection of penumbra in acute ischemic stroke. J Neuroradiol 42(5):255–260. https://doi.org/10.1016/j.neurad.2014.07.002

156. Wharton S, Bowtell R (2012) Fiber orientation-dependent white matter contrast in gradient echo MRI. Proc Natl Acad Sci 109(45):18559–18564. https://doi.org/10.1073/pnas.1211075109

157. Yablonskiy DA, Sukstanskii AL (2014) Biophysical mechanisms of myelin-induced water frequency shifts. Magn Reson Med 71(6):

1956–1958. https://doi.org/10.1002/mrm.25214

158. Acosta-Cabronero J, Milovic C, Mattern H, Tejos C, Speck O, Callaghan MF (2018) A robust multi-scale approach to quantitative susceptibility mapping. NeuroImage 183:7–24. https://doi.org/10.1016/j.neuroimage.2018.07.065
159. Lim IAL, Faria AV, Li X, Hsu JTC, Airan RD, Mori S, van Zijl PCM (2013) Human brain atlas for automated region of interest selection in quantitative susceptibility mapping: application to determine iron content in deep gray matter structures. NeuroImage 82:449–469. https://doi.org/10.1016/j.neuroimage.2013.05.127
160. Langkammer C, Pirpamer L, Seiler S, Deistung A, Schweser F, Franthal S, Homayoon N, Katschnig-Winter P, Koegl-Wallner M, Pendl T, Stoegerer EM, Wenzel K, Fazekas F, Ropele S, Reichenbach JR, Schmidt R, Schwingenschuh P (2016) Quantitative susceptibility mapping in Parkinson's disease. PLoS One 11(9):e0162460. https://doi.org/10.1371/journal.pone.0162460
161. Deistung A, Schäfer A, Schweser F, Biedermann U, Turner R, Reichenbach JR (2013) Toward in vivo histology: a comparison of quantitative susceptibility mapping (QSM) with magnitude-, phase-, and R2*-imaging at ultra-high magnetic field strength. NeuroImage 65:299–314. https://doi.org/10.1016/j.neuroimage.2012.09.055
162. Geyer S, Weiss M, Reimann K, Lohmann G, Turner R (2011) Microstructural parcellation of the human cerebral cortex – from Brodmann's post-mortem map to in vivo mapping with high-field magnetic resonance imaging. Front Hum Neurosci 5:19. https://doi.org/10.3389/fnhum.2011.00019
163. Langkammer C, Liu T, Khalil M, Enzinger C, Jehna M, Fuchs S, Fazekas F, Wang Y, Ropele S (2013) Quantitative susceptibility mapping in multiple sclerosis. Radiology 267(2):551–559. https://doi.org/10.1148/radiol.12120707
164. Lebel RM, Eissa A, Seres P, Blevins G, Wilman AH (2012) Quantitative high-field imaging of sub-cortical gray matter in multiple sclerosis. Mult Scler J 18(4):433–441. https://doi.org/10.1177/1352458511428464
165. Walsh AJ, Lebel RM, Eissa A, Blevins G, Catz I, Lu J-Q, Resch L, Johnson ES, Emery DJ, Warren KG, Wilman AH (2013) Multiple sclerosis: validation of MR imaging for quantification and detection of iron. Radiology 267(2):531–542. https://doi.org/10.1148/radiol.12120863
166. Wisnieff C, Ramanan S, Olesik J, Gauthier S, Wang Y, Pitt D (2015) Quantitative susceptibility mapping (QSM) of white matter multiple sclerosis lesions: interpreting positive susceptibility and the presence of iron. Magn Reson Med 74(2):564–570. https://doi.org/10.1002/mrm.25420
167. Taege Y, Hagemeier J, Bergsland N, Dwyer MG, Weinstock-Guttman B, Zivadinov R, Schweser F (2019) Assessment of mesoscopic properties of deep gray matter iron through a model-based simultaneous analysis of magnetic susceptibility and R2*—a pilot study in patients with multiple sclerosis and normal controls. NeuroImage 186:308–320. https://doi.org/10.1016/j.neuroimage.2018.11.011
168. Harrison DM, Li X, Liu H, Jones CK, Caffo B, Calabresi PA, van Zijl P (2016) Lesion heterogeneity on high-field susceptibility MRI is associated with multiple sclerosis severity. Am J Neuroradiol 37(8):1447–1453. https://doi.org/10.3174/ajnr.A4726
169. Li X, Harrison DM, Liu H, Jones CK, Oh J, Calabresi PA, van Zijl PCM (2016) Magnetic susceptibility contrast variations in multiple sclerosis lesions. J Magn Reson Imaging 43(2):463–473. https://doi.org/10.1002/jmri.24976
170. Pohmann R, Speck O, Scheffler K (2016) Signal-to-noise ratio and MR tissue parameters in human brain imaging at 3, 7, and 9.4 Tesla using current receive coil arrays. Magn Reson Med 75(2):801–809. https://doi.org/10.1002/mrm.25677
171. Berry AS, Shah VD, Furman DJ, White RL III, Baker SL, O'Neil JP, Janabi M, D'Esposito M, Jagust WJ (2018) Dopamine synthesis capacity is associated with D2/3 receptor binding but not dopamine release. Neuropsychopharmacology 43(6):1201–1211. https://doi.org/10.1038/npp.2017.180
172. Treit S, Naji N, Seres P, Rickard J, Stolz E, Wilman AH, Beaulieu C (2021) R2* and quantitative susceptibility mapping in deep gray matter of 498 healthy controls from 5 to 90 years. Hum Brain Mapp 42(14):4597–4610. https://doi.org/10.1002/hbm.25569
173. Lozoff B, Georgieff MK (2006) Iron deficiency and brain development. Semin Pediatr Neurol 13(3):158–165. https://doi.org/10.1016/j.spen.2006.08.004

174. Grantham-McGregor S, Ani C (2001) A review of studies on the effect of iron deficiency on cognitive development in children. J Nutr 131(2):649S–668S. https://doi.org/10.1093/jn/131.2.649S
175. McCann JC, Ames BN (2007) An overview of evidence for a causal relation between iron deficiency during development and deficits in cognitive or behavioral function. Am J Clin Nutr 85(4):931–945. https://doi.org/10.1093/ajcn/85.4.931
176. Bodnar LM, Wisner KL (2005) Nutrition and depression: implications for improving mental health among childbearing-aged women. Biol Psychiatry 58(9):679–685. https://doi.org/10.1016/j.biopsych.2005.05.009
177. Corapci F, Calatroni A, Kaciroti N, Jimenez E, Lozoff B (2010) Longitudinal evaluation of externalizing and internalizing behavior problems following iron deficiency in infancy. J Pediatr Psychol 35(3):296–305. https://doi.org/10.1093/jpepsy/jsp065
178. Insel BJ, Schaefer CA, McKeague IW, Susser ES, Brown AS (2008) Maternal iron deficiency and the risk of schizophrenia in offspring. Arch Gen Psychiatry 65(10):1136–1144. https://doi.org/10.1001/archpsyc.65.10.1136
179. Kim S-W, Stewart R, Park W-Y, Jhon M, Lee J-Y, Kim S-Y, Kim J-M, Amminger P, Chung Y-C, Yoon J-S (2018) Latent iron deficiency as a marker of negative symptoms in patients with first-episode schizophrenia spectrum disorder. Nutrients 10(11):1707. https://doi.org/10.3390/nu10111707
180. Konofal E, Lecendreux M, Arnulf I, Mouren M-C (2004) Iron deficiency in children with attention-deficit/hyperactivity disorder. Arch Pediatr Adolesc Med 158(12):1113. https://doi.org/10.1001/archpedi.158.12.1113
181. Lee H-S, Chao H-H, Huang W-T, Chen SC-C, Yang H-Y (2020) Psychiatric disorders risk in patients with iron deficiency anemia and association with iron supplementation medications: a nationwide database analysis. BMC Psychiatry 20(1):216. https://doi.org/10.1186/s12888-020-02621-0
182. Lozoff B, Jimenez E, Hagen J, Mollen E, Wolf AW (2000) Poorer behavioral and developmental outcome more than 10 years after treatment for iron deficiency in infancy. Pediatrics 105(4):e51. https://doi.org/10.1542/peds.105.4.e51
183. Oner O, Alkar OY, Oner P (2008) Relation of ferritin levels with symptom ratings and cognitive performance in children with attention deficit–hyperactivity disorder. Pediatr Int 50(1):40–44. https://doi.org/10.1111/j.1442-200X.2007.02496.x
184. Tseng P-T, Cheng Y-S, Chen Y-W, Stubbs B, Whiteley P, Carvalho AF, Li D-J, Chen T-Y, Yang W-C, Tang C-H, Chu C-S, Yang W-C, Liang H-Y, Wu C-K, Yen C-F, Lin P-Y (2018) Peripheral iron levels in children with autism spectrum disorders vs controls: a systematic review and meta-analysis. Nutr Res 50:44–52. https://doi.org/10.1016/j.nutres.2017.11.004
185. Tseng P-T, Cheng Y-S, Yen C-F, Chen Y-W, Stubbs B, Whiteley P, Carvalho AF, Li D-J, Chen T-Y, Yang W-C, Tang C-H, Chu C-S, Yang W-C, Liang H-Y, Wu C-K, Lin P-Y (2018) Peripheral iron levels in children with attention-deficit hyperactivity disorder: a systematic review and meta-analysis. Sci Rep 8(1):788. https://doi.org/10.1038/s41598-017-19096-x
186. Yao S, Zhong Y, Xu Y, Qin J, Zhang N, Zhu X, Li Y (2017) Quantitative susceptibility mapping reveals an association between brain iron load and depression severity. Front Hum Neurosci 11:442. https://doi.org/10.3389/fnhum.2017.00442
187. Daugherty AM, Raz N (2015) Appraising the role of iron in brain aging and cognition: promises and limitations of MRI methods. Neuropsychol Rev 25(3):272–287. https://doi.org/10.1007/s11065-015-9292-y
188. Bartzokis G, Lu PH, Tishler TA, Fong SM, Oluwadara B, Finn JP, Huang D, Bordelon Y, Mintz J, Perlman S (2007) Myelin breakdown and iron changes in Huntington's disease: pathogenesis and treatment implications. Neurochem Res 32(10):1655–1664. https://doi.org/10.1007/s11064-007-9352-7
189. Ghadery C, Pirpamer L, Hofer E, Langkammer C, Petrovic K, Loitfelder M, Schwingenschuh P, Seiler S, Duering M, Jouvent E, Schmidt H, Fazekas F, Mangin J-F, Chabriat H, Dichgans M, Ropele S, Schmidt R (2015) R2* mapping for brain iron: associations with cognition in normal aging. Neurobiol Aging 36(2):925–932. https://doi.org/10.1016/j.neurobiolaging.2014.09.013
190. Li K, Reichmann H (2016) Role of iron in neurodegenerative diseases. J Neural Transm 123(4):389–399. https://doi.org/10.1007/s00702-016-1508-7
191. Núñez MT, Urrutia P, Mena N, Aguirre P, Tapia V, Salazar J (2012) Iron toxicity in neurodegeneration. Biometals 25(4):761–776. https://doi.org/10.1007/s10534-012-9523-0

192. Penke L, Valdés Hernandéz MC, Maniega SM, Gow AJ, Murray C, Starr JM, Bastin ME, Deary IJ, Wardlaw JM (2012) Brain iron deposits are associated with general cognitive ability and cognitive aging. Neurobiol Aging 33(3):510–517.e2. https://doi.org/10.1016/j.neurobiolaging.2010.04.032
193. Pujol J, Junqué C, Vendrell P, Grau JM, Martí-Vilalta JL, Olivé C, Gili J (1992) Biological significance of iron-related magnetic resonance imaging changes in the brain. Arch Neurol 49(7):711–717. https://doi.org/10.1001/archneur.1992.00530310053012
194. Arnsten AFT, Goldman-Rakic PS (1998) Noise stress impairs prefrontal cortical cognitive function in monkeys: evidence for a hyperdopaminergic mechanism. Arch Gen Psychiatry 55(4):362–368. https://doi.org/10.1001/archpsyc.55.4.362
195. Williams GV, Goldman-Rakic PS (1995) Modulation of memory fields by dopamine Dl receptors in prefrontal cortex. Nature 376(6541):572–575. https://doi.org/10.1038/376572a0
196. Rodrigue KM, Daugherty AM, Foster CM, Kennedy KM (2020) Striatal iron content is linked to reduced fronto-striatal brain function under working memory load. NeuroImage 210:116544. https://doi.org/10.1016/j.neuroimage.2020.116544
197. Kalpouzos G, Garzón B, Sitnikov R, Heiland C, Salami A, Persson J, Bäckman L (2017) Higher striatal iron concentration is linked to frontostriatal underactivation and poorer memory in normal aging. Cereb Cortex 27(6):3427–3436. https://doi.org/10.1093/cercor/bhx045
198. Persson J, Garzón B, Sitnikov R, Bäckman L, Kalpouzos G (2020) A positive influence of basal ganglia iron concentration on implicit sequence learning. Brain Struct Funct 225(2):735–749. https://doi.org/10.1007/s00429-020-02032-7
199. Haber SN, Kunishio K, Mizobuchi M, Lynd-Balta E (1995) The orbital and medial prefrontal circuit through the primate basal ganglia. J Neurosci 15(7):4851–4867. https://doi.org/10.1523/JNEUROSCI.15-07-04851.1995
200. Haber SN, Knutson B (2010) The reward circuit: linking primate anatomy and human imaging. Neuropsychopharmacology 35(1):4–26. https://doi.org/10.1038/npp.2009.129
201. Darki F, Nemmi F, Möller A, Sitnikov R, Klingberg T (2016) Quantitative susceptibility mapping of striatum in children and adults, and its association with working memory performance. NeuroImage 136:208–214. https://doi.org/10.1016/j.neuroimage.2016.04.065
202. Carpenter KLH, Li W, Wei H, Wu B, Xiao X, Liu C, Worley G, Egger HL (2016) Magnetic susceptibility of brain iron is associated with childhood spatial IQ. NeuroImage 132:167–174. https://doi.org/10.1016/j.neuroimage.2016.02.028
203. Steiger TK, Weiskopf N, Bunzeck N (2016) Iron level and myelin content in the ventral striatum predict memory performance in the aging brain. J Neurosci 36(12):3552–3558. https://doi.org/10.1523/JNEUROSCI.3617-15.2016
204. Daugherty AM, Haacke EM, Raz N (2015) Striatal iron content predicts its shrinkage and changes in verbal working memory after two years in healthy adults. J Neurosci 35(17):6731–6743. https://doi.org/10.1523/JNEUROSCI.4717-14.2015
205. Biel D, Steiger TK, Bunzeck N (2021) Age-related iron accumulation and demyelination in the basal ganglia are closely related to verbal memory and executive functioning. Sci Rep 11(1):9438. https://doi.org/10.1038/s41598-021-88840-1
206. Bartzokis G, Lu PH, Tingus K, Peters DG, Amar CP, Tishler TA, Finn JP, Villablanca P, Altshuler LL, Mintz J, Neely E, Connor JR (2011) Gender and iron genes may modify associations between brain iron and memory in healthy aging. Neuropsychopharmacology 36(7):1375–1384. https://doi.org/10.1038/npp.2011.22
207. Sullivan EV, Adalsteinsson E, Rohlfing T, Pfefferbaum A (2009) Relevance of iron deposition in deep gray matter brain structures to cognitive and motor performance in healthy elderly men and women: exploratory findings. Brain Imaging Behav 3(2):167–175. https://doi.org/10.1007/s11682-008-9059-7
208. Valdés Hernández MdC, Ritchie S, Glatz A, Allerhand M, Muñoz Maniega S, Gow AJ, Royle NA, Bastin ME, Starr JM, Deary IJ, Wardlaw JM (2015) Brain iron deposits and lifespan cognitive ability. Age 37(5):100. https://doi.org/10.1007/s11357-015-9837-2
209. Burden MJ, Westerlund AJ, Armony-Sivan R, Nelson CA, Jacobson SW, Lozoff B, Angelilli ML, Jacobson JL (2007) An event-related potential study of attention and recognition memory in infants with iron-deficiency anemia. Pediatrics 120(2):e336–e345. https://doi.org/10.1542/peds.2006-2525

210. Carter RC, Jacobson JL, Burden MJ, Armony-Sivan R, Dodge NC, Angelilli ML, Lozoff B, Jacobson SW (2010) Iron deficiency anemia and cognitive function in infancy. Pediatrics 126(2):e427–e434. https://doi.org/10.1542/peds.2009-2097
211. Lozoff B, Brittenham GM, Wolf AW, McClish DK, Kuhnert PM, Jimenez E, Jimenez R, Mora LA, Gomez I, Krauskoph D (1987) Iron deficiency anemia and iron therapy effects on infant developmental test performance. Pediatrics 79(6):981–995
212. Lozoff B, Klein NK, Nelson EC, McClish DK, Manuel M, Chacon ME (1998) Behavior of infants with iron-deficiency anemia. Child Dev 69(1):24–36. https://doi.org/10.1111/j.1467-8624.1998.tb06130.x
213. Deinard AS, List A, Lindgren B, Hunt JV, Chang P-N (1986) Cognitive deficits in iron-deficient and iron-deficient anemic children. J Pediatr 108(5, Part 1):681–689. https://doi.org/10.1016/S0022-3476(86)81041-1
214. Pollitt E, Leibel RL, Greenfield DB (1983) Iron deficiency and cognitive test performance in preschool children. Nutr Behav 1(2):137–146
215. Halterman JS, Kaczorowski JM, Aligne CA, Auinger P, Szilagyi PG (2001) Iron deficiency and cognitive achievement among school-aged children and adolescents in the United States. Pediatrics 107(6):1381–1386. https://doi.org/10.1542/peds.107.6.1381
216. Pollitt E, Hathiral P, Kotchabhakdi NJ, Missell L, Valyasevi A (1989) Iron deficiency and educational achievement in Thailand. Am J Clin Nutr 50(3):687–697. https://doi.org/10.1093/ajcn/50.3.687
217. Scott SP, De Souza MJ, Koehler K, Murray-Kolb LE (2017) Combined iron deficiency and low aerobic fitness doubly burden academic performance among women attending university. J Nutr 147(1):104–109. https://doi.org/10.3945/jn.116.240192
218. Lukowski AF, Koss M, Burden MJ, Jonides J, Nelson CA, Kaciroti N, Jimenez E, Lozoff B (2010) Iron deficiency in infancy and neurocognitive functioning at 19 years: evidence of long-term deficits in executive function and recognition memory. Nutr Neurosci 13(2):54–70. https://doi.org/10.1179/147683010X12611460763689
219. Lozoff B, Beard J, Connor J, Felt B, Georgieff M, Schallert T (2006) Long-lasting neural and behavioral effects of iron deficiency in infancy. Nutr Rev 64(suppl_2):S34–S43. https://doi.org/10.1111/j.1753-4887.2006.tb00243.x
220. Wenger MJ, DellaValle DM, Murray-Kolb LE, Haas JD (2019) Effect of iron deficiency on simultaneous measures of behavior, brain activity, and energy expenditure in the performance of a cognitive task. Nutr Neurosci 22(3):196–206. https://doi.org/10.1080/1028415X.2017.1360559

Chapter 14

Dopaminergic Control of Actions and Habits

Fabien Naneix and Etienne Coutureau

Abstract

Choosing between different course of behavioral response is an essential process to survive in a complex environment. Numerous studies have demonstrated that basic processes of action control may be investigated using instrumental conditioning, as instrumental response may be dissociated in goal-directed action or habitual response depending both on different, but interacting, neuronal circuits. The dopamine system is a central element in the coordination between actions and habits. In this chapter, we describe in details the different behavioral procedures used to investigate actions and habits in rodent models, including instrumental learning, outcome devaluation, and contingency degradation. We also discuss how these procedures can be combined with other techniques to specifically investigate the role of the dopamine system in these different processes.

Key words Behavior, instrumental conditioning, reward, dopamine, outcome devaluation, contingency degradation

1 Introduction

1.1 Response Control by Actions and Habits

To survive in a changing environment, organisms have to choose by selecting actions while constantly monitoring their consequences. According to current views, instrumental conditioning in both rodents [1, 2] and primates, including humans [3, 4], may capture many of the basic cognitive mechanisms involved in choice behavior [5].

Current research considers that instrumental responding is the result of two distinctive classes of behavior. The first class is goal-directed action. Goal-directed action is a purposive behavior which depends on both the current value of the outcome that the behavior induces and the causal relationship between that behavior and its outcome [2, 6, 7]. The second class is habitual response. Contrary to goal-directed actions, habits are thought to be elicited by cues present during training and to be independent of the value of the outcome.

José Antonio Fuentealba-Evans and Pablo Henny (eds.), *Dopaminergic System Function and Dysfunction: Experimental Approaches*, Neuromethods, vol. 193, https://doi.org/10.1007/978-1-0716-2799-0_14,

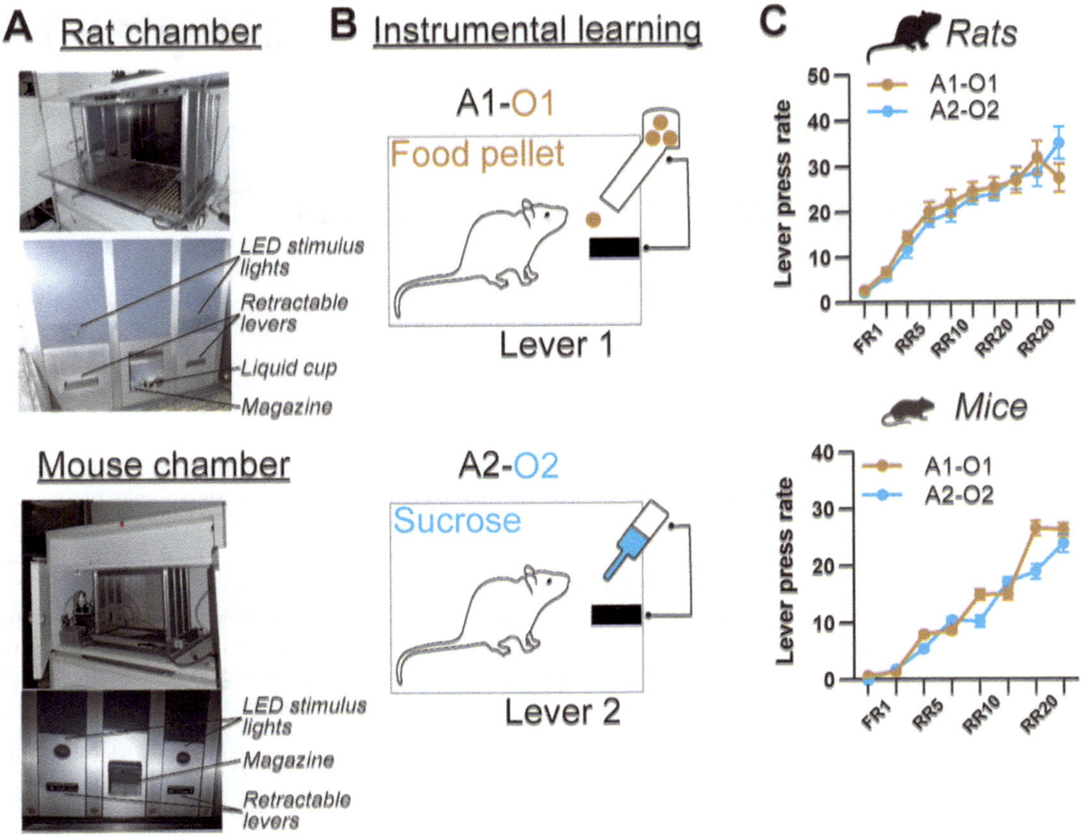

Fig. 1 General design of instrumental learning and apparatus in rodent models. (**a**) Conditioning chambers for rats (top part) and mice (bottom part). For each panel shows a general view of the apparatus from outside (top) and a zoomed view of levers on each side of a magazine where the food reward may be collected. (**b**) General design of instrumental conditioning with the same animal learning two actions (A1, A2; lever press) associated with two different outcomes (O1, food pellet; O2, drop of sucrose solution). (**c**) Representative data of the increase of instrumental performance (lever press rate) for each A–O association through different reinforcement schedules in rats (top, $n = 12$) and mice (bottom, $n = 25$)

Instrumental control by goal-directed actions or habits are best investigated using outcome devaluation and contingency degradation procedures, which represent powerful behavioral tools widely used across species. For instance, in laboratory rats moderately trained to press a lever for food (as shown in Fig. 1), reducing the appetitive value of the food (outcome devaluation, Fig. 2) or the relation between lever-press and outcome delivery (contingency degradation, Fig. 3) results in a later reduction of lever pressing as assessed in an extinction test. Thus, early in acquisition responses are flexible adaptable and goal-directed, i.e., mediated by a detailed representation of the outcome. In contrast, in overtrained rats or in rats trained with specific schedules of reinforcement, performance is sensitive neither to outcome devaluation nor to contingency degradation [8–10], thus reflecting a habitual responding. A thorough

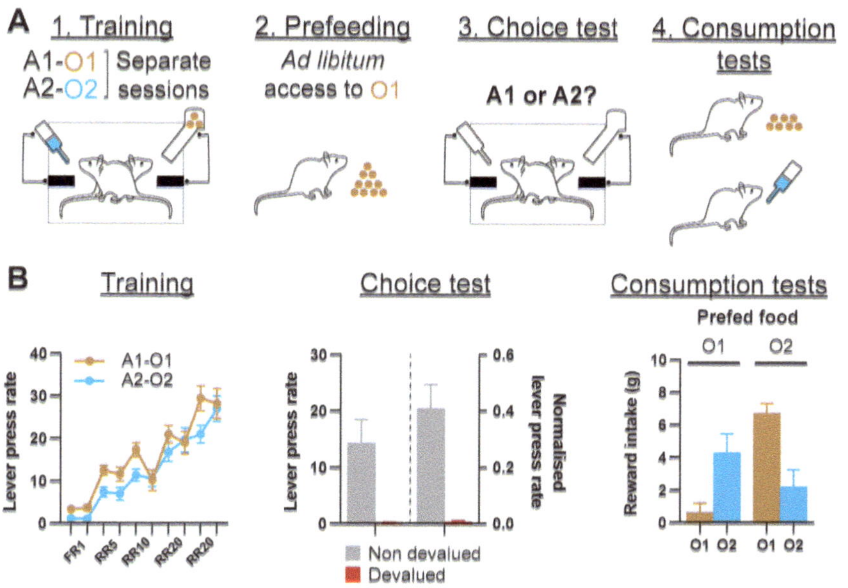

Fig. 2 Sensory-specific satiety outcome devaluation protocol. General design (**a**) and representative results (**b**) of each phase of outcome devaluation in rats ($n = 12$). After initial learning (1), animals were pre-fed with one specific food O1 for 1 h (2), before to be placed in their conditioning chamber for a choice test (3) between the two levers previously associated with O1 and O2. Goal-directed action is clearly demonstrated by a selective decrease of response on the lever associated with the pre-fed food (Devalued, red bars) compared to the other one (Non-Devalued, gray bars). After completion of this test, the selectivity of the satiety effect is tested with two consecutive consumption tests (4; one for O1 and one for O2, counterbalanced between animals) where a selective consumption of the non-pre-fed food is observed

theoretical analysis of the coordination of actions and habits is far beyond the scope of this chapter but the interested reader could refer to recent reviews.

1.2 The Dopaminergic System and Its Role in Actions and Habits

The dopamine (DA) system plays a central role in the coordination of actions and habits [11]. Dopaminergic cells in both the substantia nigra (pars compacta) and the ventral tegmental area differentially innervate the prefrontal cortex (mesocortical pathway), the striatum (nigrostriatal pathway), and the nucleus accumbens (mesolimbic pathway) [12, 13]. This broad innervation conveys various types of information such as prediction error [14, 15] or event salience and motivational processes [16, 17]. Recent work has emphasized the differential contribution of the various DA pathways to goal-directed behavior (*see* [18]). Indeed, 6-hydroxydopamine (6-OHDA) lesions of DA terminals in prefrontal region [19] alter the ability to adapt to change in contingency (but *see* also [20]), without altering behavioral adaptation to outcome devaluation. Blockade of D1/D2 DA receptors in the medial prefrontal cortex during contingency degradation similarly alter adaptation to contingency changes [19]. In contrast, microinfusions of dopamine directly into a more ventral portion of medial prefrontal cortex (infralimbic cortex) restore goal-directed responding in training

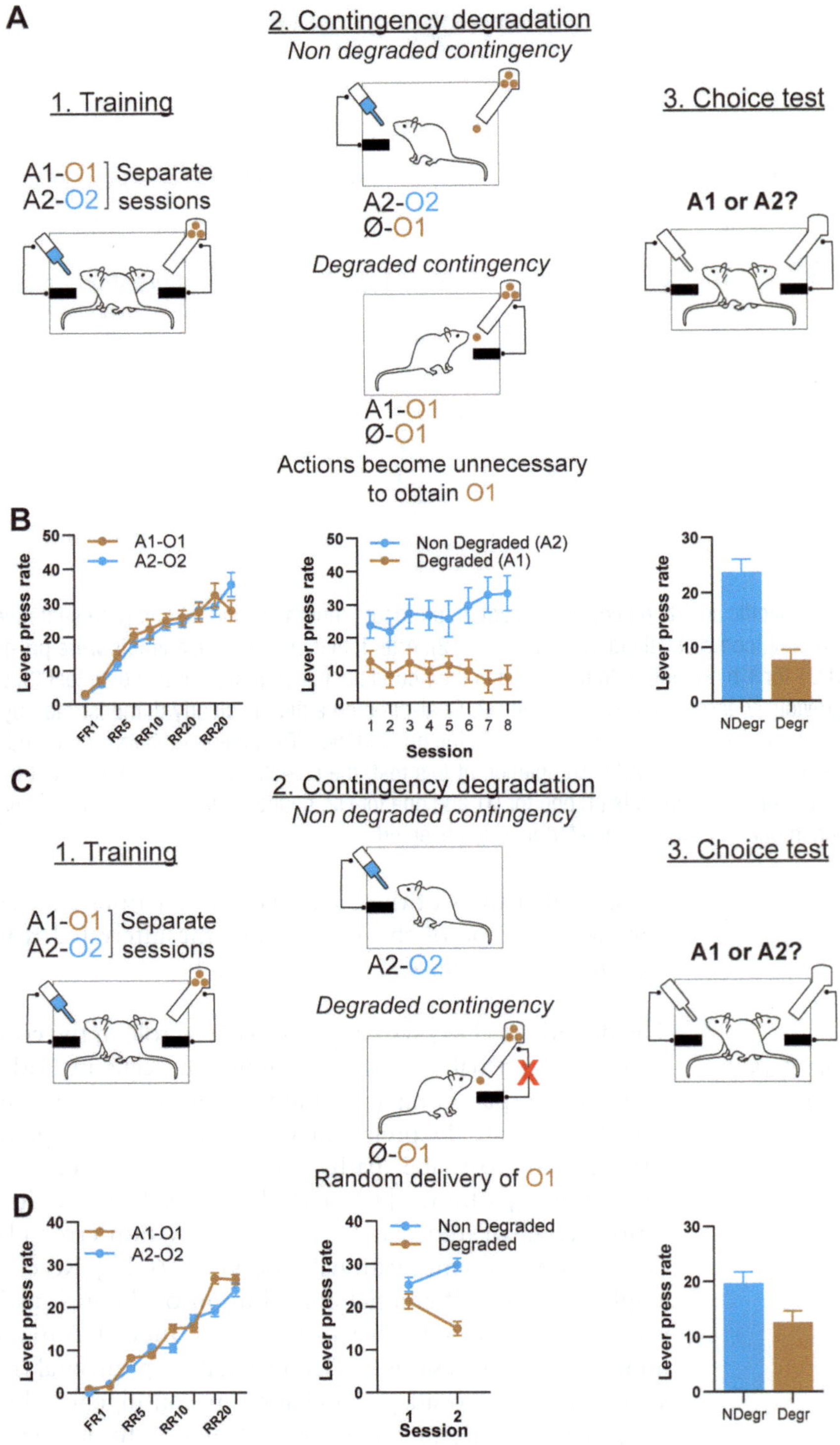

Fig. 3 Contingency degradation protocols. (**a**, **b**) General design (**a**) and representative results (B) of the contingency degradation procedure designed by Dickinson and colleagues [2, 47]. After initial learning (1), rats received two daily sessions (2) during which they received food rewards by pressing the lever but also when they do not press, with the same probability (1 pellet every 20 lever press in average or 1 pellet every 20 sec without lever press). During one session the noncontingent food is different from the food available by pressing

conditions promoting habitual responding [21]. Importantly, thorough analysis of the developmental trajectories of the DA systems has demonstrated that adult subjects' ability to execute actions according to their predicted consequences is associated with the late maturation of the DA mesocortical pathway during adolescence [22, 23]. Dopaminergic control of actions and habits is not restricted to the mesocortical system. Indeed, lesions of DA terminals in the dorsal striatum alter responding following contingency changes [20, 24] and disrupt habit formation [25]. In contrast, lesions of DA terminals in the nucleus accumbens core spare action's sensitivity to outcome devaluation [26].

Importantly, drugs acting on the DA systems dramatically impact the coordination of actions and habits. Several studies have indeed demonstrated that repeated injections psychostimulants such as amphetamine [27–29], methamphetamine [30, 31], or cocaine [32, 33] enhanced the transition of instrumental control from goal-directed to habitual, indexed by a lack of sensitivity of the instrumental response to reward devaluation (*see* [34] for a recent review on habits and drugs of abuse).

1.3 Behavioral Tests to Assess Action Control

As described above, DA systems have a profound role in the coordination of goal-directed and habitual responding. This chapter therefore describes protocol for the how-to study instrumental learning in both rats and mice. These behavioral approaches may be then combined with other in vivo techniques to investigate the dopamine system as described in other chapters, including lesions, recordings, pharmacology, circuits targeting, and selective manipulations.

2 Materials

2.1 Animals

We mainly used male naïve Long-Evans rats (Janvier Laboratories). Most of our work is conducted in adult animals weighing between 275 and 300 g [19, 35–40]. We also previously investigated goal-directed and habitual behavior in adolescent rats (3–5 weeks old) [22, 23]. We were also able to investigate this type of behavior in mice, using male naïve C57BL/6 J mice (Janvier Laboratories) weighing between 20 and 28 g.

Fig. 3 (continued) the lever (Non-Degraded condition), whereas during the other one the two foods are the same (Degraded condition). A selective decrease in instrumental performance in the Degraded condition is observed during all contingency degradation sessions and also during the subsequent choice test between the two levers (3). (**c**, **d**) General design (**c**) and representative results in mice (**d**), a simplified protocol for contingency degradation. In this case, the Non-Degraded session is similar to an instrumental learning session (A2–02, no additional reward delivery). During the other session, pressing the lever has no effect on reward delivery. Food rewards are instead delivered randomly based on each animal's delivery rate at the end of the learning phase

2.2 Housing

Rats were usually housed in pairs in polycarbonate cages (46 × 26 × 20 cm). Mice were usually housed in group of 4 standard polypropylene cages. Animals were housed in a temperature- and humidity-controlled room and maintained under a 12 h light/dark cycle (light on at 7:00 or 8:00 A.M.). Environmental enrichment is provided by tinted polycarbonate tubing elements placed into the cages. All our experiments take place during the light phase of the cycle. Before the experiments, animals are given ad libitum access to food (A04, SAFE) and water for 1 week and are handled each every day. Throughout the duration of the experiments, animals are maintained at 90% of their original weight by restricting their food intake (rats: ~12 g/day; mice: ~2 g/day). For food restriction in young animals, body weight was carefully monitored and compared to the body weight curve of control non-food-restricted animals. The amount of food provided daily was adjusted to maintain animals at 90% of their theoretical weight at this age without altering their average growth.

2.3 Behavioral Apparatus

There is an important variety of modular conditioning chambers available to train and test animals (Med Associates Inc®, Coulbourn Instruments LLC), which were successfully used by other labs. In our previous work, all experiments are conducted in identical conditioning chambers (adult rats: 40 cm wide × 30 cm deep × 35 cm high; adolescent rats/mice: 25 cm wide × 16 cm deep × 20 cm high; Imetronic, France, Fig. 1a), each located inside a sound- and light-attenuating wooden chamber (rats: 74 × 46 × 50 cm; mice: 30 × 40 × 36 cm). Each chamber has a ventilation fan producing a background noise of 55 dB and four LEDs on the ceiling for illumination of chamber during training sessions. Each chamber has two opaque panels on the right and left sides, two clear Perspex walls on the back and front sides, and a stainless steel grid floor. Left wall was equipped with a magazine connected to pellet dispensers or to a syringe pump located outside. Magazines are equipped with infrared cells to detect the animal's visits. Two retractable levels are located on each side of the magazine, controlling the delivery of the associated outcome, as a function of the current reinforcement schedule. Chambers were controlled via a computer running POLY software (Imetronic, France).

For consumption tests (*see* Outcome devaluation), we used individual cages similar to their home cage located in a separate room. A glass dish containing food pellets was fixed on the floor of each cage or a bottle with a specific solution was inserted.

2.4 Food Reward

Calibrated food pellets (rats: 45 mg; mice: 20 mg; Bio-Serv, NJ, USA) are usually used as reward outcome for instrumental behavior. In the past, we successfully used grain-based (F0165 or F0163), purified (F0021 or F0071), and sucrose pellets (F0023 and F07595). In some experiments we also used 10% sucrose solution (S9378, Merck, Germany) as reward (*see* **Note 1**).

3 Methods

3.1 Magazine Training

Initially, animals are trained for two sessions to collect food rewards during two 30-min magazine training sessions (one session for each reward, separated by approximately 2 h). Rewards are delivered on a random time 60 s schedule. The order of presentation of rewards are alternated each day and counterbalanced across animals. To decrease potential neophobia for new food, especially in mice, animals can be also exposed to the food reward directly in their home cage in addition of their daily food.

3.2 Instrumental Training

Following magazine training, all animals are trained to press a lever to obtain a reward (30 min session). We previously used the learning of either a single action–outcome relationship (A–O) [39, 40] or two simultaneously (A1–O1 and A2–O2) [19, 22, 23, 35–38] (Fig. 1b, c). When using multiple A–O learning, each session is separated by a 1–2 h break and their presentation order is counterbalanced between training days. In developing animals, conditioning schedule may go up to four sessions a day (two for each action–outcome relationships) to allow all the training and testing phase to take place in a limited developmental window (e.g., adolescence in rodent is between postnatal day 28 and 45–50).

Animals are initial trained under continuous reinforcement, fixed ratio schedule FR1, for 2–3 days (i.e., each lever press was rewarded). Animals are then shifted to variable reinforcement schedules (*see* **Note 2**). To investigate actions and goal-directed behavior, we usually used reinforcement schedule based in variable ratio/ random ratio (VR or RR), in which the reward is delivered as a probability after a defined number of action (e.g., lever presses, nose pokes) in average. For instance, an RR5 reinforcement schedule means that the animal will receive a reward every five response in average (probability to obtain outcome O doing the action A, $P(O|A) = 1/5 = 0.2$). Using this kind of schedule maintains a strong correlation between animal's performance and reward delivery, favorizing goal-directed behavior [9, 41]. Animals are usually shifted from RR5 (2 days) to RR10 (2 days, $P(O|A) = 1/10 = 0.1$) and finally RR20 (4 days, $P(O|A) = 1/20 = 0.05$). Ideally, we want to observe a progressive increase in the response rate throughout the training days and reinforcement schedules, similar for the two action–outcome relationships (*see* Fig. 1b, c).

3.3 Outcome Devaluation

Outcome devaluation is used to test the representation of action's outcome as a goal. In the literature there is two main ways to devaluate action-associated food reward: sensory-specific satiety and conditioned taste aversion (CTA). In the past we mainly used sensory-specific satiety which we will be described here (*see* **Note 3** for more details about devaluation by food aversion conditioning).

Sensory-specific satiety outcome devaluation procedure can be divided in three stages: pre-feeding phase, test phase, and consumption tests phase (Fig. 2a). We usually perform the satiety phase in individual cages, different from their home cage and the conditioning chambers. In order to familiarize the animals to the novel consumption cage, animals are placed for 30 min in an individual consumption cage after the completion of the last session of instrumental training. On the following day, animals are given free access to one type of food in these consumption cages. If a protocol with 2 actions and 2 outcomes is used, half of animals receive O1 and half of them receive O2. In the case of a single A–O learning, half of animals can receive another kind of food (e.g., chow, other type of pellets). Animals should have more food that they can eat in 60 min to be sure they are slated for this particular reward.

Immediately after the pre-feeding treatment, animals are placed in their operant chambers for a 5–15 min choice test. During this test, levers are inserted in the chamber, but no reward is delivered, meaning that instrumental performance is only dependent of animal's representation of each outcome value as a goal or not. Finally, in order to evaluate the effectiveness of pre-feeding procedure, animals are then transferred back to the consumption cages and allowed access to each of the two food rewards successively (the order of the pre-fed and the non-pre-fed reward being counterbalanced across animals) for 15 min. If the animals' actions are goal-directed, then you would observe a selective decrease in the response associated with the pre-fed food (Fig. 2b) associated with a low intake of the pre-fed food reward. In contrast, habitual behavior would be characterized by an insensitivity of animals' responses to satiety, in spite of a low intake of the pre-fed food during the consumption tests. Importantly, animals can be used as their own controls in this procedure by retraining them during 1–2 days before to perform another outcome devaluation test, using the alternative food reward during the pre-feeding.

More recently, our group conducted a series of study to investigate the importance of the delay between the pre-feeding phase and the test. We show that the effect of outcome devaluation on goal-directed behavior is observed up to 2 h after the pre-feeding. Moreover, this effect can be observed up to 5 h after the pre-feeding but only if this one is also done in the conditioning chambers, instead of in different consumption cages [37]. This important experiment shows the importance of contextual cues in the modulation of this type of behavior [42–44] (*see* **Note 4**).

3.4 Contingency Degradation

The second important parameter is the knowledge of the causal relationships between one action and its associated outcome ("*belief criteria*") [2, 18]. This is measured by the instrumental contingency, i.e., the probability to obtain the outcome by performing the action (pressing a lever; P(O|A)) compared to the probability to obtain this same outcome without performing the action (P(O|Ø)). During the training phase (*see* Subheading 3.2), the contingency is positive as subjects cannot obtain the reward without performing the action. To test the knowledge of A–O relationships, we and others used a procedure of instrumental contingency degradation in which we increase the probability to obtain the food without performing the action, making the A–O contingency null.

The most widely used protocol is based on the initial experiment by Hammond [45] revised alter by Colwill and Rescorla [46] and by Dickinson and colleagues [2, 47]. This is based on the delivery of "free" reward, noncontingent with the instrument response, in addition to the rewards an animal may obtain by its action. The reward delivery rate is based on the same probability used for the instrumental response. For instance, if the animal is trained to press the left lever (A1) to obtain a food pellet (O1) under an RR20 schedule (i.e., 1 reward every 20 responses in average, P(O1|A1) = 0.05), then the noncontingent pellet would be delivered at the same probability for each second during which the animal is not pressing the lever (P(O1|Ø) = 1/ 20 sec = 0.05). In this case the A1–O1 contingency is degraded as it is not required to perform A1 to obtain O1. However, it is important to note that in this case the additional delivery of O1 could generate a competition between the action (lever pressing) and the collection of these rewards (magazine visits), participating to the global decrease of the instrumental response. To control for this, Dickinson and colleagues [2, 47] modified the protocol to introduce two important changes. The first one is the delivery of "free reward" also during the session with the other action (right lever, A2). However, during the Non-Degraded contingency session, the "free rewards" are different from those available by performing the action (*Different condition*: P(O2|A2) = 0.01; P (O2|Ø) = 0.01; P(O1|Ø) = 0.01). In the Degraded contingency session, both the reward available by pressing the lever and by doing nothing are the same (Same condition: P(O1|A1) = 0.01; P(O1|Ø) = 0.01; Fig. 3). We previously used this kind of design in our previous studies and it is important to be sure that you have a similar level of response for the two actions before to start to contingency degradation protocol [19, 22, 23]. We used two 20-min session per day (one session per condition, Non-Degraded and Degraded) for 4–8 days. The Degraded A–O contingency is counterbalanced between animals. A second important change is the introduction of a choice test phase at the end of the degradation procedure. This is identical to the test described above for the Outcome Devaluation procedure during which animals have the choice between the two responses, without any reward delivery. If animal's behavior is goal-directed, then the knowledge of the A1–O1 relationships is updated and we would observe a lower performance of A1 (Degraded), compared to A2 (Non Degraded), and compared to A1 level before the contingency degradation (Fig. 3b).

More recently, other studies including some from our group used a more simplified procedure [35, 48]. In this procedure, one reward continues to be obtained in regular reinforcement schedule (e.g., RR20 or RR10) without additional noncontingent reward delivery (Non-Degraded condition). However, during the Degraded condition session, the other reward is delivered randomly based on animal's own reinforcement rate on the last day of training. In this condition, performing the action has no consequence, so there is no additional reward delivery compared to the other session. Here again, a goal-directed behavior is expected to exhibit a selective decrease of the response rate in the Degraded condition, which is also observed during the choice test phase as we reported in rats [35] and in mice (Fig. 3c).

3.5 Behavioral Analysis

Many behavioral parameters may be analyzed from this kind of paradigm. The main outcome is the instrumental response itself (e.g., lever press, nose poke), which may be expressed as total number or, more often, in response rate per minute. The average response during the last or two last sessions of instrumental training may be used as a measure of response baseline. Then instrumental performance during outcome devaluation and contingency degradation may be expressed in relative response based on animal's own baseline. This is a powerful way to analyze these data as it is based on the behavior of each animal, but it should not prevent the analysis of raw responses which could reveal interindividual differences. In parallel of the instrumental response itself, the analysis of approach and consummatory behaviors (magazine visits, reward earned, latencies to press or collect food, reward intake, licks) may also provide important information on other behavioral processes (Pavlovian behavior, food palatability, motivation).

Finally, feeding behavior during outcome devaluation is usually measured as total food intake during the pre-feeding phase and the consumptions tests. Consumption tests may also be analyzed as consumption index (food eaten/total food available).

3.6 Dopamine System, Action, and Habits: Future Directions

To our knowledge, all the studies which investigated the role of the DA system in action and habits have used classical tools to target dopamine circuits, including 6-OHDA lesions [19, 20, 24–26], local or systemic pharmacological approaches [19, 23], or using psychostimulants acting on DA neurons [34], which may all be combined with ex vivo analysis including immunostaining of DA neurons and terminals (e.g., tyrosine hydroxylase staining), neurochemistry analyses of dopamine and metabolites levels, or measures of dopamine receptors levels using molecular biology techniques.

With the rise of powerful genetic tools in neuroscience during the last 10 years, there is now a strong opportunity to target and manipulate the activity of specific neuronal populations and circuits during specific phases behavioral testing or with a high temporal specificity. For instance, our group recently used chemogenetic approaches to dissect the role of thalamo-cortical [35] and cortico-cortical [38] circuits in goal-directed behavior. In mice, TH-Cre and DAT-Cre lines combined with viral vectors are extensively used to target dopamine neurons [49, 50]. In rats, we also showed that such viral approaches like the use of retrograde CAV2 viruses can be used to target and manipulate different brain circuits [51], including catecholaminergic (dopamine, noradrenaline) systems [52]. However, despite numerous works using chemogenetics or optogenetics tools to manipulate DA neurons and circuits, there is no recent work so far using these approaches in the theoretical frame of goal-directed action and habits. Another exciting opportunity in these field is the recent development of optical tools like fiber photometry to record dopamine neurons activity [53] and even dopamine release [54] with a high temporal specificity, similar to what could only be previously observed with very demanding techniques such as fast scan cyclic voltammetry. This could allow now a more precise measure of dopamine signals in different part of the brains when animals express goal-directed or habit behavior.

4 Notes

1. *Food Rewards*

   We usually use calibrated food pellets. Liquid rewards are very powerful in driving instrumental learning. In that case, they are delivered inside a cup through a liquid dispenser. In rats, we used 10% sucrose solution [19, 22, 23], whereas other experiments in mice [55] used 10% sweetened condensed milk dissolved in water (3.25 kcal/g). However, it is important to make sure that they provide both similar training profile and hedonic properties as compared to solid reward when using a two actions–two outcomes procedure with one of the actions associated with solid reward and the other associated with a liquid reward.

   Liquid rewards are extensively used by numerous works for Pavlovian conditioning as it also allows to measure hedonic processes like food palatability using licking microstructure (*see* [56] for a more detailed recent review on this topic).

2. *The Role of Reinforcement Schedule*

In our experiments, we usually use reinforcement schedule based on variable ratio/random ratio (VR or RR), in which the reward is delivered as a probability after a defined number of action (e.g., lever presses, nose pokes) in average. Some other experiments use a ratio/random interval (VI or RI) schedule, in which reward is made available for procurement after a time (in sec) has elapsed where time is randomly determined after each reward [39, 40]. Rodents trained on a random ratio schedule are more likely to show goal-directed responding than rodents trained on a random interval schedule [9, 57], an effect which is thought to result from a weakened correlation between response rates and reward rates when rats are trained under interval schedule [41].

3. *Devaluation by Sensory-Specific Satiety Versus Food Aversion*

Sensory-specific satiety is a very efficient way to change the value of a reward. Since the resulting change in value is reversible, it allows successive testing [38] of instrumental performance following devaluation but also the assessment of contingency degradation in the same animals [19]. It must be noted, however, that reliable but more long-lasting devaluation can also be achieved by pairing food consumption with illness. In that case, following instrumental training, the consumption of the food which served as reward is paired with an intraperitoneal injection of lithium chloride (LiCl). The dosage and volume of LiCl can be adjusted depending on the strength of the expected aversion. Control animals receive a similar injection of vehicle solution (usually sodium chloride, NaCl). Importantly, it must be noted that goal-directed behavior is only observed when food aversion is acquired in a different cage than the conditioning chamber [40, 58].

4. *The Role of Context in Actions and Habits*

Contextual cues play a fundamental role in the coordination of actions and habits as recently reviewed [10, 44, 58]. In some studies, two different responses are trained (or overtrained) in two different distinct contexts, which allow for a within-subject assessment of actions and habits [43, 57]. This parameter remains poorly understood and, to our knowledge, no studies have investigated the role of the dopamine system in these processes.

References

1. Dickinson A (1985) Actions and habits: the development of behavioural autonomy. Philos Trans R Soc London B, Biol Sci 308:67–78
2. Balleine BW, Dickinson A (1998) Goal-directed instrumental action: contingency and incentive learning and their cortical substrates. Neuropharmacology 37:407–419
3. Valentin VV et al (2007) Determining the neural substrates of goal-directed learning in the human brain. J Neurosci 27:4019–4026

4. Tanaka SC et al (2008) Calculating consequences: brain systems that encode the causal effects of actions. J Neurosci 28:6750–6755
5. Rangel A et al (2008) A framework for studying the neurobiology of value-based decision making. Nat Rev Neurosci 9:545–556
6. Adams CD, Dickinson A (1981) Instrumental responding following Reinforcer devaluation. Q J Exp Psychol Sect B 33:109–121
7. Dickinson A (1994) Instrumental conditioning. In: Animal learning and cognition. Academic Press, San Diego, pp 45–79
8. Adams CD (1982) Variations in the sensitivity of instrumental responding to reinforcer devaluation. Q J Exp Psychol B Comp Physiol Psychol 34B:77–98
9. Dickinson A et al (1983) The effect of the instrumental training contingency on susceptibility to reinforcer devaluation. Q J Exp Psychol B Compar Physiol Psychol 35:35–51
10. Thrailkill EA, Bouton ME (2015) Contextual control of instrumental actions and habits. J Exp Psychol Anim Learn Cogn 41:69–80
11. Wickens JR et al (2007) Dopaminergic mechanisms in actions and habits. J Neurosci 27:8181–8183
12. Björklund A, Dunnett SB (2007) Dopamine neuron systems in the brain: an update. Trends Neurosci 30:194–202
13. Lammel S et al (2014) Reward and aversion in a heterogeneous midbrain dopamine system. Neuropharmacology 76 Pt B:351–359
14. Schultz W (2000) Multiple reward signals in the brain. Nat Rev Neurosci 1:199–207
15. Montague PR et al (2004) Computational roles for dopamine in behavioural control. Nature 431:760–767
16. Redgrave P, Gurney K (2006) The short-latency dopamine signal: a role in discovering novel actions? Nat Rev Neurosci 7:967–975
17. Bromberg-Martin ES et al (2010) Dopamine in motivational control: rewarding, aversive, and alerting. Neuron 68:815–834
18. Coutureau E, Parkes SL (2018) Cortical determinants of goal-directed behavior. In: Morris R et al (eds) Goal-directed decision making. Academic Press, pp 179–197
19. Naneix F et al (2009) A role for medial prefrontal dopaminergic innervation in instrumental conditioning. J Neurosci 29: 6599–6606
20. Lex B, Hauber W (2010) The role of dopamine in the prelimbic cortex and the dorsomedial striatum in instrumental conditioning. Cereb Cortex 20:873–883
21. Hitchcott PK et al (2007) Bidirectional modulation of goal-directed actions by prefrontal cortical dopamine. Cereb Cortex 17: 2820–2827
22. Naneix F et al (2012) Parallel maturation of goal-directed behavior and dopaminergic systems during adolescence. J Neurosci 32: 16223–16232
23. Naneix F et al (2013) Adolescent stimulation of D2 receptors alters the maturation of dopamine-dependent goal-directed behavior. Neuropsychopharmacology 38:1566–1574
24. Braun S, Hauber W (2012) Striatal dopamine depletion in rats produces variable effects on contingency detection: task-related influences. Eur J Neurosci 35:486–495
25. Faure A (2005) Lesion to the Nigrostriatal Dopamine system disrupts stimulus-response habit formation. J Neurosci 25:2771–2780
26. Lex B, Hauber W (2010) The role of nucleus accumbens dopamine in outcome encoding in instrumental and Pavlovian conditioning. Neurobiol Learn Mem 93:283–290
27. Nelson A, Killcross S (2006) Amphetamine exposure enhances habit formation. J Neurosci 26:3805–3812
28. Nordquist RE et al (2007) Augmented reinforcer value and accelerated habit formation after repeated amphetamine treatment. Eur Neuropsychopharmacol 17:532–540
29. Nelson AJD, Killcross S (2013) Accelerated habit formation following amphetamine exposure is reversed by D1, but enhanced by D2, receptor antagonists. Front Neurosci 7:76
30. Furlong TM et al (2017) Pulling habits out of rats: adenosine 2A receptor antagonism in dorsomedial striatum rescues methamphetamine-induced deficits in goal-directed action. Addict Biol 22:172–183
31. Furlong TM et al (2018) Methamphetamine promotes habitual action and alters the density of striatal glutamate receptor and vesicular proteins in dorsal striatum. Addict Biol 23: 857–867
32. Corbit LH et al (2014) Effects of repeated cocaine exposure on habit learning and reversal by N-acetylcysteine. Neuropsychopharmacology 39:1893–1901
33. LeBlanc KH et al (2013) Repeated cocaine exposure facilitates the expression of incentive motivation and induces habitual control in rats. PLoS One 8:e61355
34. Vandaele Y, Ahmed SH (2021) Habit, choice, and addiction. Neuropsychopharmacology 46: 689–698
35. Alcaraz F et al (2018) Thalamocortical and corticothalamic pathways differentially

contribute to goal-directed behaviors in the rat. elife 7:e32517
36. Fresno V et al (2019) A thalamocortical circuit for updating action-outcome associations. elife 8:e46187
37. Parkes SL et al (2016) A time course analysis of satiety-induced instrumental outcome devaluation. Learn Behav 44:347–355
38. Parkes SL et al (2018) Insular and ventrolateral orbitofrontal Cortices differentially contribute to goal-directed behavior in Rodents. Cereb Cortex 28:2313–2325
39. Tantot F et al (2017) The effect of high-fat diet consumption on appetitive instrumental behavior in rats. Appetite 108:203–211
40. Tran-Tu-Yen DAS et al (2009) Transient role of the rat prelimbic cortex in goal-directed behaviour. Eur J Neurosci 30:464–471
41. Yin HH, Knowlton BJ (2006) The role of the basal ganglia in habit formation. Nat Rev Neurosci 7:464–476
42. Bradfield LA et al (2020) Goal-directed actions transiently depend on dorsal hippocampus. Nat Neurosci 23:1194–1197
43. Killcross S, Coutureau E (2003) Coordination of actions and habits in the medial prefrontal cortex of rats. Cereb Cortex 13:400–408
44. Trask S, Bouton ME (2014) Contextual control of operant behavior: evidence for hierarchical associations in instrumental learning. Learn Behav 42:281–288
45. Hammond LJ (1980) The effect of contingency upon the appetitive conditioning of free-operant behavior. J Exp Anal Behav 34: 297–304
46. Colwill RM, Rescorla RA (1986) Associative structures in instrumental learning. In: Bower GH (ed) Psychology of learning and motivation. Academic Press, pp 55–104
47. Dickinson A, Mulatero CW (1989) Reinforcer specificity of the suppression of instrumental performance on a non-contingent schedule. Behav Process 19:167–180
48. Dias-Ferreira E et al (2009) Chronic stress causes frontostriatal reorganization and affects decision-making. Science 325:621–625
49. Stuber GD et al (2015) Considerations when using cre-driver rodent lines for studying ventral tegmental area circuitry. Neuron 85: 439–445
50. Lammel S et al (2015) Diversity of transgenic mouse models for selective targeting of midbrain dopamine neurons. Neuron 85:429–438
51. Morceau S et al (2019) Targeting reciprocally connected brain regions through CAV-2 mediated interventions. Front Mol Neurosci 12:303
52. Cerpa J-C et al (2020) Targeting Catecholaminergic systems in transgenic rats with a CAV-2 vector Harboring a Cre-dependent DREADD Cassette. Front Mol Neurosci 13: 121
53. Gunaydin LA et al (2014) Natural neural projection dynamics underlying social behavior. Cell 157:1535–1551
54. Labouesse MA et al (2020) GPCR-based Dopamine sensors-A detailed guide to inform sensor choice for in vivo imaging. Int J Mol Sci 21:E8048
55. Ducrocq F et al (2019) Decrease in operant responding under obesogenic diet exposure is not related to deficits in incentive or Hedonic processes. Obesity (Silver Spring) 27:255–263
56. Naneix F et al (2019) Investigating the effect of physiological need states on palatability and motivation using microstructural analysis of licking. Neuroscience. https://doi.org/10.1016/j.neuroscience.2019.10.036
57. Gremel CM, Costa RM (2013) Orbitofrontal and striatal circuits dynamically encode the shift between goal-directed and habitual actions. Nat Commun 4:2264
58. Bouton ME (2021) Context, attention, and the switch between habit and goal-direction in behavior. Learn Behav 49:349–362

Index

José Antonio Fuentealba-Evans and Pablo Henny (eds.), *Dopaminergic System Function and Dysfunction: Experimental Approaches*, Neuromethods, vol. 193, https://doi.org/10.1007/978-1-0716-2799-0,

T

U

V

W

MIX
Papier aus verantwortungsvollen Quellen
Paper from responsible sources
FSC® C105338

If you have any concerns about our products,
you can contact us on
ProductSafety@springernature.com

In case Publisher is established outside the EU,
the EU authorized representative is:
Springer Nature Customer Service Center GmbH
Europaplatz 3, 69115 Heidelberg, Germany

Printed by Libri Plureos GmbH
in Hamburg, Germany